T0107380

PUBLICATIONS OF THE ISRAEL ACADEMY

OF SCIENCES AND HUMANITIES

SECTION OF SCIENCES

FAUNA PALAESTINA

INSECTA III—ORTHOPTERA: ACRIDOIDEA

THE FAUNA PALAESTINA COMMITTEE

Pyrgodera armata, nymph

FAUNA PALAESTINA • INSECTA III

ORTHOPTERA: ACRIDOIDEA

by

LEV FISHELSON

Jerusalem 1985

The Israel Academy of Sciences and Humanities

Author's Address:
Department of Zoology
The George S. Weiss Faculty of Life Sciences
Tel Aviv University, Ramat-Aviv
69 978 Tel Aviv, Israel

ISBN 965–208–013–6
ISBN 965–208–059–4

Printed in Israel
at Keterpress Enterprises, Jerusalem

CONTENTS

PREFACE

THE FIRST STUDIES of the Acridoidea of North-East Egypt, Palestine and Syria were made during Napoleon's campaign in Egypt. The famous French scientist, J. C. Savigny, accompanied the Emperor in Palestine and Egypt in 1798–1801. The material collected by him included several species of grasshoppers from the Holy Land, described later as: *Eremippus savignyi* (Krauss), *Dociostaurus genei* (Ocskay), *D. maroccanus* (Thunberg), *Duroniella laticornis* (Krauss), *D. fracta* (Krauss), *Ramburiella truchmana* (Fischer-Waldheim), *Prionosthenus galericulatus* Stål, and *Sphingonotus angulatus* Uvarov.

At the end of the nineteenth century, several expeditions visited this region, most notably those of the botanist H. C. Hart in 1890–91, and of the zoologists Th. Barrois (in 1890) and E. Festa (in 1893). Their intensive collecting yielded a number of new species of Acridoidea, among them *Dociostaurus hauensteini* (Bolivar) and *Prionosthenus bethlemita* (Bolivar). Another important contribution was made by A. Kneucker. The material he collected in Palestine during 1904 was identified by H. A. Krauss (1909).

The most comprehensive publications on the taxonomy and ecology of the Acridoidea of Palestine are those of P. A. Buxton and B. P. Uvarov, extending from the early 1920's to the mid-1930's. R. Ebner collected specimens in Palestine in 1928. At about this time, F. S. Bodenheimer began his work. Bodenheimer was the founder of scientific entomology in Palestine, and later his paper "Ökologisch-Zoogeographische Untersuchungen über die Orthopterenfauna Palästinas" (1935) was the first and, to the present day, the only work to summarize the knowledge on the Acridoidea of Israel.

The research on Acridoidea in Israel is continuing at the Hebrew University of Jerusalem and Tel Aviv University, with emphasis on environmental biology, physiology, behaviour, endocrinology and genetics of the various genera. Some of the resulting publications are, e.g., by Shulov (1952), Pener & Shulov (1960), Fishelson (1960, 1969), Shulov & Pener (1961), Pener (1966), Broza & Pener (1969, 1972), Blondheim & Shulov (1972), Blondheim (1978), Orshan & Pener (1979), Pener & Orshan (1980), Pener et al. (1981).

All the Israeli material dealt with in this volume is deposited in the Entomological Collection of the Department of Zoology of Tel Aviv University, Ramat Aviv.

All figures have been drawn from material collected in Israel and Sinai.

The numbers in parentheses which appear after the locations refer to the geographical areas on the map of Israel and Sinai, at the end of the book. Spelling of names of localities in Israel and Sinai is according to the maps published by the Survey of Israel.

ACKNOWLEDGEMENTS

I am grateful to all those colleagues who placed their collections of Acridoidea at my disposal. I am particularly indebted to the following colleagues and friends who helped me with advice and the identification of material : Prof. H. Bytinski-Salz ; the late Prof. B. P. Uvarov ; the late Dr. V. M. Dirsh ; Dr. D. Hollis ; Prof. M. P. Pener ; Prof. A. Shulov ; and Prof. Y. Wahrman.

Acknowledgement is due to Mr. S. Schäfer for the fine illustrations and Mr. A. Shoob for photography.

Special thanks are due to Prof. O. Theodor and Prof. F. D. Por for their critical reading of the manuscript, and to Ms. I. Ferber for her thoughtful editorial work and preparation of the manuscript for publication.

The manuscript was edited for the printer by Mr. R. Amoils. The book was seen through the press by Mr. S. Reem, Director of the Publications Department of the Israel Academy of Sciences and Humanities.

INTRODUCTION

Morphology and Biology of Acridoidea

The superfamily Acridoidea forms a well-defined taxonomical unit of the order Orthoptera, with the following morphological characters: hind legs saltatorial, with strong femora; unsegmented short cerci at the posterior end of the abdomen; forewings (tegmina), if present, narrow, not folded, covering the triangular hind wings; pronotum covering the thorax dorsally and laterally; mandibles dentate; tympanal organ, if present, situated on the sides of the first abdominal segment.

External Morphology

Head: Strong, hypognathous with mouth parts directed ventrally, or opisthognathous with mouth parts directed ventrally and posteriorly; head inside view oval, acute, obtuse or oblique (Figs. 1, 2). Frons usually with a median vertical frontal ridge, which may be flat, with parallel or diverging margins, raised or obliterated (Fig. 3). Three simple ocelli on the frons: one in the middle of the frontal ridge, and two higher up, close to the margin of the compound eyes. A subocular groove extends from the eye to the clypeus, dividing the frons from the genae (Figs. 4, 5). The vertex is situated between and anterior to the eyes, being usually flat, horizontal, sometimes depressed and usually with lateral and small median carinae; two depressions (foveolae) usually present on the lateral carinae of the vertex, their form being triangular, oblong or quadrangular (Figs. 6–8). The occiput is usually swollen, smooth or with a pattern of small ridges. Antennae filiform, leaf-shaped or scalpeliform, usually shorter than half the body length.

Thorax: The thorax consists of pro-, meso- and metathorax. The prothorax is slightly separated from the mesothorax; the pronotum is large, its dorsal surface flat or roof-shaped, with or without two lateral carinae and a median longitudinal carina (Figs. 9, 10), and one, two or three transverse sulci or grooves: the first sulcus divides the pronotum into an anterior part, the prozona, and a posterior part, the metazona. The prosternal plate is slighly sunken, flat, or bears a spine or tubercle between the forelegs (Figs. 11, 12). According to Rowell (1961), this tubercle bears sensilla important for coordination of leg motion. Mesosternal and metasternal plates with their sclerites are separated only indistinctly by transverse sulci of varying form, and are important as taxonomic characters (Figs. 13, 14).

Wings: The wings are usually longer than the abdomen, rarely shorter or absent. Forewings are narrow, opaque and with dense venation. Hind wings large, transparent or with a colour pattern. The main veins of the wings are as follows (Figs. 15, 16): costa (C); subcosta (Sc); radius (R), usually with a radius-sector (Rs) which

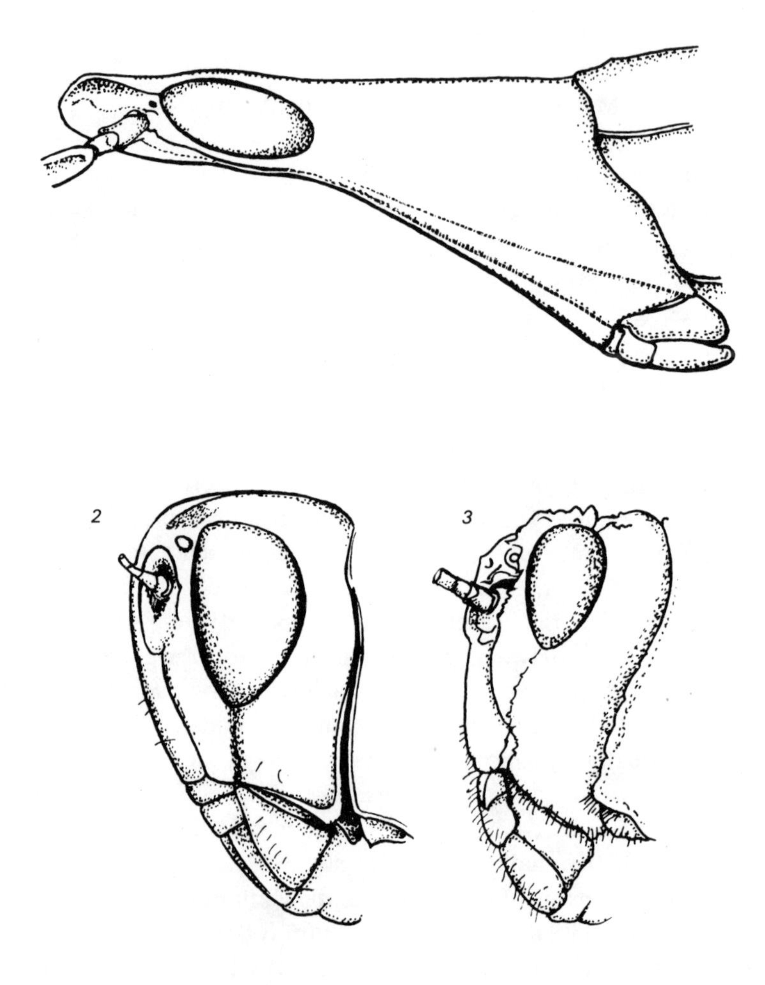

Figs. 1–3 : Head profiles
1. *Acrida bicolor* (acute) ; 2. *Eyprepocnemis plorans* (opisthognathous) ;
3. *Tmethis pulchripennis* (hypognathous)

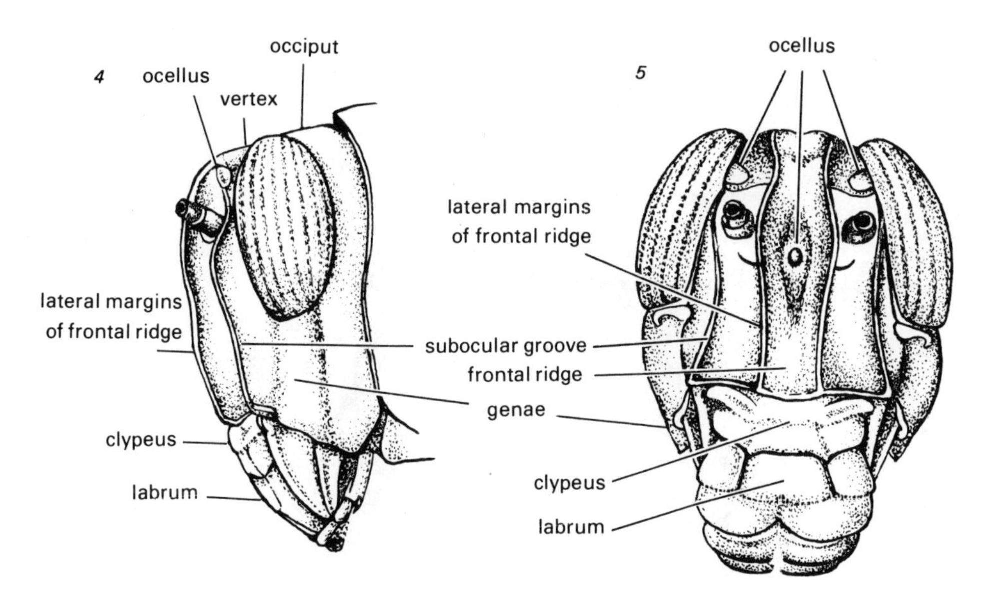

Figs. 4–5 : Head of *Anacridium aegyptium*
4. lateral ; 5. frontal

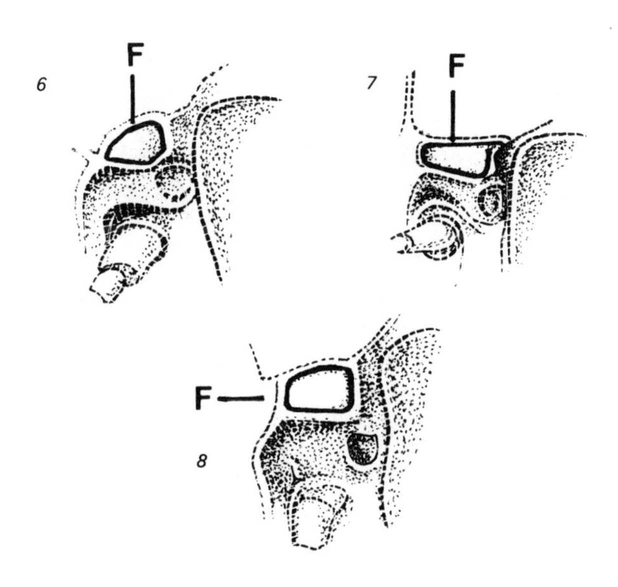

Figs. 6–8 : Foveolae (F) of vertex
6. triangular in *Oedipoda* ; 7. oblong in *Aiolopus* ; 8. square in *Notostaurus*

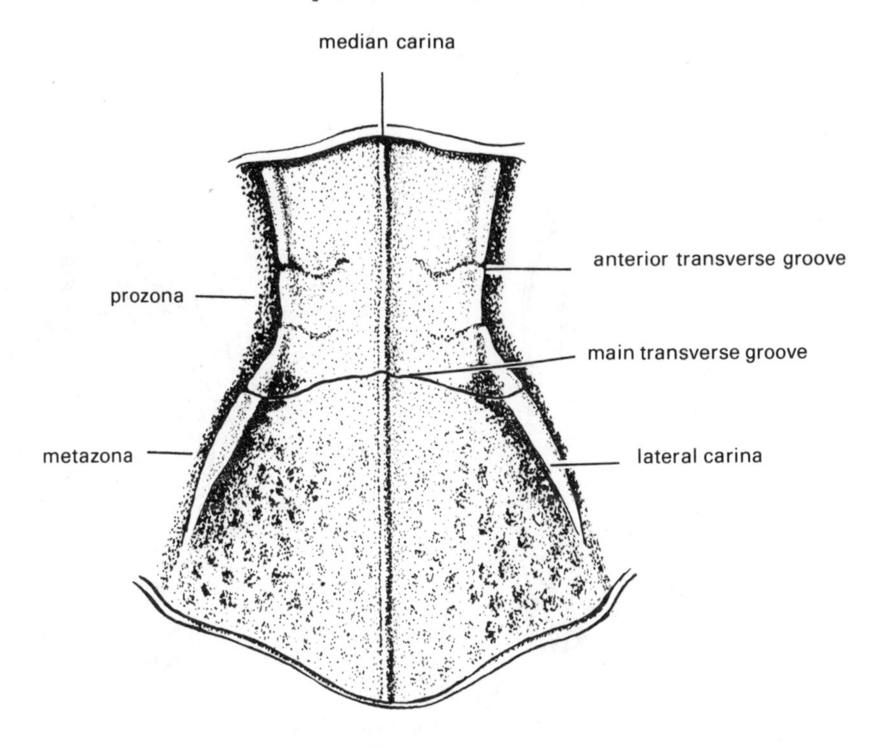

Fig. 9 : Pronotum of *Chorthippus peneri*, dorsal

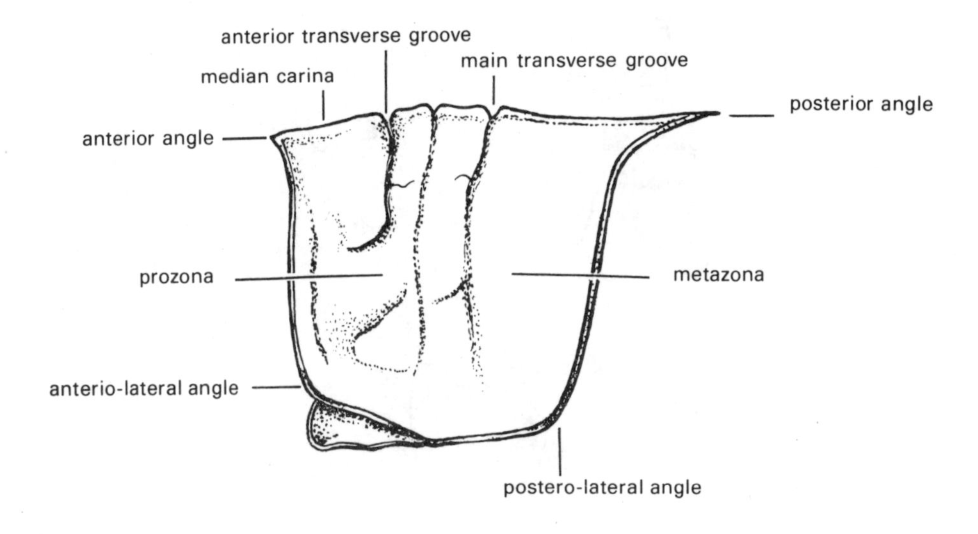

Fig. 10 : Pronotum of *Anacridium aegyptium*, lateral

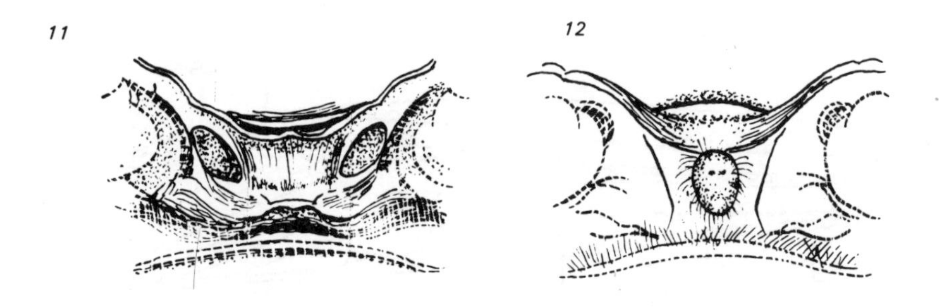

Figs. 11–12: Prosternal plates
11. without process *(Tmethis pulchripennis)*; 12. with a tubercle *(Anacridium aegyptium)*

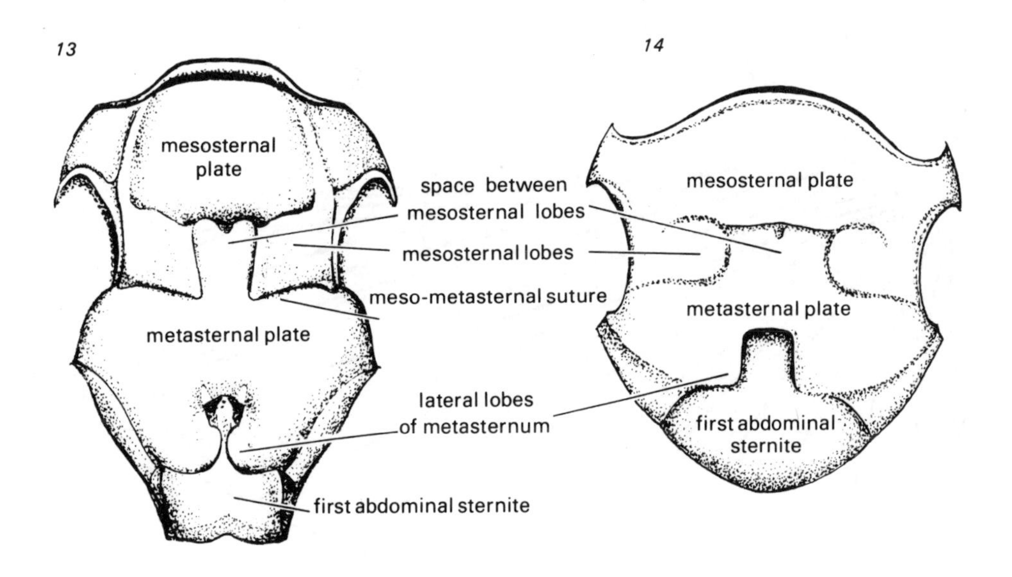

Figs. 13–14: Sternal plates
13. *Anacridium aegyptium*; 14. *Chorthippus peneri*

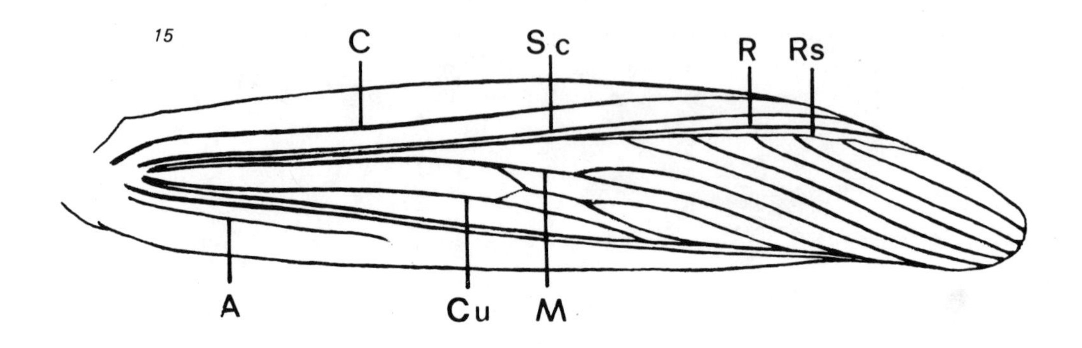

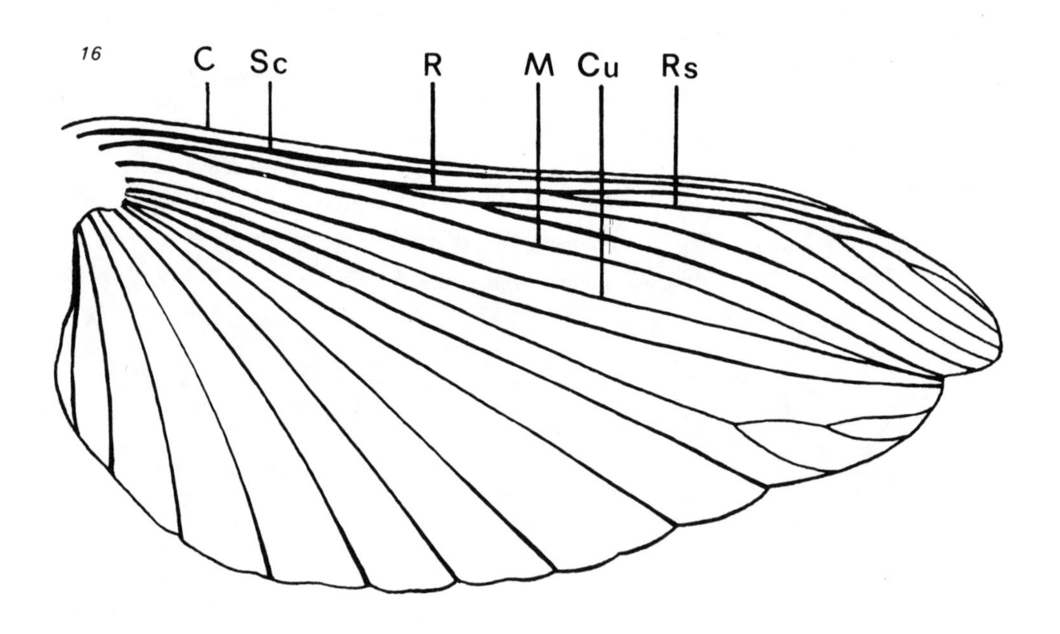

Figs. 15–16 : Wings of *Schistocerca gregaria* with the main veins
15. tegmen; 16. wing
C–costa ; Sc–subcosta ; R–radius ; Rs–radius-sector ;
A–anal ; Cu–cubitus ; M–media

forms the base of R_{2-5}; media (M); cubitus (Cu); and two anal veins (A). On the hind wings there are also jugal veins behind the anal veins, along which the wing-plate is folded (Fig. 16). These main veins divide the wings into the following fields : the precostal field — bordered by the costal vein ; the costal field — behind the costal vein; the subcostal field — bordered by the subcosta and radius; the radial field — between radius and media; the median field — between media and cubitus. The wing development is apparently phenotypically controlled in some species, and in such cases, e.g., in the genus *Cyclopternacris*, there are macropterous, normal forms in some geographical regions, and brachypterous populations in others.

Legs : The legs are simple; forelegs and mid legs are similar; the hind legs are large and saltatorial (Fig. 17). The femora are strong, their outer surface covered with two rows of symmetrical plates or with tuberculate ridges. Hind tibiae long, with two rows of spines at their dorsal margins, and with two pairs of movable spurs at the apex (Figs. 18, 19). Tarsus three-segmented, with a pair of claws, with or without an arolium.

Abdomen : Tergite 10 bears two non-segmented, curved or spine-like cerci in both sexes (Figs. 20–22). Sternite 9 of male forms a subgenital plate which encloses the genitalia (Fig. 21). The ovipositor of the female consists of six valves, of which four — two dorsal and two ventral — are strongly armoured and curved (Fig. 20), forming an effective organ for digging.

INTERNAL ANATOMY

Digestive System : The intestine forms a more or less uniform tube, extending from the buccal cavity to the anus. The stomodaeum is lined with a cuticle, usually with numerous rows of sharp denticles of specific form and pattern in the different genera. A strong gizzard at the end of the stomodaeum is separated from the midgut by six cardiac valves. The ventricle, which follows, bears six gastric coeca on the outer side, each divided into an anterior and posterior part. Posteriorly, the midgut bears numerous malpighian tubes and six pyloric valves. The proctodaeum is divided into three not distinctly separated parts. Salivary and rectal glands are also present.

Respiratory System : This very complicated system is represented by branched tracheae and air-sacs, which are connected with the outside by lateral spiracles, one pair on each segment. The spiracles are connected to the spiracular trachea, the central air-carrying duct. This tube has branches, the most important of which are those connecting the pair of dorsal tracheae that extend laterally along the heart, and the pair of ventral, abdominal tracheae extending along the intestine. All these tracts have numerous branches to the integument, intestine, neural cord and the fat body. The air-sacs are usually enlarged parts of the tracheae, and there are cephalic, thoracic and metathoracic sacs. The sacs in the abdomen are smaller but very numerous. All these air-sacs usually extend latero-dorsally and are connected by small tracheae. The number and distribution of the sacs vary in different species and they are apparently adapted to the environment. Strong flyers have large air-sacs. Recent studies (personal observation) have shown that typical desert forms also show the same pattern

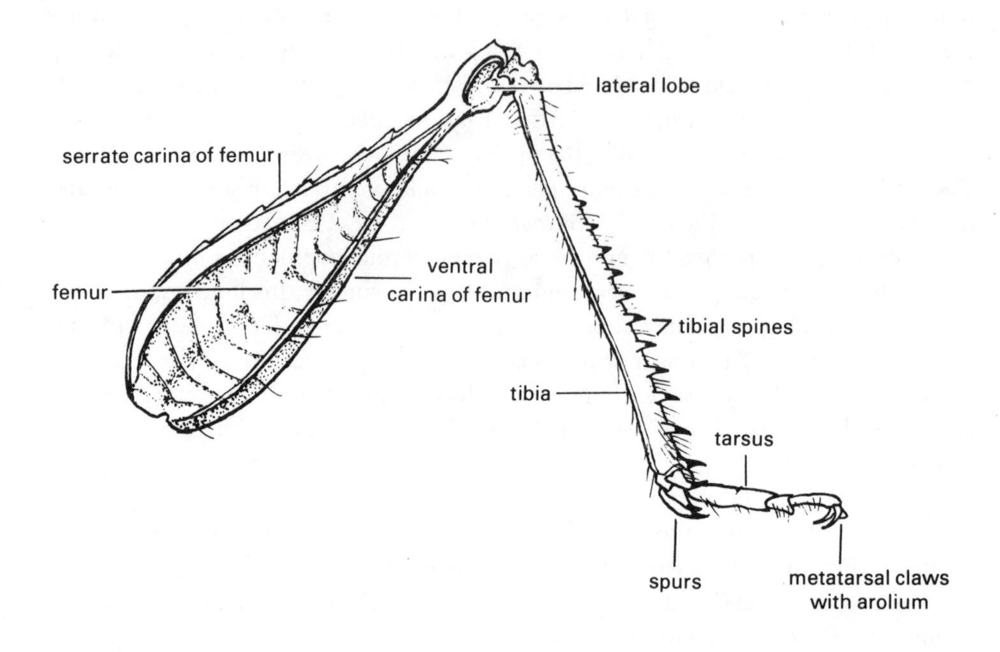

Fig. 17 : Hind leg of *Eyprepocnemis plorans*

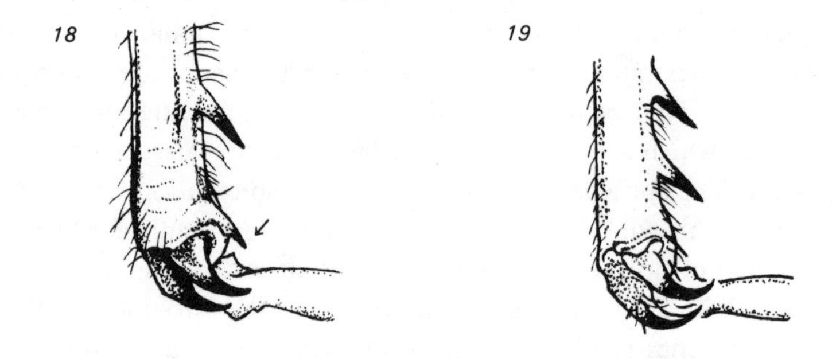

Figs. 18–19 : Distal part of tibia
18. with apical spine ; 19. without apical spine

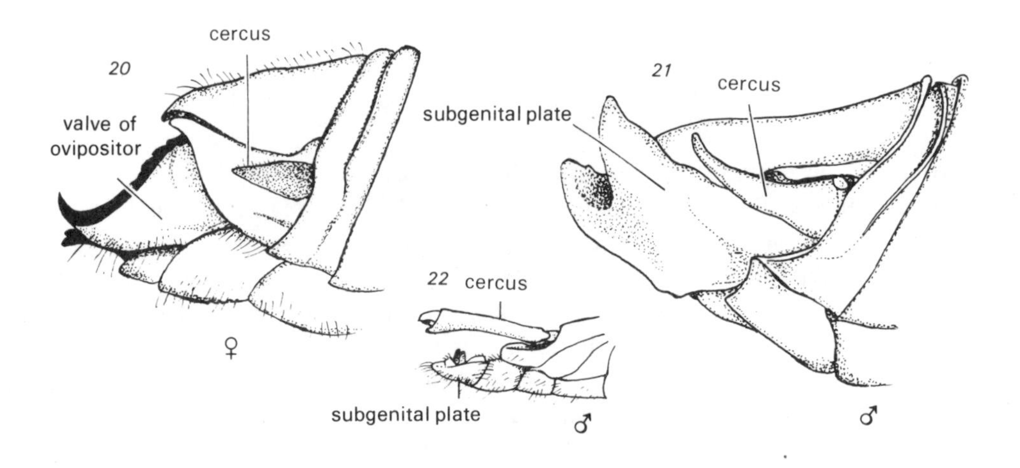

Figs. 20–22: Types of abdomen
20. *Anacridium aegyptium*, female; 21. *A. aegyptium*, male; 22. *Calliptamus barbarus*, male

of development of the air-sacs, even if they are wingless, e.g., *Acinipe zebratus.* This phenomenon is apparently in accordance with thermal regulation, which is very important for inhabitants of hot and dry deserts.

Circulatory System: The circulatory system consists of a contracting tube-like vessel attached to the dorsal part of the abdomen, the heart and two cephalic ampullae situated close to the base of the antennae. The haemolymph is usually colourless or slightly green and contains various types of amoebocytes.

Excretory System: The malpighian tubes, the main organs of excretion, are often numerous. They are also used to store vitamin B. The endocrine system is formed by a pair of corpora cardiaca situated above the oesophagus and a pair of corpora allata situated close to the fore intestine.

Reproductive System: Reproduction is usually sexual, but parthenogenesis has also been observed (e.g., in *Schistocerca gregaria*).

Male Genitalia: The male reproductive organs consist of an unpaired testis and two vasa deferentia which become united and open into an ejaculator. Attached to the end part of this duct are the accessory glands that form the spermatophore and the aedeagus. The form and structure of the genitalia are important taxonomically (Dirsh, 1957, 1973).

Female Genitalia: The female reproductive organs consist of two ovaries, two oviducts and an unpaired uterus with a tubular spermatheca. The number of ovarioles varies markedly, from 6–8 in small species to 140–180 in larger species (see also Waloff, 1950; Fishelson, 1960). The number of eggs which are produced by a female varies accordingly. *Eremogryllus hammadae*, a very small species, deposits 10–15 eggs in each oviposition, while the large *Poekilocerus bufonius* (Plate V : 6) or *Acinipe zebratus* deposit 150–200 eggs in one oviposition (Fishelson, unpublished).

Reproduction and Development

Fertilization takes place after copulation during which the male transfers a spermatophore into the female genital opening. The form and size of the spermatophore differ in various taxonomic units (Dirsh, 1957); it is usually bottle-shaped, the neck-part being introduced into the female genital opening. The reproductive behaviour begins with courtship, the most characteristic manifestation of which is stridulation, in most species produced by friction of the hind femur over the protruding veins of the forewings, or over a lateral plate of the abdominal tergite, e.g., in the genus *Tmethis*. The sounds produced are very specific and act as an intraspecific coding system. Many species of the genera *Sphingonotus, Morphacris* and *Chorthippus* may also produce sounds during flight. Receptive females react to these specific sounds by producing their own sounds or by approaching the stridulating male. In some species, e.g., *Sphingonotus rubescens*, the males approach the stridulating female, aggregating around her and performing a mating dance. In *Acrida bicolor*, the stridulating male remains in one place and the females move to him. Confronting a female, the male touches her with his antennae and, if she is receptive, he climbs upon her, clinging to her back by means of his claws. Copulation may last from several minutes to four or even eight hours. Oviposition begins shortly afterwards. The female sometimes begins to lay eggs with the male still in the copulating position.

Eggs: The eggs are elongate, 3–10 mm long. They are usually slightly curved, and covered by a thin, yellowish chorion which becomes dark brown some hours after oviposition. The outer surface of the chorion has a characteristic and species-specific structure of ridges and tubercles. Below the chorion, there is a vitelline membrane which protects the egg against dehydration and pressure. Micropyles, in groups or single, are situated at one end of the egg. During oviposition, the female inserts the ovipositor into the soil, sometimes, as in *Poekilocerus bufonius*, to a depth of 14–16 cm. The eggs are laid layer upon layer, each surrounded by proteinous foam secreted by the accessory glands. Particles of soil are cemented together by the drying foam and an egg-pod is formed. The egg-pod of most of the desert-inhabiting species or species with prolonged development is very strong. In species that develop without a diapause, e.g., *Schistocerca gregaria* and *Morphacris fasciata*, the foam only forms a delicate porous protecting cup (Plate V : 6). While extracting her abdomen from the soil, the female continues to produce foam and thus a column of foam is also formed between the eggs and the surface of the soil. Some species, e.g., of the genus *Docio-staurus*, cover the foam with a lid of soil. The eggs of forms with a prolonged diapause, e.g., *Acinipe zebratus* or *Ocneropsis bethlemita*, are well protected by the outer envelope. Especially thick walls are present in egg-pods that remain buried in the dry sand throughout the summer and from which the hoppers hatch only after the first rains.

Embryonic Development and Hatching : In most species embryonic development continues for six to 10 months, so that there is only one generation per year. An embryonic diapause has been described in species that hatch in the spring because of low temperatures in winter, and in species that hatch in autumn because of dehydration and high temperatures in summer (Shulov, 1952). Hatching from each egg-pod is synchronous and the hatched larvae immediately begin to make their way upwards through the dry foam cover. Experiments with different species have shown that hatching can be brought on by simply wetting the eggs. In *Poekilocerus bufonius*, the larvae hatch two hours after wetting, in *Acinipe zebratus* after four hours, and in *Leptopternis maculata* after one hour (Fishelson, 1960, and unpublished). This again shows that all the larvae in a single egg-pod reach a certain stage of development and that hatching is induced by an external factor (see also Shulov & Pener, 1961). Being negatively geotactic, the larvae always climb upwards, using the cervical ampullae as a digging organ. Upon reaching the surface the larvae moult, shedding the embryonic envelope, which forms a whitish aggregate, and the first instars emerge. These are usually 3–7 mm long, resembling the adult, except in size, in the less numerous antennal segments and in the absence of wings. The hoppers soon begin to feed and the integument hardens. Development usually passes through five moults, so that there are five stages which differ not only in size, but also in additional segments of the antennae and in the appearance and development of wings in the alate species (see also Dirsh, 1968). The different stages are identified as follows:

Key to Larval Stages

1. Primordia of wings absent; if present they are situated laterally on the thorax — 2
- Primordia of wings visible, situated dorsally behind the pronotum — 4
2. Antennae with 13 or 14 segments. Primordia of wings, if present, in the form of protuberances on the metanotum. — **Stage I**
- Antennae with 15–22 segments. Primordia of wings visible, on mesonotum and metanotum, with veins on them — 3
3. Antennae with 15–19 segments; only single veins on the primordia. — **Stage II**
- Antennae with 17–22 segments. Primordia of wings distinctly projecting, with several veins. — **Stage III**
4. Primordia of wings as long as or shorter than pronotum; inner pair (tegmina) much shorter than external (wings). Antennae with 20–22 segments. — **Stage IV**
- Primordia of wings as long as or longer than pronotum. Both pairs of primordia of more or less equal length. Antennae with 22–26 segments. — **Stage V**

In apterous species the differences between the stages are less distinct, and the number of segments of the antennae and the structures of head, pronotum and genital plates should be checked for each species. In hoppers of alate species the wing primordia are contiguous dorsally, whereas in adults of apterous species the wing rudiments are situated laterally and are not contiguous dorsally (Figs. 23, 24).

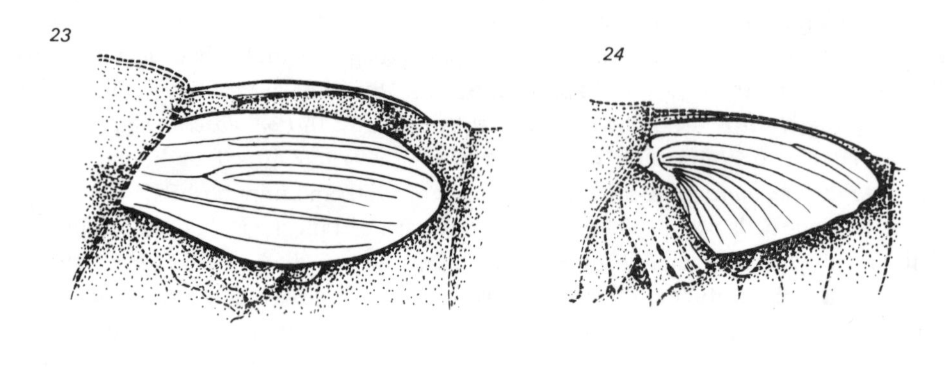

Figs. 23–24: Vestigial wings
23. wing rudiment of *Pareuprepocnemis syriaca*; 24. wing primordium of a hopper

Soon after the last moult, the insects are ready for reproduction. In general, the reproductive cycle in most species begins with oviposition towards winter, and hatching in spring. The adults then usually die. Specimens of some species, e.g., *Poekilocerus bufonius, Ochrilidia tibialis, Anacridium aegyptium, Ocneropsis bethlemita* and *Acrotylus insubricus*, may spend the winter as hoppers or adults.

Development is markedly influenced by environmental factors, especially temperature. Thus, for example, the development of *Poekilocerus bufonius* from oviposition to imago lasts three months at 30°C, whereas hoppers of the same species live for nine months without moulting at 20°C. The same was observed in *Prionosthenus galericulatus* (Fishelson, unpublished). The temperature control of the development plays a decisive role in the annual distribution pattern of the various species of grasshoppers (Fig. 25).

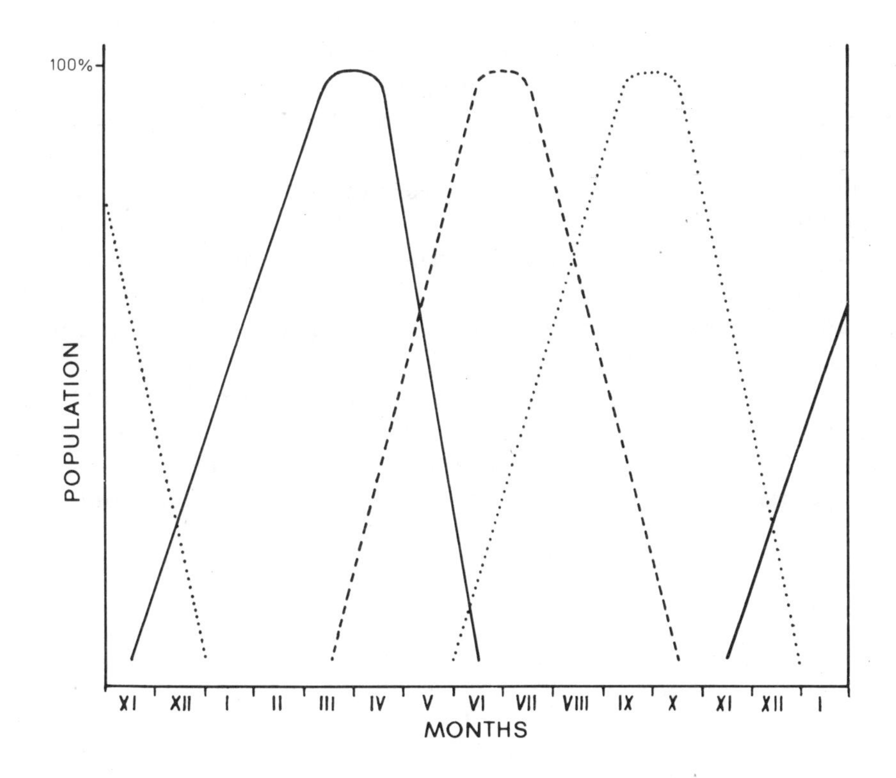

Fig. 25 : Annual cycles of three species of Acridoidea in one habitat (Tel Aviv)
——— *Duroniella laticornis*; – – – – *Leptopternis gracilis*; · · · · · · *Pyrgomorphella granosa*

Distribution and Composition of the Fauna

The fauna of Acridoidea of the eastern Mediterranean includes elements from faunas and regions presently contiguous with or which had connections with our region in the past. The fauna consists of Ethiopian, North African, Arabian, South-Eastern and North-Eastern Mediterranean, South-Eastern European, Transcaucasian, and Middle and South Asiatic elements, as well as endemic species. The fauna of Acridoidea occurring here consists mainly of xerophilous forms characteristic of semi-arid and arid regions. According to the definition of Uvarov (1948, 1977), there are groups of genera such as *Tmethis*, *Eremotmethis* and *Utubius* that form the 'Archeremic' stock, which is distributed from Africa to India. Other genera of the same group are found in Central-South America and Australia. Another group forms the 'Atlantidian' stock, of which the major genus, *Sphingonotus* (including 110 species), is found in the deserts of Asia, Africa and India, as well as Mexico, Galapagos and Australia, but not in South America (Mishchenko, 1936). The genus *Sphingonotus* is also very rich in species in our region. Yet another group of genera, probably of later origin, is that forming the 'Paneremic' stock, found in all deserts of the Old World. The typical genera in the fauna are those related to the genus *Sphingonotus*: *Leptopternis, Hyalorrhipis* and *Helioscirtus. Sphodromerus* and *Ochrilidia* also belong to this stock (Uvarov, 1977).

The Ethiopian stock includes forms distributed in the arid and subtropical savannas of East Africa, which also dominate the grasslands in Israel. Most numerous are species of the genera *Heteracris, Truxalis, Acrida, Pyrgomorpha* and *Stenohippus.*

The Mediterranean stock includes hygrophilous genera such as *Sphenophyma, Acinipe, Ocneropsis* and *Ramburiella.* The Transcaucasian fauna is represented here by the species of the genus *Chorthippus* and its relatives.

Species which entered the eastern Mediterranean from regions with a different climate show varying annual cycles of reproduction and occurrence, in accordance with their humidity and temperature requirements. This causes seasonal changes in the species-spectrum of the Acridoidea of this region (Fig. 26). In April and May, about

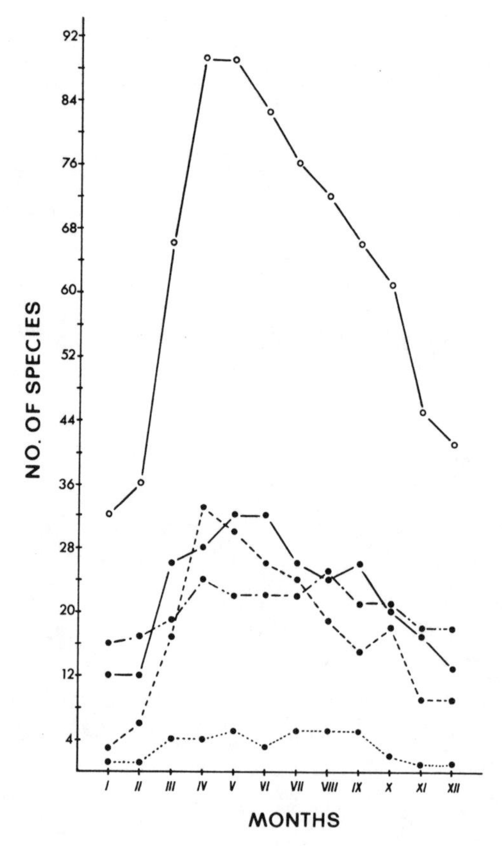

Fig. 26: Annual changes of the Acridoidea fauna in Israel, according to the faunal elements ······· Asiatic; ---- Afroeremic; ——— Mediterranean; —·— Ethiopian; ○—○ Total

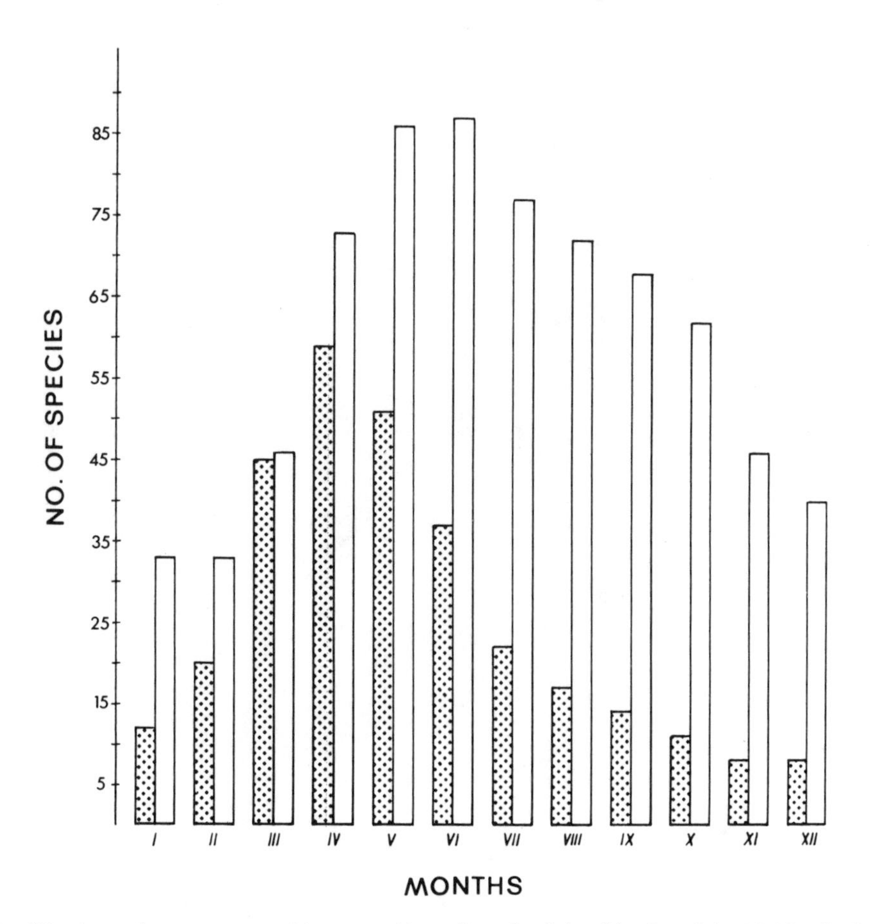

Fig. 27: Annual occurrence of hoppers (dotted) and adults (blank) of Acridoidea in Israel

90 species were collected, comprising about 85% of all the species known from our region. The frequent occurrence of hoppers in spring and the development of 74–86 species during the hot months of the year (Fig. 27), even in the northern part of the region, clearly point to the dominance of thermophilous forms in the fauna of Acridoidea.

Other important factors which influence the spectrum of grasshopper species are the geomorphology and the climatic conditions in specific habitats of the region, especially in the northern localities which are cooler and have a higher rainfall than the dry desert region.

Table 1 summarizes the distribution of acridoid species along a profile from the Mediterranean shore through the Coastal Plain (A) and up the Mount Carmel Ridge (C). This is represented schematically in Fig. 28. Three zones can be distinguished in this profile: (A) the Coastal Plain with various types of sandy soils; (B) the slopes of mountains covered with bushes and grasses forming the 'batha' and 'garigue'; and (C) the hilltops. Typical psammobiotic species (e.g., *Sphingonotus*

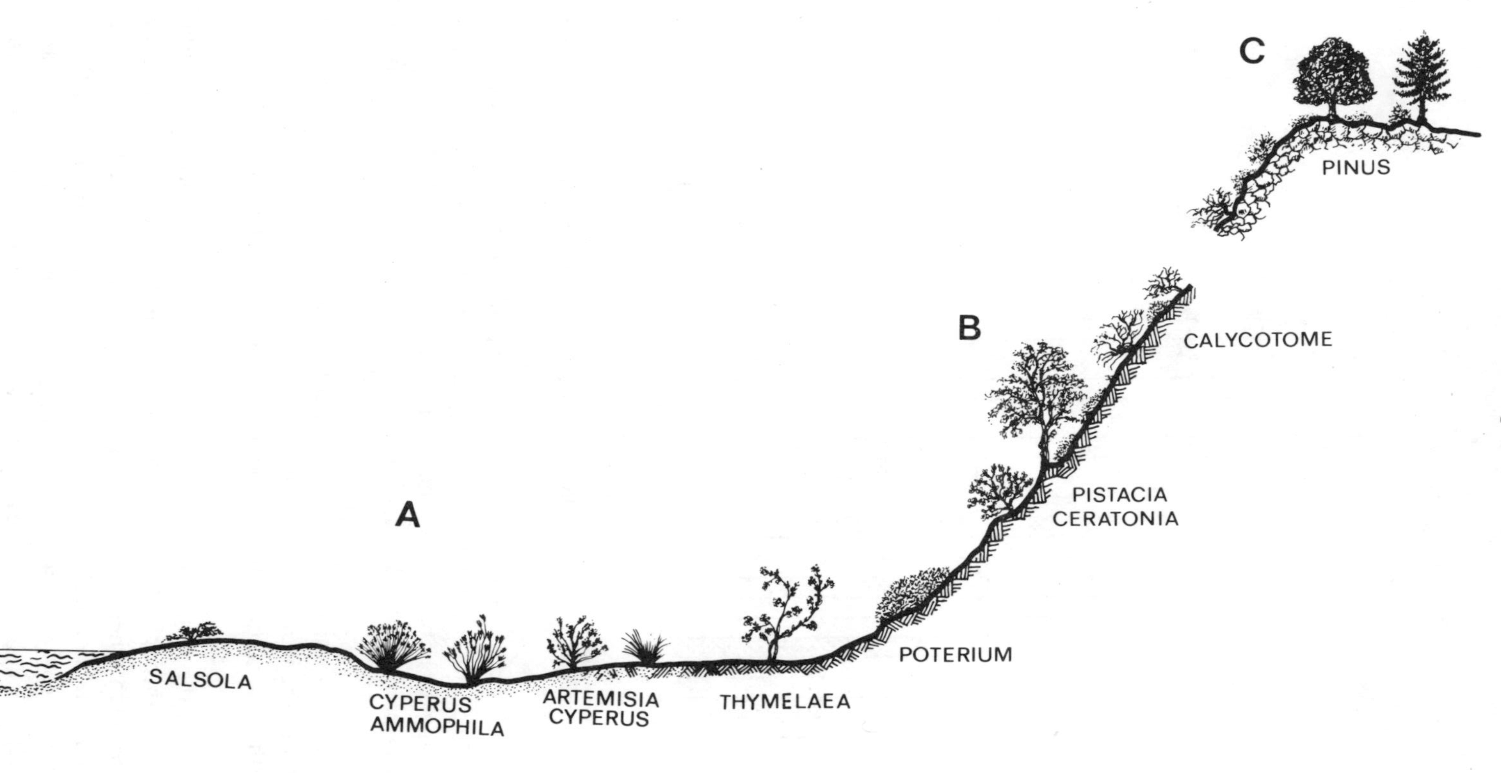

Fig. 28 : Geo-floristic schematic section from the shoreline to the top of Carmel Ridge
A — the Coastal Plain with various types of sandy soils
B — the slopes covered with bushes and grasses forming the 'batha' and 'garigue'
C — the hilltops (explanation in text)

angulatus, Ochrilidia geniculata and *Leptopternis maculata*) are found only on sand; other species (*Dociostaurus curvicercus, Pyrgomorphella granosa* and *Oedipoda miniata*) are found both on the plain and on the slope; while yet other species (*Pareupre-pocnemis syriaca, Pezotettix judaica, Chorthippus dorsatus* and *Acinipe hebraeus*) are encountered only on the highest part of the hills. Altogether 33 species were found along this profile, among which the geophilous forms constitute the minority (Table 2).

A similar profile in a more southern, arid habitat, e.g., in the Central Negev (17), features 28 species, of which 50% are geophilous. Deeper into the desert these species form the dominant population. In the mountains of Sinai along the Gulf of 'Aqaba (Elat), 11 of the 14 species collected were geophilous. Thus, from the north to the south, two phenomena are observed: a general drop in the number of species, together with a rise in the percentage of geophilous forms (Fig. 29). This is closely connected with the decrease in vegetation cover from north to south, which is the result of the decrease in rainfall.

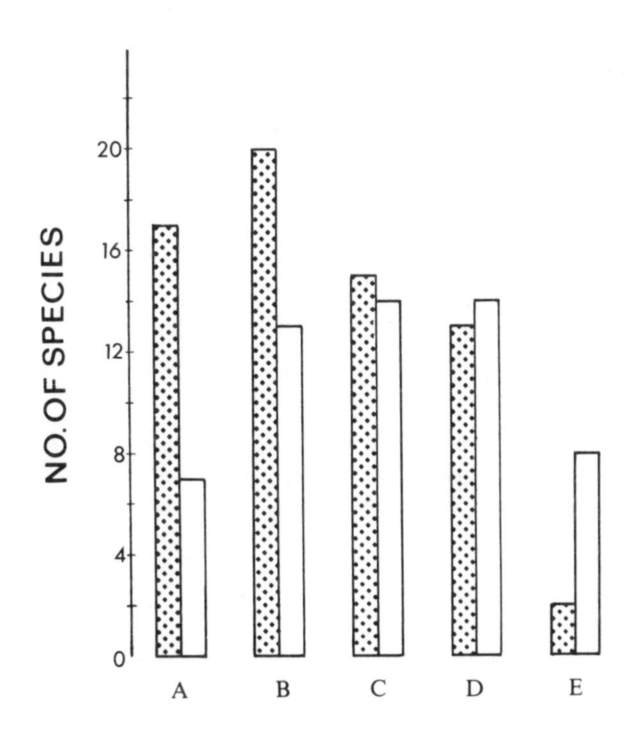

Fig. 29: Geophilous (blank) and phytophilous (dotted) Acridoidea in different localities
A — Upper Galilee B — Carmel Ridge
C — Central Negev D — Judean Hills
E — Southern Negev

TABLE 1
Distribution of grasshoppers along the profile illustrated in Fig. 28 (Mount Carmel)

Zone	*A*			*B*		*C*
	Dunes	*Sand*	*Hamra*	*Batha*	*Garigue*	*Hilltops*
Sphingonotus angulatus	+	+	−	−	−	−
Morphacris fasciata	−	+	+	+	+	−
Leptopternis gracilis	+	+	−	−	−	−
L. maculata	−	+	+	−	−	−
Acrotylus patruelis	−	+	+	−	−	−
Acrotylus insubricus	−	−	+	+	−	−
Calliptamus barbarus des.	−	+	+	−	−	−
C. palaestinensis	−	+	+	+	−	−
Oedipoda miniata	−	−	+	+	−	−
O. aurea	−	−	−	+	−	−
O. caerulescens	−	−	−	+	−	−
Sphenophyma rugulosa	−	−	−	+	+	−
Ochrilidia geniculata	+	+	+	−	−	−
Tropidopola longicornis	+	+	+	−	−	−
Heteracris littoralis	−	+	−	−	−	−
H. annulosus	−	+	+	+	+	−
Pyrgomorpha conica	−	+	+	+	+	−
Aiolopus strepens	−	+	+	−	−	−
A. thalassinus	−	−	+	+	+	+
Eyprepocnemis plorans	−	−	−	+	+	−
Pareuprepocnemis syriaca	−	−	−	−	+	+
Anacridium aegyptium	−	−	+	−	−	−
Acrida bicolor	−	−	+	+	−	−
Truxalis grandis	−	+	+	−	−	−
Pyrgomorphella granosa	−	−	+	+	−	−
Dociostaurus curvicercus	−	+	+	+	−	−
Notostaurus anatolicus	−	−	−	+	−	−
Pezotettix judaica	−	−	−	+	+	−
Prionosthenus galericulatus	−	−	−	+	+	+
Duroniella laticornis	−	−	−	+	+	+
Chorthippus dorsatus	−	−	−	+	+	+
Xerohippus savignyi	−	−	−	+	−	−
Acinipe hebraeus	−	−	−	−	+	+

TABLE 2
Number of phytophilous and geophilous forms along the profile illustrated in Fig. 28 (Mount Carmel)

Zone	*Geophilous*	*Phytophilous*	*Total number of species*
A	9	12	21
B	6	16	22
C	1	5	6

Ecological Associations of Grasshoppers

A habitat, even the simplest one, involves numerous factors, each of them acting on the organisms inhabiting it. Some of these factors, such as the physical and chemical parameters, are obvious and easy to detect. Others, such as the biological interactions among the neighbouring species, are more cryptic and difficult to estimate. The modelling of the physiognomy of grasshoppers found in various habitats seems to be controlled by two dominant parameters — the nature of the plant cover and the type of soil (substrate). In accordance with this, two main ecotypes are found among the grasshoppers in Israel: one is phytophilous and the other is geophilous. The phytophiles are grasshopper species that use the plants not only as a food-source, but also as a resting place and hideout. Their morphological adaptations to this habitat are: large arolia serving for attachment, streamlining and smooth body structures; heads at an acute or sub-acute angle (opisthognathous); long and delicate colourless wings; and usually green or greenish-grey or brown coloration. The behavioural adaptations are as follows: resting along branches or stems and, if flying up, moving from plant to plant. When disturbed, many species crawl around the stem or branch, remaining on the side opposite to the intruder. Such behaviour was observed in *Ochrilidia* spp., *Anacridium aegyptium* (colour plate: 1), *Tropidopola longicornis* (colour plate: 2), *Aiolopus simulatrix* (colour plate: 3), *Derycoris albidula* and *Pamphagulus bodenheimeri* (colour plate: 4). Other phytophiles, such as *Poekilocerus bufonius, Eyprepocnemis plorans, Duroniella lucasi, Heteracris* spp. and *Thisoicetrinus pterostichus*, drop from the plant, digging among the dense branches close to the ground.

The geophilous grasshoppers are inhabitants of the ground surface; for them plants serve as a food-source, but when not feeding they seek refuge on the bare or almost bare soil. Morphologically this group differs from the former one in the following features: regression of the arolia; stronger saltatorial legs; stronger and more granulate body; shorter wings, in many cases with various colours and markings;

head hypognathous and oblique in profile. Body colours usually match the colour of the soil on which they are living. This is especially prominent in *Eremotmethis carinatus* and *Tmethis pulchripennis* (colour plate : 7), in which the successive colours of nymphs are in accordance with the rocks on which the instars develop. Most Israeli grasshoppers belong to this group of species, so heavily represented in the arid parts by species of *Sphingonotus* and in northern Israel by *Oedipoda* spp. (colour plate : 5) and *Acrotylus* spp. (colour plate : 6).

With regard to behaviour, the geophilous forms tend, when disturbed, to remain *in situ*, crawling into a crevice or towards a stone, and only flying off at the 'last minute'. When at rest on the ground, they usually settle 'head to sun', thus casting the smallest shadow. The forms living on sand (*Hyalorrhipis calcarata* [colour plate : 8], *Leptopternis* spp., *Eremogryllus hammadae*) are also diggers, throwing the sand behind them with their hind legs. They are thus able to cover themselves, leaving only the upper part of the head and the antennae exposed above the surface. Such specialized adaptation of species from different taxonomic groups has produced many examples of convergent development, e.g., the typical psammobiotic forms : *Tenuitarsus angustus* (Pyrgomorphinae), *Hyalorrhipis calcarata* (Oedipodinae) and *Eremogryllus hammadae* (Acridinae).

Associations of Species

If faunas of Acridoidea are compared, the following associations of geophilous forms, which demonstrate a great affinity to the type of soil, are found:

Dunes : This habitat is formed by typical arid enclaves along the Mediterranean shore (4, 8, 9), ending in the north near the border of Lebanon. Further east, dunes penetrate into Central Sinai (21), the Central Negev (17) and the 'Arava Valley (14). This habitat, which is very extreme, has a very characteristic psammophilous population. On the dunes in the south, the typical species are *Tenuitarsus angustus* and *Hyalorrhipis calcarata* (Fig. 30). Along the shore and in the Central Negev, *Hyalorrhippis* is accompanied by *Eremogryllus hammadae* and *Leptopternis gracilis*, and by *Sphingonotus angulatus*. Near Haifa, *Sphingonotus angulatus* is found together with *Leptopternis maculata* and further north with *Sphingonotus eurasius*.

Mountains: From the south throughout the central part of Sinai (22, 21, 20), the mountains continue on both sides of the 'Arava Valley (14) and towards the north and in the Galilee (1, 2), on the Golan Heights (18) and Mount Hermon (19). The southern part of this habitat, which almost reaches the northern end of the Dead Sea (13), is very dry, mainly bare or with very sparse vegetation. The typical species here are *Scintharista notabilis, Cyclopternacris cincticollis* and species of the genera *Sphingonotus* and *Sphodromerus* (Fig. 31). Of the species occurring in the south, only *Sphingonotus rubescens* continues northwards, penetrating into regions which receive more rain and have a richer vegetation. The dominance of brachypterous forms begins here : species of *Pezotettix, Pareuprepocnemis syriaca, Acinipe hebraeus,*

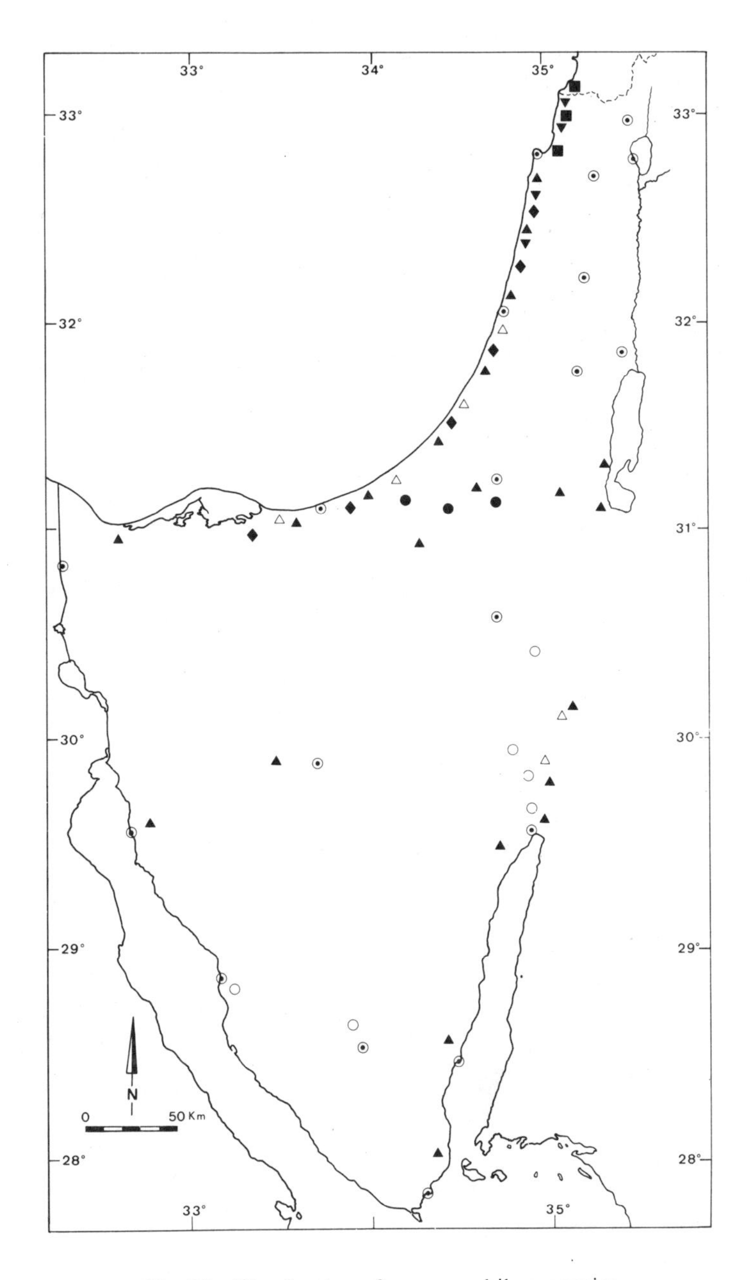

Fig. 30 : Distribution of psammophilous species

▲ *Leptopternis gracilis*
△ *Hyalorrhipis calcarata*
▼ *Leptopternis maculata*
○ *Tenuitarsus angustus*
● *Eremogryllus hammadae*
■ *Sphingonotus eurasius*
◆ *Sphingonotus angulatus*

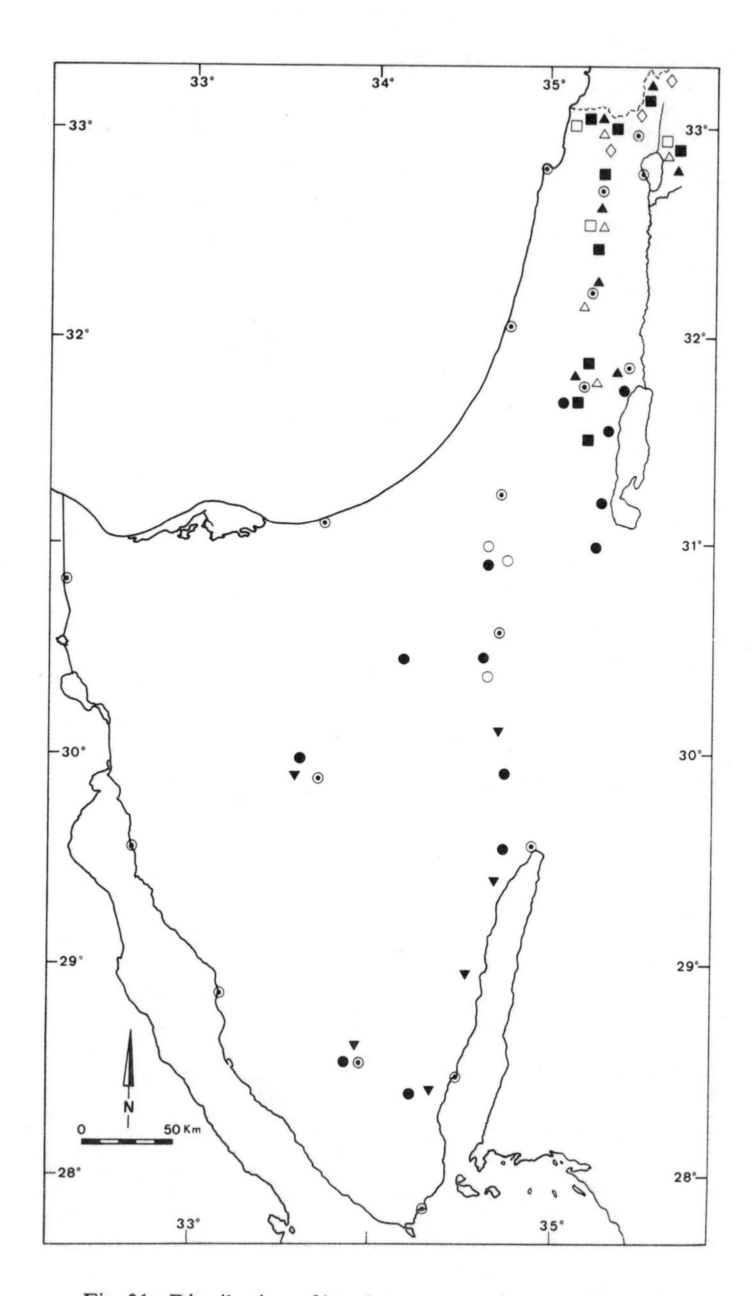

Fig. 31 : Distribution of brachypterous and mountain species

●	*Sphodromerus pilipes*	▼	*Scintharista notabilis*
○	*Acinipe zebratus*	■	*Prionosthenus galericulatus*
▲	*Pezotettix judaica*	△	*Pareuprepocnemis syriaca*
□	*Acinipe hebraeus*	◇	*Sphenophyma rugulosa*

Sphenophyma rugulosa — ending on the Golan Heights with *Ocneropsis kneuckeri, O. lividipes, Notostaurus cephalotes* and *Chorthipus dirshi.*

Coastal Plain : This consists of the area between the sand dunes influenced by winds and sea, and the mountain slopes. It is cultivated in the greater part and was also cultivated in the past. In the southern part of the region, the most common species are those of *Sphingonotus.* In Northern Sinai (20), the Gaza Strip (9) and the western part of the Negev (15), in areas with an annual rainfall exceeding 100 mm, *Acrotylus insubricus* predominates, accompanied in the north by *Acrotylus patruelis, Oedipoda miniata* and species of *Calliptamus.*

Phytophilous species form other types of associations. In this case there is a strong connection between special types of vegetation and their fauna of Acridoidea.

The *Tropidopola–Ochrilidia* association is found on *Phragmites, Typha* and *Juncus.* Various species of these plants occur from the north to the south and with them a typical group of Acridoidea. Thus, in the northern part of the Jordan Valley (7) and along the streams in the Golan Heights (18), the most common species are *Locusta migratoria, Anacridium aegyptium, Eyprepocnemis plorans* and *Tropidopola longicornis.* The same plants in the Coastal Plain (4, 8, 9) and the Sinai Peninsula (20, 21, 22) are inhabited by *Ochrilidia tibialis, Tropidopola longicornis, Eyprepocnemis plorans* and *Aiolopus thalassinus.* Towards the hotter, more arid parts of the region the dominance of the genus *Ochrilidia* increases (Fig. 32). It is represented by several species on all the plants of the genera *Juncus* and *Ammophila. Tropidopola longicornis* penetrates into the desert along the Dead Sea (13) solonchaks and into oases in the Sinai Peninsula (20, 21, 22).

The *Heteracris–Eyprepocnemis* association shows a great affinity for bushes in the northern part of the region, especially for *Polygonum microphyllum* and *Artemisia monosperma*, and for *Retama rotem, Artemisia herba-alba* and *Haloxylon* spp. in the southern part. In the north, the most common species of this group in the Jordan Valley (7) are *Thisoicetrinus pterostichus, Heteracris littoralis* and *Eyprepocnemis plorans.* Towards the Coastal Plain the group is accompanied by *H. adspersus* and *H. annulosus.* Of these species, *H. littoralis* and *H. annulosus* are the typical bush-dwelling grasshoppers in the Coastal Plain, accompanied by *Pyrgomorpha conica.* The bushes of the Negev desert and the Sinai Peninsula are mainly inhabited by *H. annulosus* and *Pamphagulus bodenheimeri* (Fig. 32).

The *Dociostaurus–Morphacris* association is found in summer mainly in areas with low, usually dry, ephemeral grasses. Of these species, *Morphacris fasciata* is the most common, its distribution extending from the Syrian Plateau to the southern desert areas of Israel, where it is sometimes accompanied by *Mioscirtus wagneri.*

Other members of this association are species of the *Dociostaurus* group, of which the brachypterous forms, e.g., *Notostaurus cephalotes* and *Dociostaurus hauensteini*, are characteristic of highlands in the north. On slopes and drier places they are accompanied by *D. genei.* Towards the Coastal Plain the dominant species in these grass habitats are *Notostaurus anatolicus* in the north and *Dociostaurus curvicercus, D. genei* and species of Egnatiinae in the south.

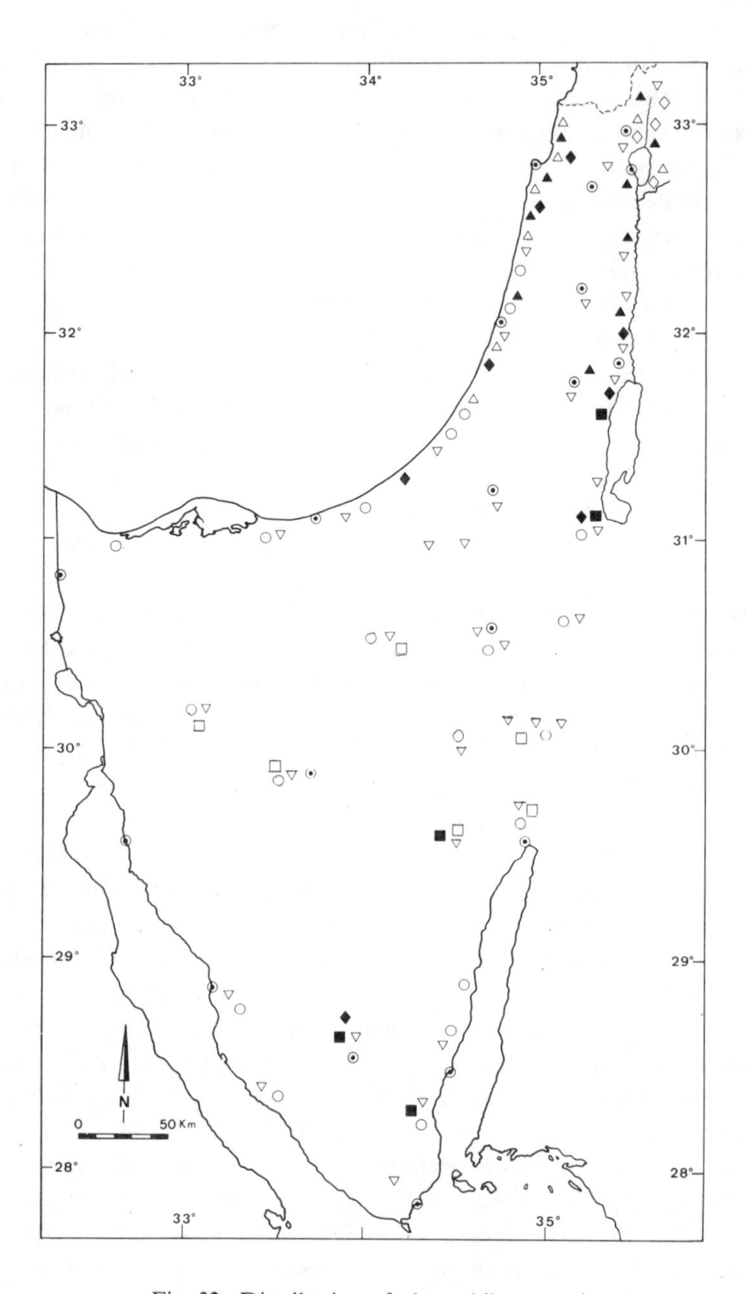

Fig. 32 : Distribution of phytophilous species

▽ *Heteracris annulosus*
▲ *Eyprepocnemis plorans*
□ *Pamphagulus bodenheimeri*
◆ *Tropidopola longicornis*
△ *Heteracris littoralis*
■ *Dericorys millierei*
○ *Ochrilidia* spp.
◇ *Thisoicetrinus pterostichus*

SYSTEMATIC PART

TAXONOMY

The publication of Ander (1949) established the basic taxonomy of the Acridoidea and they are considered as a superfamily of the suborder Caelifera. Later came the revisions of Bei-Bienko & Mishchenko (1951), Beier (1955) and Dirsh (1961). The most recent revisions of Acridoidea were published by Dirsh (1974) and Jago (1977). Dirsh, like several authors in the past, raises the taxonomic rank of various units, using few and very small differences in structure of sclerites, mostly genital ones. The dangers of such an approach can be exemplified by the fact that, according to Dirsh, the genera *Acrida* and *Truxalis* belong to two different subfamilies, notwithstanding the fact that it is very difficult to separate these genera on the basis of their morphology and anatomy. The classification of Bei-Bienko & Mishchenko (1951) and Mishchenko (1952) is used here, not because it is the most natural, but because it is the simplest and most useful, with the minimum pretension to a phylogenetic clarification. At present, 1,527 genera of Acridoidea are known, and their adaptability is so high that the use of external or internal anatomical characters for phylogenetic hypotheses and evolutionary analysis can only result in more confusion.

According to the definition of Bei-Bienko & Mishchenko (1951), the Acridoidea are a superfamily which contains two families: the Tetrigidae and the Acrididae. Both families are divided into subfamilies.

The synonymy includes selected references to publications dealing with species from Israel and surrounding regions. Full synonymy in Johnston (1956; 1968) and Bei-Bienko & Mishchenko (1951).

Key to the Families of Acridoidea in Israel and Sinai

1. Pronotum very long, covering most or all abdominal segments (Fig. 34). **Tetrigidae**
– Pronotum short, covering only the thorax and bases of wings (Fig. 36). **Acrididae**

Family TETRIGIDAE

Diagnosis: Small species, body usually narrowing posteriorly.

Head: Hypognathous or slightly opisthognathous. Frontal ridge below median ocellus, with a single median carina.

Thorax: Pronotum long, with a posterior process which covers the entire abdomen (Fig. 33). Tegmina rudimentary, strongly abbreviate, forming platelets.

Wings: Usually fully developed, long, covered by process of pronotum.

Legs: Hind femur with Brunner's organ; tympanal organ absent. Front and mid tarsi with two segments, hind tarsus with three segments; arolium absent.

Distribution: Numerous species are found in tropical and subtropical areas; in humid places, among low and dense vegetation.

Orthoptera: Acridoidea

Some authors (Ander, 1949; Beier, 1955; Dirsh, 1961) consider the Tetrigidae as a suborder Tetrigoidea, that includes also the families Tridactylidae and Cylindrachetidae.

Key to the Genera of Tetrigidae in Israel and Sinai

1. Vertex much wider than diameter of eye. Median carina of pronotum roof-shaped anteriorly (Fig. 33). **Tetrix** Latreille
- Vertex narrower than or as wide as eye. Median carina of pronotum linear, only slightly raised (Fig. 34). **Paratettix** Bolivar

Genus TETRIX Latreille, 1802

Histoire naturelle générale et particulière des Crustacées et des Insectes. Orthoptera, Acrididae, 3: 284.

Acrydium Geoffrey E. L., 1762, *Histoire abrégée des Insectes qui se trouvent aux Environs de Paris*, 1: 390.
Bulla Schrank H., 1781, *Enumeratio Insectorum Austriae Indigenorum,* p. 242.
Acridium Schrank H., 1801, *Fauna Boica*, Nürnberg, 2: 32.
Tettix Charpentier T. de, 1841, *Germars Z. Ent.*, 3: 315.

Type Species: *Tetrix subulata* (Linnaeus, 1761).
Diagnosis: Body usually rugose. Head moderately hypognathous, lower than the median carina of pronotum. Frontal ridge low, projecting only between antennae. Margins of vertex raised, vertex wider than eye. Antennae inserted below ventral margin of eye, not reaching margins of lateral lobes of pronotum posteriorly. Median carina of pronotum partly roof-shaped (Fig. 33). Wings covered entirely by pronotum.
Distribution: Throughout the world, also in temperate regions.
Two species in Israel.

Key to the Species of Tetrix in Israel

1. Median carina of pronotum continuing posteriorly without becoming lower; surface of pronotum more or less smooth. **T. subulata** (Linnaeus)
- Median carina of pronotum roof-shaped anteriorly, distinctly lower above and between hind legs; surface of pronotum rugose (Fig. 33). **T. depressa** (Brisout)

Tetrix subulata (Linnaeus, 1761)

Type Locality : 'Europe'.

Gryllus subulatus Linnaeus C., 1761, *Fauna Svecica*, 2nd ed., p. 236.
Tetrix subulata —. Finot A., 1890, *Faune de France, Insects Orthoptères*, Paris, p. 167, pl. 8, fig. 116.
Tettix subulatus —. Zacher F., 1917, *Die Geradflügler Deutschlands und ihre Verbreitung*, Berlin, p. 267.
Acridium subulatum —. Buxton P. A. & B. P. Uvarov, 1923, *Bull. Soc. R. ent. Égypte*, p. 207.
Acrydium subulatum —. Bodenheimer F. S., 1935, *Arch. Naturgesch.*, 4(2) : 209.
Tetrix subulata —. Ramme W., 1951, *Mitt. zool. Mus. Berl.*, 27 : 383.

Robust, short. Head hypognathous. Frontal ridge usually straight ; vertex obtuse or rounded apically, dorsally smooth or granulate. Pronotum with prozona narrower than metazona, rugose laterally, almost smooth dorsally ; median carina strongly raised anteriorly, gradually lowered posteriorly. Wings entirely covered by pronotum. Hind femur narrow, its maximum length more than 3.0–3.2 times its width.
Coloration : 'Soil-like' with pale or dark dorsal parts.
Measurements (mm) : Body ♀ 14.2, ♂ 11.0; hind femur ♀ 6.6, ♂ 5.0.
Distribution : Siberia to Central Asia, Central Europe, North America.
Israel : Upper Galilee (1), Central Coastal Plain (8).

Tetrix depressa (Brisout, 1848)
Fig. 33

Type Locality : 'Syria'.

Tettix depressa Brisout de Barneville, L., 1848, *Annls Soc. ent. Fr.* (2), 6 : 424.
Tetrix acuminata Brisout de Barneville, L., 1851, *ibid.* (2), 8 : 113
Tettix charpentieri Fieber F. X., 1853, *Lotos*, 3 : 141.
Tetrix elevata Fieber F. X., 1853, *ibid.*, 3 : 144.
Tettix rudis Walker F., 1871, *Catalogue of the specimens of Dermaptera Saltatoria in the Collection of the British Museum*, London, Suppl., Part V, p. 87.
Acrydium depressum —. Chopard L., 1922, *Faune de France, 3. Orthoptères et Dermaptères*, Paris, p. 138, fig. 318.
Tetrix depressa —. Ramme W., 1951, *Mitt. zool. Mus. Berl.*, 27 : 383.

Body depressed, short, rugose. Head almost hypognathous. Face densely tuberculate ; frontal ridge slightly projecting between bases of antennae. Interocular space 1.5 times as wide as eye. Vertex almost straight apically, projecting in middle, its median carina raised, continuing posteriorly to carina of pronotum. Pronotum very rugose, carinate anteriorly, constricted laterally ; median carina roof-shaped to the middle, distinctly lower posteriorly (Fig. 33). Prosternal collar entirely covering the mouth parts. Wings much shorter than pronotum. Hind femur 2.5 times as long as wide. Dorsal and ventral margins of mid and hind femurs undulate. Hind tibia with five spines on the outer line

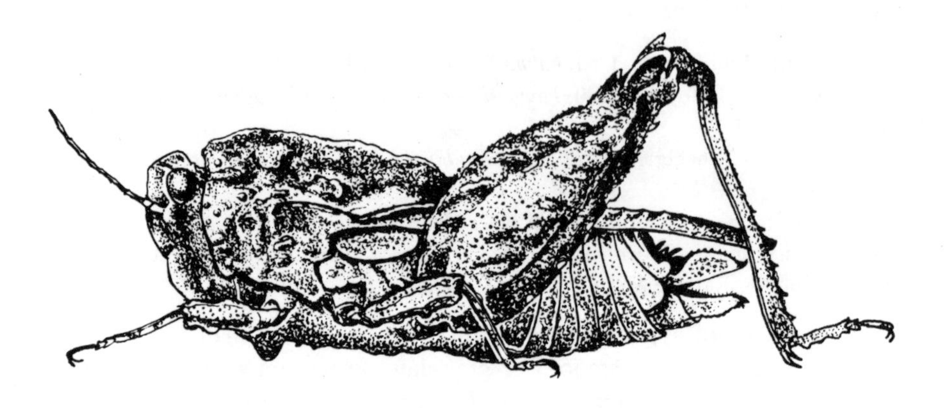

Fig. 33 : *Tetrix depressa* (Brisout, 1848), female

and minute dense dentations. Arolium absent on hind tarsus.
Coloration : Dark grey, with light fields on prozona.
Measurements (mm) : Body ♀ 9.5 ; hind femur ♀ 6.0.
Distribution : Crimea, Caucasus, Central Asia, Afghanistan, North and West Iran, Southern Europe, North Africa.
Israel : Lower Galilee (2; Tiv'on, 16 March, 1954; among dense vegetation).

Genus PARATETTIX Bolivar, 1887

Annls Soc. r. ent. Belg., 31 : 270

Type Species : *Paratettix meridionalis* (Rambur, 1838).
Diagnosis : Slender. Head hypognathous, slightly projecting above pronotum ; vertex narrower than or as wide as eye ; with or without median carina. Antennae stout, inserted below eye margins or between them. Pronotum compressed anteriorly, extending posteriorly beyond middle of hind tibia ; median carina extending to posterior margin. Prozona with short lateral carinae anteriorly ; ventral posterior part of lateral lobes of pronotum raised laterally. Wings extending beyond apex of pronotum (Fig. 34).
Distribution : Tropics of the Old World.
Two species in Israel.

Paratettix

Key to the Species of Paratettix in Israel

1. Interocular distance much narrower than vertical diameter of eye. Antenna inserted at level of ventral margin of eye (Fig. 34). **P. meridionalis** (Rambur)
– Interocular distance as long as or slightly longer than vertical diameter of eye. Antennae inserted distinctly below ventral margin of eye. **P. ocellatus** Uvarov

Paratettix meridionalis (Rambur, 1838)

Figs. 34, 35

Type Locality : 'Europe'.

Tetrix meridionalis Rambur J. P., 1838, *Faune entomologique de l'Andalousie. Orthoptera*, 2 : 65.
Tettix dohrnii Fieber F. X., 1853, *Lotos*, 3 : 142.
Tettix ophthalmica Fieber F. X., 1853, *ibid.*, 3 : 145.
Paratettix meridionalis —. Finot A., 1890, *Faune de France, Insectes Orthoptères*, Paris, p. 170.
Paratettix meridionalis —. Buxton P. A. & B. P. Uvarov, 1923, *Bull. Soc. R. ent. Égypte*, p. 207.
Paratettix meridionalis —. Bodenheimer F. S., 1935, *Arch. Naturgesch.*, 4(2) : 209.
Paratettix meridionalis —. Bodenheimer F. S., 1935, *Animal Life in Palestine*, Jerusalem, p. 321.
Paratettix meridionalis —. Ramme W., 1951, *Mit. zool. Mus. Berl.*, 27 : 423.

Slender, smooth or with dense minute tubercles. Head hypognathous or, in males, slightly opisthognathous ; face with numerous tubercles, wider ventrally . Frontal ridge convex to the median ocellus, then depressed, and continuing ventrally in a carina. Vertex not projecting before eyes, much narrower than diameter of eye, concave, sparsely granulate ; median carina present ; length of antennal segments 3–5 times their width. Pronotum with three carinae on prozona, the median one low or obliterated anteriorly ; median carina on metazona distinct along its entire length ; lateral carinae diverging anteriorly, gradually converging posteriorly ; pronotum usually covered with minute dentations. Margins of front and mid femurs undulate.
Coloration : Brown-grey or dark grey, soil-like, sometimes pale with a prominent black transverse band on the level of articulation of hind legs.
Measurements (mm) : Body ♀ 9.0–12.5, ♂ 6.5–8.5 ; head and pronotum ♀ 11.5–14.5, ♂ 10.0–15.5 ; hind femur ♀ 5.0–5.7, ♂ 4.3–5.0.
Distribution : Mediterranean countries.
Israel : Upper and Lower Galilee (1, 2), Jordan Valley (7), Coastal Plain (4, 8, 9), Judean Hills (11), Northern Negev (15).
Dense populations are found in Israel in irrigated fields of the Central Coastal Plain (8), often 50 specimens per square metre. Among the samples collected throughout the year were copulating adults, females laying eggs and hoppers at various stages of development. Specimens were seen jumping into the water, swimming expertly, 'paddling' with their hind legs.

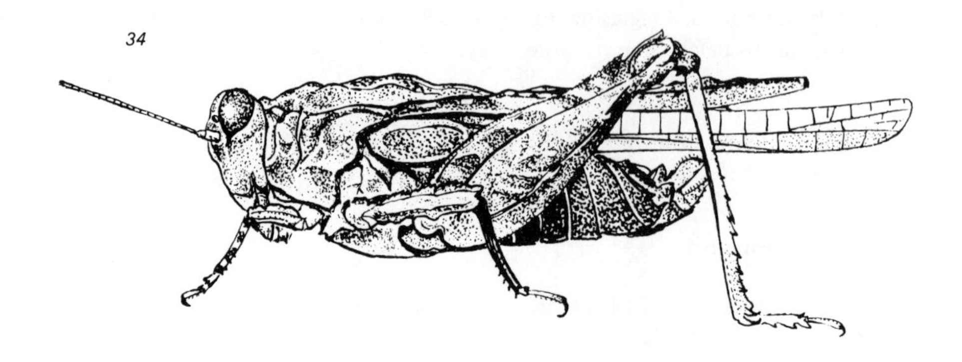

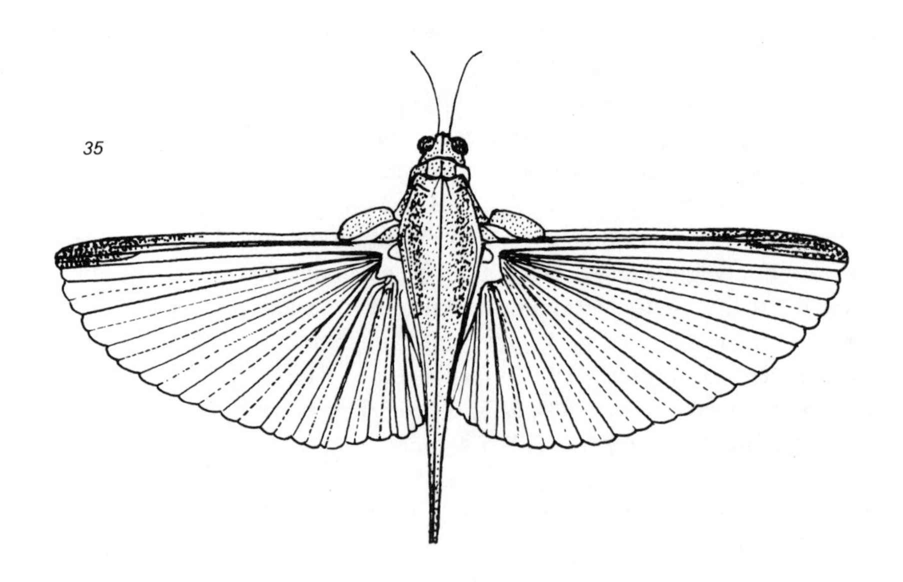

Figs. 34–35 : *Paratettix meridionalis* (Rambur, 1839)
34. female ; 35. wings open

Paratettix ocellatus Uvarov, 1936

Type Locality : 'Oman'.

Paratettix ocellatus Uvarov B. P., 1936, *J. Linn. Soc.* (Zool.), 39(268) : 522.

Rugose, especially laterally ; antennae inserted below ventral margins of eyes ; length of segments 2–3 times their width. Vertex with raised median carina, its width equal to vertical eye diameter. Pronotum narrow anteriorly, with sharply raised carinae, its dorsal surface slightly concave, with numerous dense transverse carinulae and rugulae posteriorly; median carina distinctly lower beyond articulation of hind legs. External surface of femur rugose ; dorsal and ventral margins undulate.
Coloration : Brownish-grey with two black spots, forming a transverse band at the point where median carina of pronotum becomes lower.
Measurements (mm) : Body ♀ 7.5–8.5 ; head and pronotum ♀ 10.5–11.8 ; hind femur ♀ 4.5–4.8.
Distribution : Oman, East and North Africa, Eastern Mediterranean.
Israel : Golan Heights (8, near water bodies).

Family ACRIDIDAE

Key to the Subfamilies of Acrididae in Israel and Sinai

1. Prosternum with a conical process or a sharp spine between the forelegs (Fig. 51), or anterior margin of prosternum raised, forming a collar-like extension ventrally which covers the mouth parts (Fig. 91) — 2
– Prosternum without process or spine between forelegs — 4
2. Prosternal process conical (Fig. 71); hind femur serrate dorsally (Fig. 59), its external median field covered by two rows of symmetrical plates. — **Catantopinae**
– Prosternal process sharp ; external median field of hind femur granulate or with irregular lines — 3
3. Head opisthognathous, frons at an acute angle ; apical fastigial areas present ; foveolae usually situated close together or contiguous (Fig. 85). — **Pyrgomorphinae**
– Head hypognathous or slightly prognathous ; apical fastigial areas absent ; foveolae distant from each other (Fig. 99). — **Pamphaginae**
4. Transverse suture of mesosternum curved posteriorly (Fig. 104). — **Egnatiinae**
– Transverse suture of mesosternum straight or slightly curved (Fig. 72) — 5
5. Head opisthognathous or slightly hypognathous ; foveolae square, oblong, or absent. Antennae usually ensiform (Fig. 108) or filiform ; arolium large, usually as long as or smaller than the claws. — **Acridinae**
– Head obtuse, hypognathous or slightly opisthognathous, pointed posteriorly (*Aiolopus*, Fig. 216). Foveolae triangular, small, rarely trapezoidal, or absent. Antennae filiform ; arolium small or absent. — **Oedipodinae**

Subfamily CATANTOPINAE

Diagnosis : Size variable, usually medium sized or large, rarely small and apterous. Head hypognathous, slightly opisthognathous, except in the genus *Tropidopola* Stål, where it is acutely cylindrical (Fig. 70). Foveolae indistinct, often absent. Antennae filiform. Pronotum with three carinae, the median one often strongly elevated (Fig. 73), the lateral ones usually continuous, sometimes interrupted or indistinct either at the posterior or anterior end (Figs. 42, 50). Prosternum with a distinct conical or cylindrical process between the forelegs (Fig. 71). External median field of hind femur with two rows of symmetrical plates (Fig. 61).
Cosmopolitan, especially in the tropics, including geophilous and phytophilous genera. Several of the most important gregarious forms, such as *Schistocerca* and *Calliptamus*, belong to this subfamily.

Key to the Genera of Catantopinae in Israel and Sinai

1. Hind tibia without external apical spine, leaving a space betwen the spines and the spurs — 2
- Hind tibia with external apical spine (Fig. 76) — 13

2. Tegmina fully vestigial, small, oblong, extending over the first or second abdominal segment (Figs. 37, 38) — 3
- Tegmina fully developed, covering the entire abdomen or its greater part — 5

3. Body hairy ; foveolae distinct, quadrangular to oblong, protruding on the frons. Vestigial tegmina narrow at the base, widening apically (Fig. 37). Hind tibia dark grey. — **Sphenophyma** Uvarov
- Body smooth ; foveolae absent. Vestigial tegmina oblong — 4

4. Large, over 20 mm in length (Fig. 38). Vestigial tegmina oblong, 6.0–9.0 mm long. Hind tibia reddish. — **Pareuprepocnemis** Brunner-Wattenwyl
- Small, less than 20 mm in length. Vestigial tegmina 2.0–2.5 mm long. Hind tibia pale. — **Pezotettix** Burmeister

5. Head opisthognathous (Figs. 39, 43) — 6
- Head hypognathous or slightly prognathous (Fig. 48) — 8

6. Arolium of hind legs large, 1.5 times longer than claws (Fig. 41). Hind tibia with 9–11 spines on outer margin. — **Eyprepocnemis** Fieber
- Arolium not longer than claws. Hind tibia with 13–16 spines on outer margin — 7

7. Antennae very long, in male twice, in female 1.5 times as long as head and pronotum together (Fig. 42). Lateral carinae of pronotum indistinct. — **Thisoicetrinus** Uvarov
- Antennae shorter ; in male 1.5 times as long as, in female as long as head and pronotum together (Figs. 43, 45, 46). Lateral carinae of pronotum distinct along their entire length. — **Heteracris** Walker

8. Prosternal process laterally compressed (Fig. 51). Mesosternal lobes narrow and long, distinctly longer than maximum width — 9

- Prosternal process different. Mesosternal lobes wide and short, as long as or shorter than maximum width 10
9. Median carina of pronotum distinct along its entire length, slightly raised, convex, intersected by three deep transverse sulci (Fig. 48). Posterior angle of pronotum obtuse. Wings with a dark band. **Anacridium** Uvarov
- Median carina of pronotum not raised ; prozona rounded, much narrower than metazona (Fig. 50) ; posterior margin of pronotum rounded. Wings colourless. **Schistocerca** Stål
10. Pronotum with three carinae which are distinct along their entire length (Fig. 57) 11
- Carinae of pronotum partly obliterated 12
11. Internal spur of hind tibia 1.5–2.0 times longer than external spur. Apical ventral lobe of male cercus without distal dentations. Wings usually with dark venation (Fig. 52). **Metromerus** Uvarov
- Internal spur of hind tibia only slightly longer than external spur. Apical ventral lobe of male cercus with two distal dentations (Fig. 55). **Calliptamus** Serville
12. Hind femur short and wide, expanded beyond the middle (Fig. 61). Wings with dark venation. **Sphodromerus** Stål
- Hind femur slender, narrow, without expansion beyond the middle (Fig. 63). Wings transparent. **Cyclopternacris** Ramme
13. Apterous or with very abbreviated vestigial tegmina 14
- Wings fully developed, posteriorly reaching the hind knee or advancing beyond it 15
14. Rugose and without a trace of wings. Prozona acutely raised anteriorly (Fig. 64). **Pamphagulus** Uvarov
- Smooth, with residual tegmina situated laterally on the thorax (Fig. 65). Pronotum without elevation. **Pezotettix** Burmeister
15. Head cylindrical, acute, opisthognathous. Pronotum rounded, smooth (Fig. 70). **Tropidopola** Stål
- Head hypognathous, high. Pronotum with a helmet-like elevation anteriorly (Fig. 73). **Dericorys** Serville

Genus SPHENOPHYMA Uvarov, 1934

Eos, Madr., 10 : 114

Platyphyma Stål C., 1876, *K. svenska VetenskAkad. Handl.*, 4 (5) : 17.

Type Species : *Platyphyma rugulosa* Stål, 1876.
Diagnosis : Body short, very hairy, surface rough. Head slightly opisthognathous. Frontal ridge flat, slightly depressed near median ocellus. Fastigial foveolae punctate, quadrangular. Vertex slightly concave ; lateral margins raised ; longitudinal carina extending onto rugose occiput. Pronotum wide ; median carina slightly raised and intersected by three transverse sulci. Lateral carinae usually obliterated in middle ; posterior margin of pronotum curved inwards.
Vestigial tegmina widening towards apex (Fig. 37). Prosternal process wide and wedge-shaped, tuberculate and hairy dorsally. Dorsal carina of hind femur delicately serrate.

Monotypic.
Distribution : Syria, Lebanon, Israel.

Sphenophyma rugulosa (Stål, 1876)
Figs. 36, 37

Type Locality : 'Syria'.

Platyphyma rugulosa Stål C., 1876, *K. svenska VetenskAkad. Handl.*, 4(5) : 18.
Sphenophyma rugulosa —. Uvarov B. P., 1934, *Eos, Madr.*, 10 : 116.

Coloration : Uniform brown to pale brown, or with dark band on body, especially on dorsal and inner side of hind femur.
Measurements (mm) : Body ♀ 19.0–23.0, ♂ 15.5–16.5 ; tegmina ♀ 3.1–3.7, ♂ 2.9–3.5.
Sphenophyma rugulosa inhabits mountain areas (up to 1,900 m altitude) with low, sparse or dense vegetation ; blends in well with the environment.
Distribution, Israel : Upper and Lower Galilee (1, 2), Mount Hermon (19), Golan Heights (18), Jordan Valley (7).
A comparison of the population of Mount Meron with that of Mount Hermon shows a great variation, especially of the rugosity of the body and serration of the femurs.

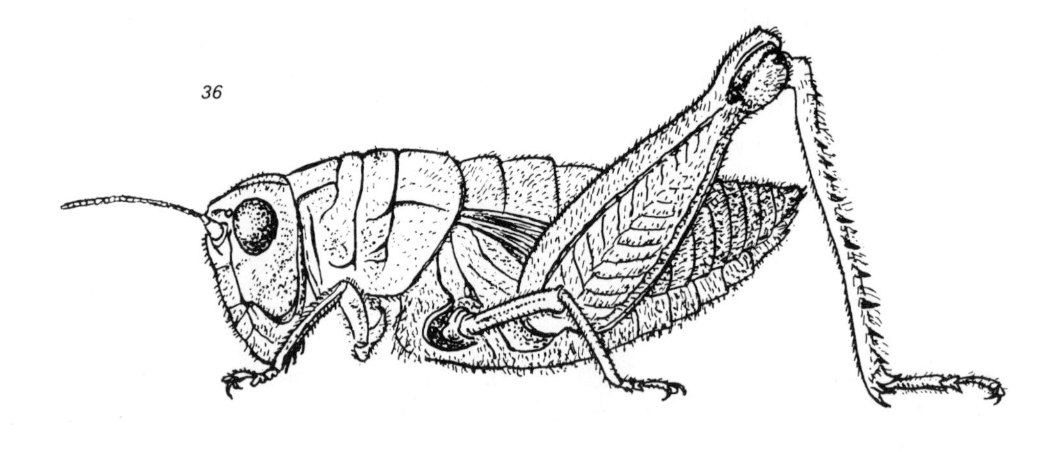

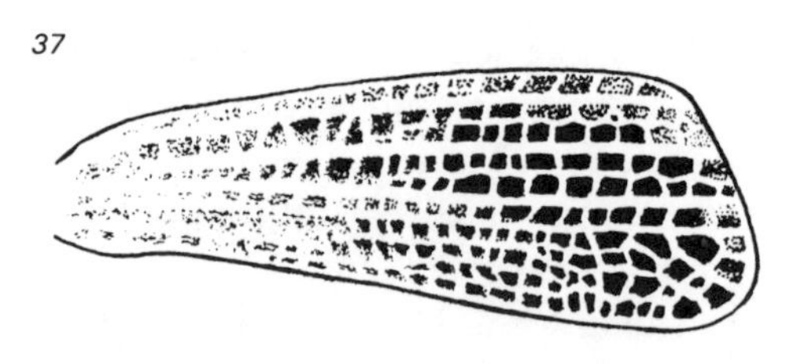

Figs. 36–37 : *Sphenophyma rugulosa* (Stål, 1876)
36. female ; 37. tegmen

Genus PAREUPREPOCNEMIS Brunner von Wattenwyl, 1893

Annali Mus. civ. Stor. nat. Genova, (2) 13 : 48

Type species : *Pareuprepocnemis syriaca* Brunner-Wattenwyl, 1893.
Diagnosis : Medium sized, body smooth, hairs sparse. Head slightly opisthognathous, wider than pronotum. Frontal ridge flat, sparsely punctate, obliterated below median ocellus. Eyes elliptical, twice as long as wide. Vertex sloping ; fastigium pentagonal ; median carina present. Foveolae obliterated. Antennae shorter than head and pronotum together. Pronotum with three transverse sulci; lateral carinae obliterated in metazona. Metazoa densely punctate, its length equal to that of anterior part of prozona. Posterior and anterior margins of pronotum rounded. Mesosternal interspace narrow and long ; margins diverging gradually. Metasternal interspace square or with contiguous lobes. Prosternal process cylindrical. Vestigial wings elongate, extending only on two abdominal segments, usually pointed apically, with prominent longitudinal veins at the apex. Hind femur slender, longer than tibia. Arolium shorter than claws.
Coloration : Uniformly dark brown or pale brown, usually with a pattern of pale lines on head and lateral carinae of pronotum. Two dark wide bands present on dorsal and external surface of hind femur. Inner and ventral sides of hind femur blackish-blue with a white ring before knee. Hind tibia and tarsus red ; tibia with a black ring before knee.
Monotypic.

Pareuprepocnemis syriaca Brunner-Wattenwyl, 1893

Fig. 38 ; Plate I : 1

Type Locality : Jerusalem.

Pareuprepocnemis syriaca Brunner von Wattenwyl C., 1893, *Annali Mus. civ. Stor. nat. Genova*, (2) 13 : 48.
Caloptenus festae Giglio-Tos E., 1892, *Boll. Musei Zool. Anat. comp. R. Univ. Torino*, 8 (164) : 18.
Pareuprepocnemis festae Giglio-Tos E., 1894, *ibid.*, 9 (191) : 2.
Pareuprepocnemis giglio-tosi Bolivar I., 1914, *Trab. Mus. Cienc. nat. Madr.* (ser. zool.), no. 20, p. 18.
Pareuprepocnemis syriaca —. Buxton P. A. & B. P. Uvarov, 1923, *Bull. Soc. R. ent. Égypte*, p. 207.
Pareuprepocnemis syriaca —. Bodenheimer F. S., 1935, *Animal Life in Palestine*, Jerusalem, pp. 38, 79, 86, 87, 89, 311, 320, 323.
Pareuprepocnemis syriaca —. Bodenheimer F. S., 1935, *Arch. Naturgesch.*, 4 (2) : 207.

Measurements (mm) : Body ♀ 29.0–30.0, ♂ 21.0–23.0 ; tegmina ♀ 7.0–9.5, ♂ 5.0–5.5.
Distribution : Syria, Lebanon, Israel.
Israel : Jordan Valley (7), Coastal Plain (4, 8), Samaria (6), Judean Hills (11), Golan Heights (18).

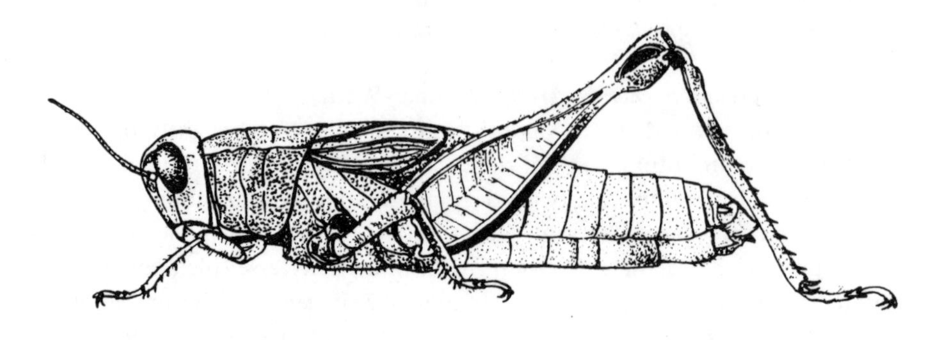

Fig. 38 : *Pareuprepocnemis syriaca* Brunner-Wattenwyl, 1893, female

Inhabits mountain tops and slopes with rich bushy vegetation. Together with *Acinipe hebraeus, Pezotettix judaica* and *Prionosthenus galericulatus* it forms the most typical community of apterous grasshoppers in this region. When in danger *P. syriaca* jumps away, hiding among branches and spines of *Putorium spinosum*.

Genus EYPREPOCNEMIS Fieber, 1853
Lotos, 3 : 98

Euprepocnemis Stål C., 1873, *Recensio Orthopterorum*, Stockholm, 1 : 75.
Eyprepocnemis —. Rehn J. A. G., 1901, *Proc. Acad. nat. Sci. Philad.*, 53 : 377.
Eyprepocnemis —. Uvarov B. P., 1943, *Proc. Linn. Soc. Lond.*, 1942–3 : 25.

Type Species : *Gryllus plorans* Charpentier, 1825.
Diagnosis: Medium sized or large. Head large, slightly opisthognathous. Eyes oval, elongate. Frontal ridge flat, sparsely punctate, obliterated below median ocellus. Vertex smooth, with signs of carinulae. Pronotum densely punctate posteriorly; prozona and metazona of equal length or, metazona shorter in male; posterior margin rounded. Median carina distinct along its entire length, intersected by three transverse sulci. Lateral carinae distinct in prozona, obliterated in metazona (Fig. 40). Prosternal process large, conical, slightly curved posteriorly. Hind tibia slightly shorter than hind femur, its apex with a short spine. Arolium much longer than claw (Fig. 41).
Distribution : The genus contains about 25 species distributed throughout the tropics and subtropics of the Old World. One species in our region.

Eyprepocnemis plorans (Charpentier, 1825)
Figs. 39–41

Type Locality : 'Lusitania'.

Gryllus plorans Charpentier T. de, 1825, *Horae entomologicae*, Wratislaviae, p. 134.
Acridium plorans —. Costa O. G., 1836, *Fauna del regno di Napoli, Ortotteri*, Naples, p. 7, pl. 1, fig. 1.
Caloptenus plorans —. Fischer L. H., 1853, *Orthoptera Europaea*, Leipzig, p. 376.
Eyprepocnemis plorans —. Fieber F. X., 1853, *Lotos*, 3 : 98.
Heteracris plorans —. Walker F., 1870, *Catalogue of the specimens of Dermaptera Saltatoria in the Collection of the British Museum*, London, Part IV, p. 655.
Cyrtacanthacris ornatipes Walker F., 1870, *ibid.*, Part III, p. 575.
Heteracris consobrina Walker F., 1870, *ibid.*, Part IV, pp. 673, 674.
Pezotettix (Euprepocnemis) plorans —. Stål C., 1873, *Recensio Orthopterorum*, Stockholm, 1 : 76.
Euprepocnemis plorans —. Stål C., 1876, *K. svenska VetenskAkad. Handl.*, 4 (5) : 16.
Cyrtacanthacris ornatipes —. Hart H. C., 1891, *Fauna and Flora of Sinai, Petra and Wadi Arabah*, London, p. 183.
Cyrtacanthacris ornatipes —. Giglio-Tos E., 1893, *Boll. Musei Zool. Anat. comp. R. Univ. Torino*, 8 (164) : 8.
Euprepocnemis ornatipes —. Kirby W. F., 1902, *Trans. R. ent. Soc. Lond.*, p. 114.
Euprepocnemis consobrina —. Kirby W. F., 1910, *A synonymic catalogue of the Orthoptera*, Vol. III, Orthoptera Saltatoria, Part II, London, p. 561.
Euprepocnemis plorans —. Uvarov B. P., 1921, *Trans. R. ent. Soc. Lond.*, 1921 : 110, 119.
Euprepocnemis plorans —. Buxton P. A. & B. P. Uvarov, 1923, *Bull. Soc. R. ent. Égypte*, p. 207.
Euprepocnemis plorans —. Innes W., 1929, *Mém. Soc. R. ent. Égypte*, 3 (2) : 143.
Euprepocnemis plorans —. Bodenheimer F. S., 1935, *Animal Life in Palestine*, Jerusalem, p. 320.
Euprepocnemis plorans —. Bodenheimer F. S., 1935, *Arch. Naturgesch.*, 4 (2) : 207.
Euprepocnemis plorans —. Ramme W., 1951, *Mitt. zool. Mus. Berl.*, 27 : 428.
Euprepocnemis plorans —. Uvarov B. P. & V. D. Dirsh, 1952, *Verh. Natur. Ges.*, Basel, 33 : 14.

Medium sized or large, slender, sparsely hairy. Head slightly opisthognathous ; vertex sloped ; fastigium slightly protruding. Frontal ridge convex, punctate ; margins diverging, disappearing towards clypeus. Eyes large, oblong, vertical diameter nearly twice as long as horizontal, four times longer than subocular groove. Occiput slightly globular, with or without a carina. Antennae filiform, in male longer than head and pronotum together, in female shorter ; proximal segments square. Metazona of pronotum densely punctate, with obliterated converging lateral carinae ; median carina slightly raised, intersected by three transverse sulci, the posterior one situated either in the middle of the pronotum or slightly beyond it. Lateral carinae in metazona indistinct, converging (Fig. 40). Prosternal process rounded apically ; length of metasternal interspace more than twice its width. Metasternal lobes usually contiguous, covering the interspace.

Tegmina narrow, reaching to or extending over hind knee, slightly rounded apically. Radius-sector with two branches ; costal field with a straw-colored stripe. Hind femur longer than tibia ; hind tibia with 10 inner spines and 10 or 11 outer ones.

Coloration: Brown or pale brown with a prominent dark dorsum of pronotum, marginate by pale lines (Fig. 40). Tegmina pale anteriorly, brown posteriorly, with scattered anterior blotches. Wings transparent, yellowish. Inner surface of hind femur yellow; outer surface with a dark line in the middle. Basal part of hind tibia bluish, apex red; tarsus red.

Measurements (mm): Body ♀ 30.5–45.0, ♂ 23.0–31.5; tegmina ♀ 22.5–34.5, ♂ 18.5–27.0.

Distribution: Western Europe, Africa, Asia.

Israel: Upper and Lower Galilee (1, 2), Jordan Valley (7), Coastal Plain (4, 8, 9), Carmel (3).

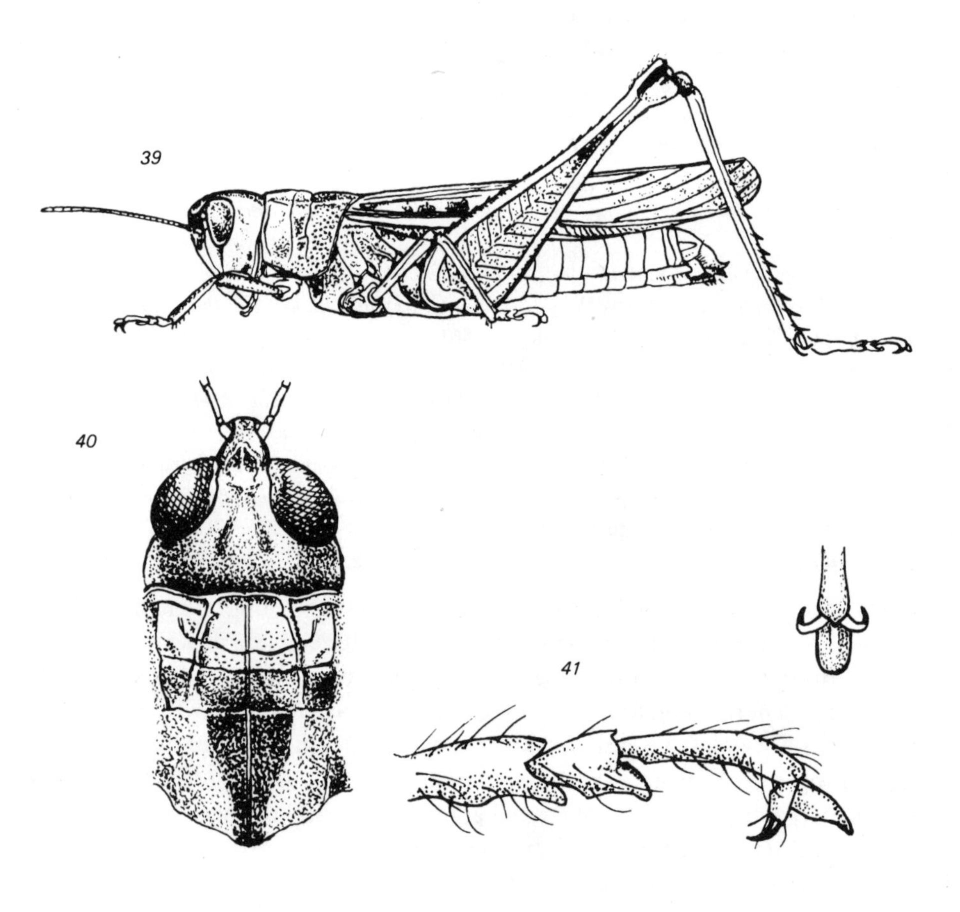

Figs. 39–41: *Eyprepocnemis plorans* (Charpentier, 1825)
39. female; 40. head and pronotum; 41. tarsus and arolium

Usually found on bushes of *Polygonum* and *Artemisia* and on stalks of high Gramineae. Hoppers appear from April to September and are easily recognized, even in the early stages of development, by the brown dorsum of the pronotum.

Genus THISOICETRINUS Uvarov, 1921

Trans. R. ent. Soc. Lond., 1921 : 128

Calliptamus —. Fischer de Waldheim G., 1846–1849, *Orthoptera Imperii Russici,* Moscow, p. 240.
Eyprepocnemis Fieber F. X., 1853, *Lotos*, 3 : 98.
Euprepocnemis Stål C., 1876, *K. svenska VetenskAkad. Handl.*, 4 (5) : 14.
Thisoicetrus (partim) Brunner von Wattenwyl C., 1893, *Annali Mus. civ. Stor. nat. Genova*, (2) 13 : 150.
Thisoecetrus (partim) Jacobson G. G. & V. L. Bianchi, 1902, *Orthoptera and Odonata of the Russian Empire*, St. Petersburg, pp. 174, 205, 318 [in Russian].
Thisoicetrinus Uvarov B. P., 1921, *Trans. R. ent. Soc. Lond.*, 1921 : 128.
Thisoicetrinus —. Uvarov B. P., 1925, *Locusts and Grasshoppers of the European part of the U.S.S.R. and Western Siberia*, Moscow, pp. 87, 92 [in Russian].

Type Species : *Oedipoda pterosticha* Fischer-Waldheim, 1833.
Diagnosis : Medium sized to large, smooth. Head large, slightly opisthognathous. Frontal ridge flat ; margins parallel, slightly diverging above median ocellus. Vertex protruding, hexagonal, with distinct median carina. Eyes oval. Antennae of male long, filiform, more than twice as long as head and pronotum together; 1.5 times as long in the female. Pronotum oblong, intersected by three transverse sulci, the posterior one situated beyond the middle. Median carina distinct along its entire length, intersected by three transverse sulci ; lateral carinae usually obliterated. Metazona deeply punctate. Prosternal process sharp, curved backwards. Hind femur narrow and long.
Coloration : Usually greenish-yellowish with a black band extending from between the eyes via pronotum to middle of tegmina. Tegmina speckled with small black spots. Wings greenish, transparent. Hind tibia reddish.
Distribution : Russia, Asia Minor, Central Asia.
Monotypic.

Thisoicetrinus pterostichus (Fischer-Waldheim, 1833)
Fig. 42

Type Locality : 'Caucasus'.

Oedipoda pterosticha Fischer de Waldheim G., 1833, *Bull. Soc. Imp. Nat. Moscou*, 6 : 384.
Acridium dorsatum Fischer de Waldheim G., 1839, *ibid.*, 12 : 301.

Calliptamus dorsatus —. Fischer de Waldheim G., 1846–49, *Nouv. Mém. Soc. Imp. Nat. Moscou*, p. 240.

Eyprepocnemis fischeri Fieber F. X., 1853, *Lotos*, 3 : 98.

Euprepocnemis dorsata —. Stål C., 1876, *K. svenka VetenskAkad. Handl.*, 4 (5) : 17.

Thisoecetrus pterostichus —. Jacobson G. G. & V. L. Bianchi, 1902, *Orthoptera and Odonata of the Russian Empire*, St. Petersburg, p. 205 [in Russian].

Thisoecetrinus pterostichus —. Uvarov B. P., 1925, *Locusts and Grasshoppers of the European part of the U.S.S.R. and Western Siberia*, Moscow, p.93, Figs. 107, 108 [in Russian].

Thisoecetrinus pterostichus —. Bodenheimer F. S., 1935, *Arch. Naturgesch.*, 4 (2) : 207.

Thisoecetrinus pterostichus —. Uvarov B. P., 1939, *Novit. zool.*, 4 : 1.

Thisoicetrinus pterostichus —. Bei-Bienko G. Ya. & L. L. Mishchenko, 1951, *Locusts and Grasshoppers of the U. S. S. R. and Adjacent Countries*, Moscow, I : 264 [in Russian].

Thisoecetrinus pterostichus —. Ramme W., 1951, *Mitt. zool. Mus. Berl.*, 27 : 428.

Smooth, slender, females large, males medium sized. Head opisthognathous, especially in male, slightly raised above pronotum. Frontal ridge flat, sparsely punctate, slightly widening above median ocellus. Vertex flat, hexagonal, with distinct median carina extending onto the globular occiput. Antennae much longer than head and pronotum together ; most of their segments 2–3 times longer than wide. Pronotum slightly raised along median carina, intersected by all three transverse sulci. Prozona 1.2–1.5 times longer than metazona ; lateral carinae only slightly marked. Prosternal process cylindrical, long, curved towards mesosternum. Margins of mesosternal interspace in female diverging posteriorly ; in male almost parallel in middle, diverging posteriorly and anteriorly ; twice as long as its minimum width. Length of metasternum twice its width. Metasternal lobes contiguous in middle or only slightly separated.

Tegmina extending beyond hind knee ; venation regular, not dense, with a network of elongate cells at the apex. Wings triangular, narrow. Hind femur slender, its length 4.5–6.0 times its maximum width. Hind tibia almost as long as hind femur ; 14–16 spines on outer margin and 12–16 on inner one. Arolium large, almost three-fifths of the claws' length.

Coloration : Greenish-brown, brownish-yellow, usually with a black band extending from vertex to posterior margin of pronotum and continuing onto the posterior (dorsal) margin of tegmina, the latter more distinct in females. Tegmina with a few scattered small black dots (Fig. 42). Wings greenish, transparent. Outer, upper and inner surfaces of hind femur with dark cross-bands which are usually obliterated in the greenish forms. Hind tibia entirely reddish, sometimes with black and pale rings near knee.

Measurements (mm) : Body ♀ 35.0–55.4, ♂ 20.0–22.3 ; tegmina ♀ 23.6–37.5, ♂ 16.5–24.2.

Distribution, Israel : Upper Galilee (1), along the Jordan River system.

Usually found in and among bushes along water bodies, hiding on the branches.

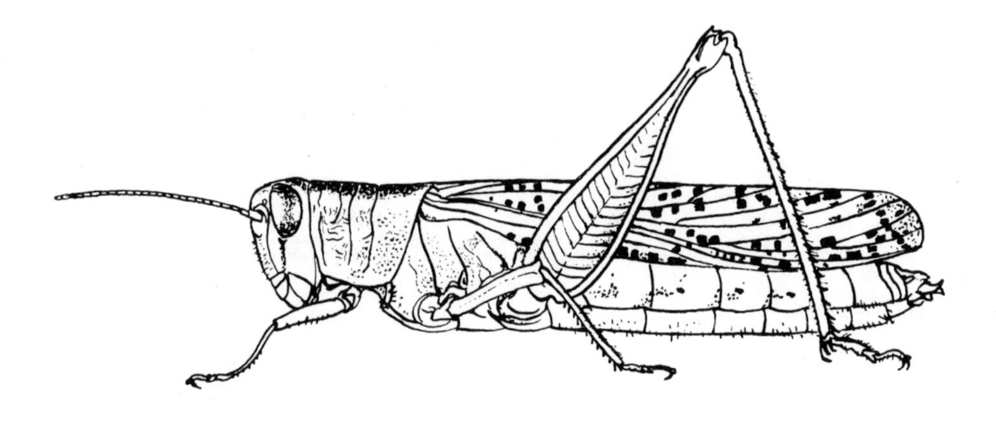

Fig. 42 : *Thisoicetrinus pterostichus* (Fischer-Waldheim, 1833), female

Genus HETERACRIS Walker, 1870

Catalogue of the specimens of Dermaptera Saltatoria in the Collection of the British Museum, London, Part IV, p. 655

Thisoecetrus Jacobson G. G. & V. L. Bianchi, 1902, *Orthoptera and Odonata of the Russian Empire*, St. Petersburg, pp. 174, 205, 318 [in Russian].
Heteracris —. Bolivar I., 1915, *Trab. Mus. Cienc. nat. Madr.* (ser. zool.), no. 20, pp. 7, 31.
Heteracris —. Uvarov B. P., 1921, *Trans. R. ent. Soc. Lond.*, p. 133.

Type Species : *Acridium herbaceum* Serville, 1838.
Diagnosis : Head slightly opisthognathous ; frontal ridge convex, smooth, flat in profile. Vertex slightly depressed, with very indistinct median carina. Eyes oval, elongate. Antennae filiform and longer than or as long as head and pronotum together. Pronotum widening slightly towards posterior margin ; prozona longer than metazona. Median carina distinct, intersected by three distinct transverse sulci ; lateral carinae distinct, sometimes blurred posteriorly. Metazona diffusely or densely punctate, posterior margin rounded. Prosternum with cylindrical or slightly conical processes. Cerci of male compressed laterally, curved, extending beyond apex of supraanal plate. Valves of ovipositor short, straight, without apical dentations; ventral valves with a distinct preapical notch. Tegmina fully developed, with rounded apex.
Distribution : About 30 species around the Mediterranean, Africa and Asia.
Three species in Israel, all typical phytophilous forms, inhabitants of bushes.

Key to the Species of Heteracris in Israel and Sinai

1. Subgenital plate of male with two separated tubercles at apex (Fig. 44). Hind femur very slender, its length about 5–6 times its maximum width. Tegmina with scattered small brown blotches. **H. adspersus** (Redtenbacher)
- Subgenital plate of male without apical tubercles, rounded (Fig. 47). Hind femur thicker and shorter, less than 5 times longer than wide. Tegmina with large brown blotches usually forming transverse bands 2
2. Hind tibia with 12–14 spines on outer line and 10–12 on inner line (Fig. 45). **H. annulosus** Walker
- Hind tibia with 15–18 spines on outer line and 12–14 on inner line (Fig. 46). **H. littoralis similis** (Brunner-Wattenwyl)

Heteracris adspersus (Redtenbacher, 1889)
Figs. 43, 44

Type Locality : 'Transcaspia' (Vienna Museum).

Euprepocnemis adspersa Redtenbacher J., 1889, *Wien. ent. Ztg.*, 8 : 30.
Euprepocnemis littoralis —. Finot, A., 1896, *Annls Soc. ent. Fr.*, 65 : 541, 543.
Thisoecetrus adspersus —. Jacobson G. G. & V. L. Bianchi, 1902, *Orthoptera and Odonata of the Russian Empire*, St. Petersburg, pp. 205, 319 [in Russian].
Thisoicetrus adspersus —. Werner F., 1905, *Sber. Akad. Wiss. Wien*, 114 (1) : 360, 361, 366, 426.
Thisoecetrus adspersus —. Buxton P. A. & B. P. Uvarov, 1923, *Bull. Soc. R. ent. Égypte*, p. 207.
Thisoicetrus adspersus —. Innes W., 1929, *Mém. Soc. R. ent. Égypte*, 3 (2) : 146, 149.
Thisoecetrus adspersus —. Bodenheimer F. S., 1935, *Arch. Naturgesch.*, 4 (2) : 206.
Thisoicetrus adspersus —. Uvarov B. P., 1939, *Novit. zool.*, 41 : 377, 378.
Euprepocnemis littoralis —. Chopard L., 1943, *Faune Emp. Franç.*, Paris, 1 : 409 (partim).
Thisoicetrus adspersus —. Bei-Bienko G. Ya. & L. L. Mishchenko, 1951, *Locusts and Grasshoppers of the U. S. S. R. and Adjacent Countries*, Moscow, I : 266 [in Russian].
Thisoicetrus adspersus —. Ramme W., 1951, *Mitt. zool. Mus. Berl.*, 27 : 428.

Medium sized or small. Head opisthognathous, especially in male. Eyes projecting laterally; frontal ridge flat, widening apically, with margins gradually converging towards clypeus. Fastigium sloping; vertex pentagonal, flat; occiput globular with slightly marked carina. Antenna filiform, in male half the length of, in females as long as or slightly longer than head and pronotum together; number of segments 24–27, punctate, length of median segment twice its width. Pronotum slightly raised in metazona, punctate; lateral carinae slightly converging in prozona, usually obliterated in metazona. Median carina intersected by three transverse sulci, the first situated beyond the middle of pronotum. Prosternal process conical, short, hairy, slightly curved towards sternal plate. Mesosternal lobes square; mesosternal interspace 1.5 times as long as wide, its margins diverging. Metasternal interspace in female square, in male either absent or very narrow. Subgenital plate of male with two apical tubercles (Fig. 44). In female upper valves of ovipositor with transverse ridges on the surface.

Tegmina reaching hind knee or extending slightly beyond it ; base opaque, intercalary vein present in median field. Wings elongate, transparent. Hind femur 6–6.2 times longer than maximum width, narrowing apically. Hind tibia slender, with 13–15 spines along inner margin and 15–17 on outer margin.
Coloration : Brown or pale brown, with a black dorsal field from the occiput and continuing on the discus of pronotum, usually with lighter margins. Numerous small brown blotches on tegmina. Two oblique dark lines on inner and outer surface of hind femur ; knee dark with white lobes. Hind tibia almost reddish, with light and dark areas near knee.
Measurements (mm) : Body ♀ 23.0–36.0, ♂ 16.5–22.5 ; tegmina ♀ 20.5–26.3, ♂ 15.5–17.5.
Distribution : Central Asia, through Persia, Israel, Saudi Arabia, North Africa, Spain. Israel : Northern Coastal Plain (4), Upper Galilee (1), Jordan Valley (7).

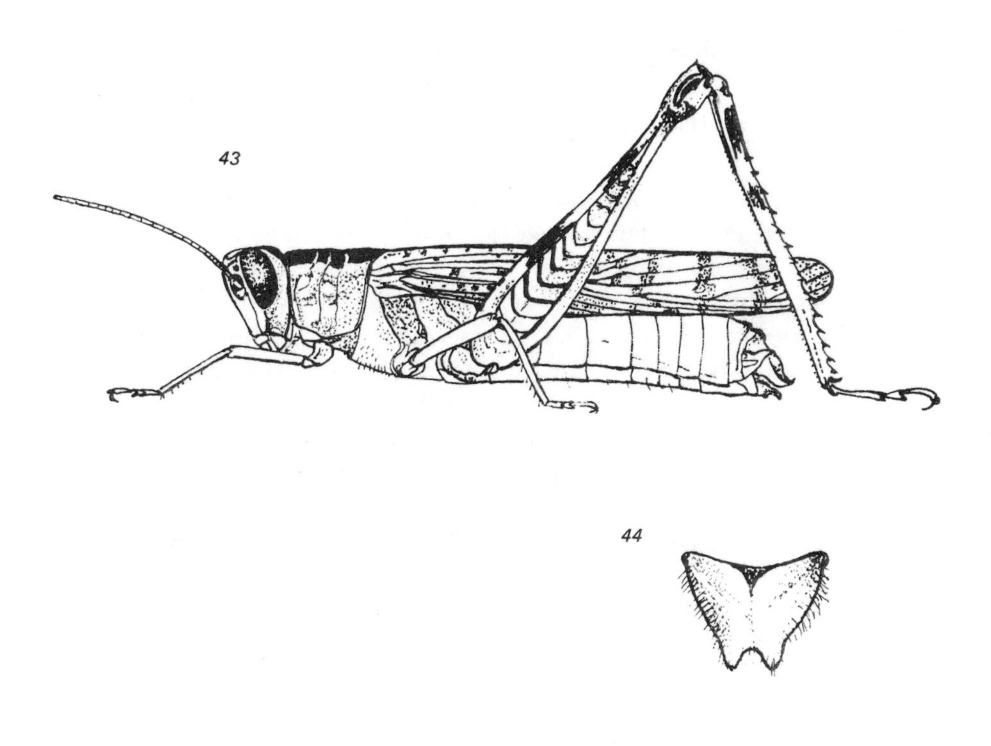

Figs. 43–44 : *Heteracris adspersus* (Redtenbacher, 1889)
43. female ; 44. subgenital plate of male

Heteracris annulosus Walker, 1870

Fig. 45

Type Locality unknown (British Museum).

Heteracris annulosa Walker F., 1870, *Catalogue of the specimens of Dermaptera Saltatoria in the Collection of the British Museum*, London, Part IV, pp. 673, 674.
Euprepocnemis littoralis Bonnet E. & A. Finot, 1884, *Revue Sci. nat. Montpellier* (3), 4 : 232.
Thisoicetrus littoralis —. Krauss H. A. & J. Vosseler, 1896, *Zool. Jb.* (Syst.), 9 : 542.
Euprepocnemis annulosa —. Kirby W. F., 1910, *A synonymic catalogue of the Orthoptera*, Vol. III, Orthoptera Saltatoria, Part II, London, p. 560.
Thisoicetrus charpentieri Bolivar I., 1914, *Méms R. Soc. esp. Hist. nat.*, 8 (5) : 209.
Thisoicetrus littoralis charpentieri —. Uvarov B. P., 1923, *Novit. zool.*, 30 : 75.
Thisoicetrus theodori Uvarov B. P., 1929, in : *Ergebnisse der Sinai-Expedition*, Leipzig, p. 102.
Pezotettix (Euprepocnemis) charpentieri —. Sjöstedt Y., 1932, *Ark. Zool.*, 24, A (1) : 44.
Thisoecetrus continuus Walker. Bodenheimer F. S., 1935, *Animal Life in Palestine*, Jerusalem, pp. 322, 323.
Thisoecetrus continuus —. Bodenheimer F. S., 1935, *Arch. Naturgesch.*, 4 (2) : 206.
Thisoicetrus annulosus —. Uvarov B. P., 1939, *Novit. zool.*, 41 : 379.

Medium sized, robust. Head slightly opisthognathous; frontal ridge flat, sparsely punctate, its margins parallel, obliterated before clypeus. Occiput short, slightly globular, with or without median carina. Antennae of female as long as head and pronotum together, in male much longer; number of segments 25–27. Pronotum slightly raised in prozona along median carina; lateral carinae distinct along their entire length or obliterated in metazona. Prosternal process cylindrical, straight. Mesosternal lobes in female square, in male longer than wide; mesosternal interspace twice as long as wide; metasternal interspace in female twice as long as wide, in male interspace absent.
Tegmina extending beyond hind knee; intercalary vein absent. Hind femur swollen at the base, its length 3.7–4.8 times its maximum width. Hind tibia much shorter than femur, with 12–14 spines on outer line and 10–12 on inner line.
Coloration: Brown or pale brown, usually with two pale stripes extending from occiput, over both lateral carinae of pronotum, merging and extending along median line of tegmen. Hind femur usually without dark band on outer surface, with two bands on inner surface, the basal one usually sending a black stripe towards point of articulation. Hind tibia pale reddish or red, with or without black ring before middle.
Measurements (mm): Body ♀ 29.0–35.5, ♂ 19.0–22.0; tegmina ♀ 23.0–30.5, ♂ 16.0–18.0; hind femur ♀ 17.0–23.0, ♂ 13.0–13.5.
Distribution, Israel: Coastal Plain (4, 8, 9), Judean Hills (11), Jordan Valley (7), Dead Sea Area (13), Negev (15, 16, 17), 'Arava Valley (14), Central Sinai (21).

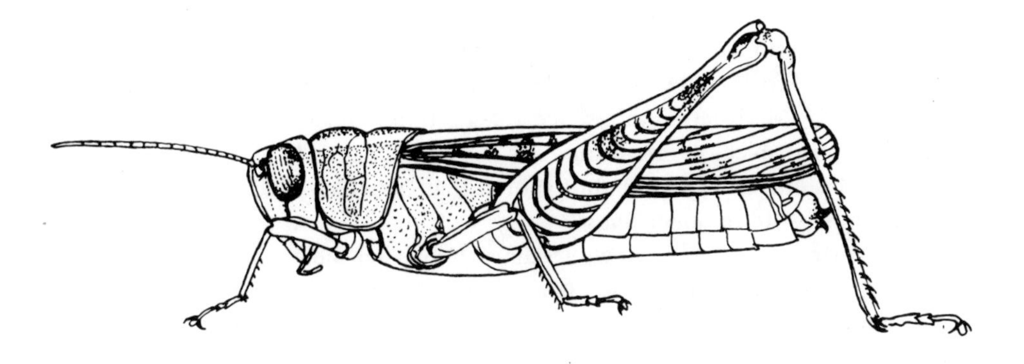

Fig. 45 : *Heteracris annulosus* Walker, 1870, female

Heteracris littoralis similis (Brunner-Wattenwyl, 1861)
Figs, 46, 47 ; Plate I : 2

Type Locality : 'Egypt'. Type lost.

Gryllus littoralis Rambur J. P., 1938, *Faune entomologique de l'Andalousie. Orthoptera*, Paris, 2 : 78.
Caloptenus similis Brunner von Wattenwyl C., 1861, *Verh. zool.-bot. Ges. Wien*, 11 : 224.
Cyrtacanthacris notata Walker F., 1870, *Catalogue of the specimens of Dermaptera Saltatoria in the Collection of the British Museum*, London, Part IV, p. 574.
Cyrtacanthacris notata —. Hart H. C., 1891, *Fauna and Flora of Sinai, Petra and Wadi Arabah*, London, p. 183.
Thisoecetrus littoralis —. Krauss H. A., 1909, in: Kneucker A., *Verh. naturw. Ver. Karlsruhe,* 21 : 37.
Thisoicetrus similis —. Kirby W. F., 1910, *A synonymic catalogue of the Orthoptera*, Vol. III, Orthoptera Saltatoria, Part II, London, p. 559.
Caloptenus similis —. Uvarov B. P., 1923, *Entomologist's mon. Mag.*, 59 : 86.
Thisoecetrus littoralis —. Buxton P. A. & B. P. Uvarov, 1923, *Bull. Soc. R. ent. Égypte*, p. 206.
Thisoecetrus littoralis —. Bodenheimer F. S., 1935, *Animal Life in Palestine*, Jerusalem, pp. 320, 322.
Thisoecetrus littoralis asiaticus Uvarov. Bodenheimer F. S., 1935, *Arch.Naturgesch.*, 4(2) : 205.
Thisoicetrus littoralis similis —. Uvarov B. P., 1939, *Novit. zool.*, 41 : 378, 381, Fig. 297 S.
Thisoicetrus littoralis similis —. Bei-Bienko G. Ya. & L. L. Mishchenko, 1951, *Locusts and Grasshoppers of the U. S. S. R. and Adjacent Countries*, Moscow, I : 266 [in Russian].
Thisoicetrus littoralis —. Ramme W., 1951, *Mitt. zool. Mus. Berl.*, 27 : 428.

Large or medium sized. Head hypognathous or slightly opisthognathous ; frontal ridge very wide, narrowing below fastigium, its maximum width greater than interorbital distance. Vertex rounded, with median carina continuing onto occiput. Anten-

nae slightly longer than head and pronotum together; distal segment elongate. Pronotum with distinct carinae; lateral carinae diverging posteriorly; median carina slightly raised. Prosternal process cylindrical, slightly constricted in the middle, swollen apically. Mesosternal interspace 1.2–1.5 times longer than wide.
Tegmina rounded apically; intercalary vein absent. Length of hind femur 4.5–5 times its maximum width. Hind tibia slightly shorter than hind femur, with 15–18 spines on outer line and 12–14 on inner line.
Coloration: Mostly dark brown, with yellowish lines on the dorsal side and numerous brown cross-bands on tegmina. Hind tibia reddish or reddish-violet. Inner surface of femur yellowish or grey with a basal black blotch and two black cross-bands; 2–3 elongate dark blotches on outer surface.
Measurements (mm): Body ♀ 26.4–50.5, ♂ 20.6–29.5; tegmina ♀ 31.0–37.5, ♂ 19.5–24.5; hind femur ♀ 19.0–28.0, ♂ 14.0–17.0.
Distribution: Pakistan, Central Asia, Israel, Cyprus, North Africa.
Israel: Upper Galilee (1), Coastal Plain, (4, 8, 9), Jordan Valley (7).
This is the most common species of *Heteracris* in the Coastal Plain and near inland water bodies in the autumn and early winter months.

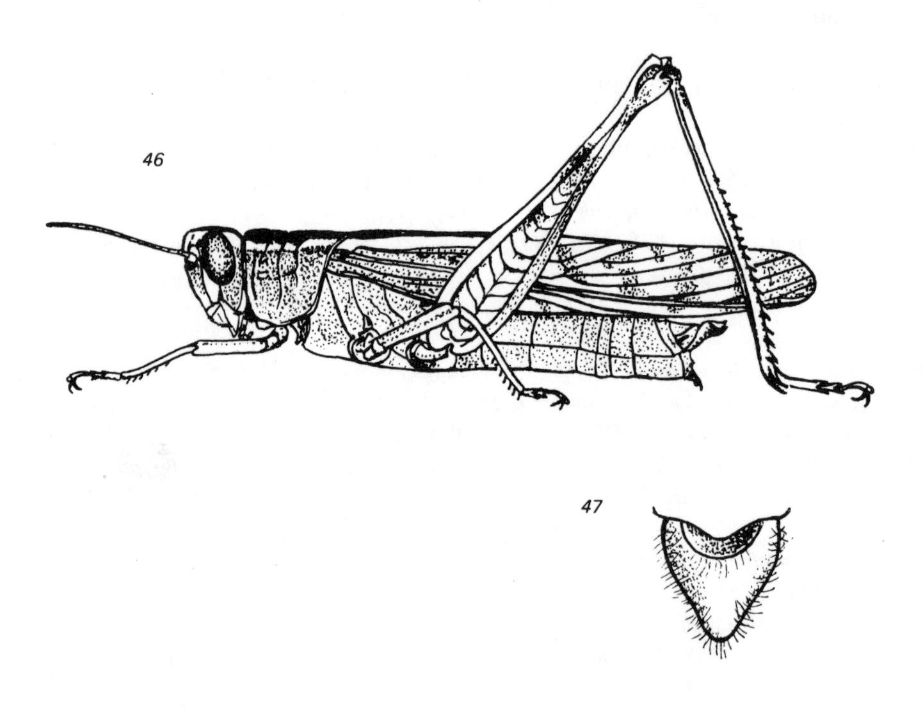

Figs. 46–47: *Heteracris littoralis similis* (Brunner-Wattenwyl, 1861)
46. female; 47. subgenital plate of male

Genus ANACRIDIUM Uvarov, 1923

Ann. Mag. nat. Hist., I. (9) 11 : 141 ; II. (9) 11 : 485

Type Species : *Gryllus (Locusta) aegyptius* Linnaeus, 1764.
Diagnosis : Large, slightly rugose and hairy. Head short, slightly opisthognathous ; face hairy, punctate ; frontal ridge depressed, its margins parallel, converging apically. Vertex pentagonal, depressed; occiput slightly globular ; median carina absent. Antennae filiform, longer than head and pronotum together. Eyes protruding, oval ; vertical axis twice as long as horizontal axis. Pronotum compressed, raised along the median carina, distinctly intersected by three transverse sulci, the posterior one situated beyond the middle of the pronotum. Prosternal process large and straight. Mesosternal interspace longer than wide.
Tegmina long, almost reaching distal half of hind tibia. Wings with dark base and dark venation. Dorsal and ventral carinae of hind femur serrate. Arolium as long as or longer than claws.
Distribution : Five species distributed throughout South Asia, Africa, the Red Sea Islands and Southern Europe.
One species in our region.

Anacridium aegyptium aegyptium (Linnaeus, 1764)

Fig. 48 ; colour plate : 1

Type Locality : 'Egypt'.

Gryllus (Locusta) aegyptius Linnaeus C., 1764, *Museum S. R. M. Ludovicae Ulricae*, p. 138.
Gryllus lineola Fabricius J. C., 1781, *Species Insectorum*, 1 : 365.
Acrydium lineola —. Olivier G. A., 1791, *Encyclopédie Méthodique, Histoire Naturelle*, 6 : 221.
Gryllus nubecula Thunberg C. P., 1815, *Mém. Acad. Sci. St.-Pétersb.*, 5 : 238.
Gryllus rubecula —. Thunberg C. P., 1824, *ibid.*, 9 : 416.
Podisma companum Costa O. G., 1836, *Fauna del regno di Napoli. Ortotteri*, Naples, p. 47, pl. 4, fig. 5A, b–d.
Podisma appulum Costa O. G., 1836, *ibid.*, p. 44, pl. 4, figs. 3a–b, 4.
Podisma lineola —. Costa O. G., 1836, *ibid.*, p. 4, pl. 1, figs. 2a–d.
Acridium indecisum Walker F., 1870, *Catalogue of the specimens of Dermaptera Saltatoria in the Collection of the British Museum*, London, Part III, p. 585 ; Part IV, p. 623.
Acridium albidiferum Walker F., 1870, *ibid.*, Part IV, p. 627.
Acridium aegyptium —. Stål C., 1873, *Recensio Orthopterorum*, Stockholm, 1 : 63.
Cyrtacanthacris aegyptius —. Karsch F., 1893, *Berl. ent. Z.*, 38 : 88, 89.
Acridium aegyptium —. Giglio-Tos E., 1893, *Boll. Musei Zool. Anat. comp. R. Univ. Torino*, 8 (164) : 8.
Locusta (Locusta) aegyptia —. Karny H., 1907, *Sber. Akad. Wiss. Wien*, 116 : 307.
Orthacanthacris aegyptia —. Kirby W. G., 1910, *A synonymic catalogue of the Orthoptera*, Vol. III, Orthoptera Saltatoria, Part II, London, p. 444.
Orthacanthacris indecisa —. Kirby W. G., 1910, *ibid.*, p. 445.

Orthacanthacris aegyptia —. Navas L., 1911, *Revista Monserratina*, Barcelona, p. 2.
Anacridium aegyptium —. Buxton P. A. & B. P. Uvarov, 1923, *Bull. Soc. R. ent. Égypte*, p. 204.
Anacridium aegyptium —. Uvarov B. P., 1927, *Bull. Soc. Sci. nat. Maroc.*, 7 : 213.
Anacridium aegyptium —. Innes W., 1929, *Mém. Soc. R. ent. Égypte*, 3 (2) : 130.
Anacridium aegyptium —. Bodenheimer F. S., 1930, *Die Schädlingsfauna Palästinas*, Berlin, pp. 316, 360, 378.
Anacridium aegyptium —. Bodenheimer F. S., 1935, *Animal Life in Palestine*, Jerusalem, pp. 79, 320, 321, 323, 356.
Anacridium aegyptium —. Bodenheimer F. S., 1935, *Arch. Naturgesch.*, 4 (2) : 202.
Flamiruizia stuardoi Liebermann S., 1943, *Revta Soc. ent. Argent.*, 11 : 401.
Anacridium aegyptium —. Ramme W., 1951, *Mitt. zool. Mus. Berl.*, 27 : 428.

Large and strong, compressed laterally, hairy and slightly rugose, especially the pronotum. Head short, slightly opisthognathous. Frontal ridge with raised, almost parallel margins, converging below vertex, reaching clypeus ventrally. Vertex depressed, rhomboidal-pentagonal. Eyes oblong, their vertical diameter 1.2–1.5 times longer than subocular groove. Antennae filiform, longer than head and pronotum together. Pronotum laterally compressed ; median carina raised, deeply intersected by three transverse sulci. Metazona as long as prozona ; posterior angle of pronotum obtuse. Sternum hairy and slightly rugose, longer than wide. Mesosternal interspace 1.5 times longer than wide. Metasternal interspace very narrow or absent.
Tegmina extending beyond middle of hind tibia (if legs extended), their length 5.6–7.0 times their maximum width. Venation at base dense, irregular ; radius-sector with 4–5 branches. Wings triangular. Length of hind femur 4.5–5.2 times its maximum width ; all carinae serrate. Hind tibia longer than femur. Arolium of male longer than claws, in female as long as or shorter than claws.
Coloration : Brown, pale brown or dark brown, with numerous darker blotches. Head and pronotum usually with a pattern of light fields. Antennae usually black. Median carina of pronotum usually pale or orange-yellow. Tegmina opaque ; wings with black venation and a dark band on basal third that often widens to cover entire basal half. Dorsal and ventral external carinae of hind femur with black parts. Ventral part of femur mostly reddish. Hind tibia brown-grey, bluish-grey, or dark bluish ; spines bluish with black apices. Hoppers green or greenish-brown, with a typical raised median carina of pronotum.
Measurements (mm) : Body ♀ 53.6–78.6, ♂ 48.0–53.6 ; tegmina ♀ 58.2–72.5, ♂ 46.5–50.5.
Distribution : Southern U.S.S.R., Central Asia, Caucasus, western Afghanistan, eastern Mediterranean, North Africa.
Israel : Upper and Lower Galilee (1, 2), Carmel (3), Coastal Plain (4, 8, 9), Judean Hills (11), Jordan Valley (7), Dead Sea Area (13), Central Negev (17).
Found on citrus trees and grape vines. In desert areas found on *Acacia, Retama*, in the north also on *Phragmites, Tamariscus* and tobacco plants. In captivity, development of eggs at 27°C takes about 35 days. Adults live through the winter ; hoppers found from April till August.

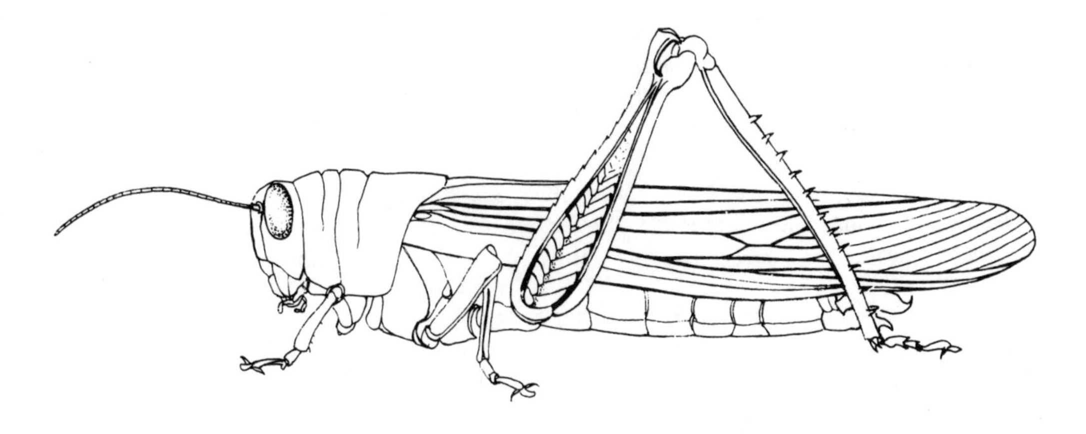

Fig. 48 : *Anacridium aegyptium aegyptium* (Linnaeus, 1764), female

Genus SCHISTOCERCA Stål, 1873

Recensio Orthopterorum, Stockholm, 1 : 64

Acridium (Schistocerca) Stål C., 1873, *Recensio Orthopterorum*, Stockholm, 1 : 64.
Schistocerca —. Stål C., 1876, (2) *K. svenska VetenskAkad. Handl.*, 4 (5) : 11.

Type Species : *Gryllus gregarius* Forskål, 1775.
Diagnosis : Large. Head hypognathous, high and narrow. Frontal ridge depressed around median ocellus ; carinae indistinct. Vertex short, widened before eyes, punctate ; median carina absent. Antennae long, as long as or longer than head and pronotum together. Pronotum anteriorly cylindrical, intersected by three transverse sulci ; median carina obliterated (Fig. 50). Prozona narrower than metazona. Posterior angle of pronotum obtuse. Mesosternal interspace long and narrow. Sternum hairy. Prosternal process rounded apically (Fig. 51).
Tegmina fully developed, extending beyond hind knee. Wings transparent ; bands absent. Male genital plate with a deep apical incision.
Distribution : Numerous species in Central and South America. Only one species in Africa and South-West Asia. Occurs in two phases : ph. *solitaria* and ph. *gregaria.*

Schistocerca gregaria (Forskål, 1775)
Figs. 49–51 ; Plate II : 1

Type Locality : Cairo.

Gryllus gregarius Forskål P., 1775, *Descriptiones animalium, avium, amphibiorum, etc. quae in itinere orientali observavit Petrus Forskål*, p. 81.

Acridium peregrinum Olivier G. A., 1804, *Voyage dans l'Empire Ottoman, l'Egypte et la Perse, fait par ordre du Gouvernement pendant les six premières années de la République*, 2 : 425 footnote ; 4 : 388 footnote.

Acrydium tartaricum Latreille P. A., 1804, *Histoire naturelle générale et particulière des Crustacées et des Insectes. Orthoptera, Acrididae*, 12 : 150.

Gryllus rufescens Thunberg C. P., 1815, *Mém. Acad. Sci. St.-Pétersb.*, 5 : 245.

Acridium flaviventre Burmeister H., 1838, *Handbuch der Entomologie*, Berlin, 2 (2) : 631, 1014.

Acridium (Schistocerca) peregrinum —. Stål C., 1873, *Recensio Orthopterorum*, Stockholm, 1 : 65.

Schistocerca peregrina —. Bolivar I., 1892/93, *Rev. biol. nord Fr.*, 5 : 486.

Schistocerca peregrina —. Giglio-Tos E., 1893, *Boll. Musei Zool. Anat. comp. R. Univ. Torino*, 8 (164) : 8.

Schistocerca gregaria —. Krauss H. A., 1907, *Denkschr. Akad. Wiss. Wien*, 71 (2): 12 (separate).

Schistocerca tatarica —. Kirby W. F., 1910, *A synonymic catalogue of the Orthoptera*, Vol. III, Orthoptera Saltatoria, Part II, London, p. 459.

Schistocerca gregaria —. Bodenheimer F. S., 1930, *Die Schädlingsfauna Palästinas*, Berlin, pp. 94–109.

Schistocerca gregaria —. Bodenheimer F. S., 1932, *Biol. Zbl.*, 52 : 598.

Schistocerca gregaria —. Bodenheimer F. S., 1935, *Animal Life in Palestine*, Jerusalem, pp. 82, 85, 281, 282, 349, 350, 351, 356, 361.

Schistocerca gregaria —. Bodenheimer F. S., 1935, *Arch. Naturgesch.*, 4 (2) : 208.

Schistocerca gregaria —. Ramme W., 1951, *Mitt. zool. Mus. Berl.*, 27 : 428.

Characters as mentioned in description of genus in general. Wings long, extending posteriorly to end of tibia ; median field of tegmina closed ; radius-sector with two branches. Mesosternum swollen ; lobes long and narrow. Metasternal interspace of female almost square, in male very narrow.

Coloration : Mature specimens of ph. *gregaria* lemon-yellow (males) and pinkish-brown (females) ; ph. *solitaria* pale greenish or grey. Tegmina with brown blotches. Reddish coloration common especially during low winter temperatures. In our region only specimens of ph. *gregaria* are found, remnants of previous invasions.

Measurements (mm) : Body ♀ 50.7–61.0, ♂ 45.8–55.3 ; tegmina ♀ 52.9–63.8, ♂44.6–60.5.

Distribution, Israel : Carmel (3), Coastal Plain (4, 8, 9), Northern and Central Negev (15, 17), 'Arava Valley (14), Dead Sea Area (13).

There are two large regions of breeding grounds which produce swarms of *S. gregaria.* The main breeding ground is in Africa, especially south of the Sahara and in the arid regions of Ethiopia (e.g., Ogaden). Others are situated in Asia : Persia, Pakistan, and part of India. From these localities, *S. gregaria* invades Southern Europe, the Middle East, Central Asia and Southern Russia. Since this species breeds in arid, semi-desert zones, far from centres of human civilization and cultivation, it is very difficult to control. Great efforts are being made by the Anti-Locust Research Centre (London) to predict and prevent possible outbreaks.

In one of the recent publications of Dirsh (1974), the African population of *S. gregaria*

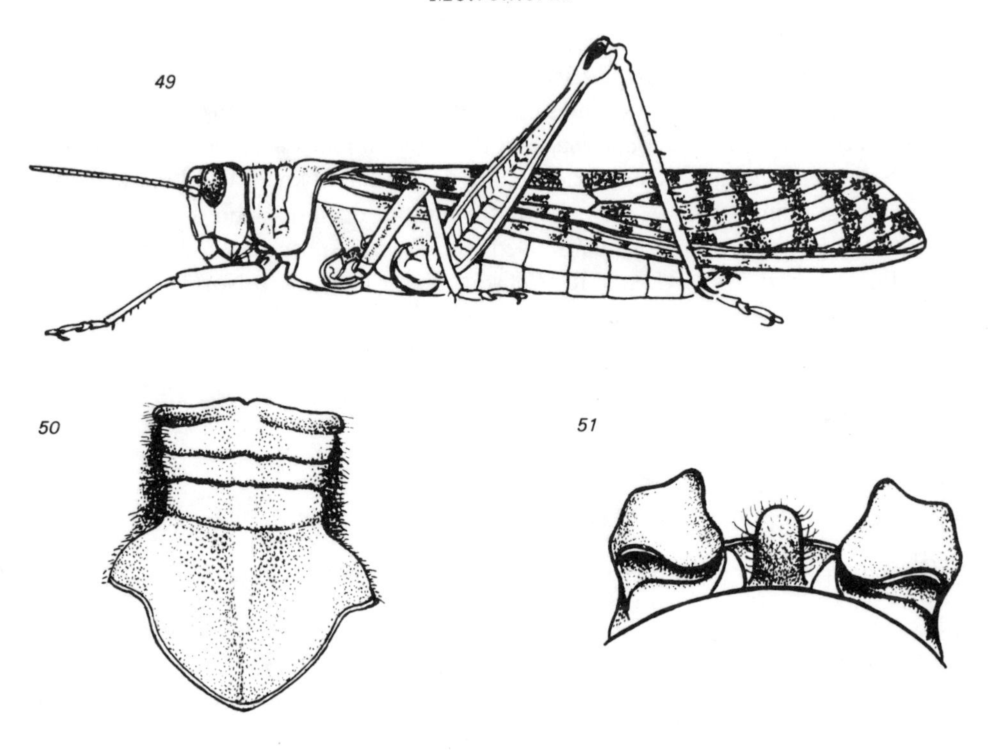

Figs. 49–51 : *Schistocerca gregaria* (Forskål, 1775)
49. female ; 50. pronotum ; 51. prosternal process

is regarded as a subspecies of *S. americana.* As *S. gregaria* has never been found on the American continent, and as its solitary phase forms stable populations reproducing in Africa and Asia, it should be regarded as a well-defined Old World species. The origin of this invader seems to be from an American ancestor, probably closely related to *S. americana.*

Genus METROMERUS Uvarov, 1938

Ann. Mag. nat. Hist., 11 (1) : 379

Kripa (partim) Uvarov B. P., 1922, *Trans. R. ent. Soc. Lond.*, 1922, pp. 118, 124.

Type Species : *Caloptenus coelesyriensis* Giglio-Tos, 1893.

Diagnosis : Medium sized or small, hairs sparse. Head hypognathous or slightly opisthognathous. Frontal ridge flat, punctate, obliterated before clypeus. Vertex

concave, with rounded margins; foveolae absent. Antennae filiform with 23–25 segments. Pronotum with three distinct carinae, intersected by three transverse sulci, the posterior one situated in the middle of pronotum. Posterior angle of metazona obtuse. Sternum square; mesosternal and metasternal lobes wide, separated. Wings fully developed, almost reaching or extending beyond hind knee. Mid tibia with two distinct longitudinal grooves. Cerci of males without dentations on distal lobe.
Distribution: A monotypic genus divided into several subspecies found in South-West Europe, part of the U. S. S. R., Central Asia, Caucasus to North Africa. Only one subspecies in our region.

Metromerus coelesyriensis coelesyriensis (Giglio-Tos, 1893)
Fig. 52

Type Locality: 'Syria' (Turin Museum).

Caloptenus coelesyriensis Giglio-Tos E., 1893, *Boll. Musei Zool. Anat. comp. R. Univ. Torino*, 8 (164): 10, fig. 4.
Kripa coelesyriensis —. Uvarov B. P., 1922, *Trans. R. ent. Soc. Lond.*, 1922: 125, pl. 1, fig. 2.
Calliptamus coelesyriensis —. Buxton P. A. & B. P. Uvarov, 1923, *Bull. Soc. R. ent. Égypte*, p. 205.
Kripa coelesyriensis —. Bodenheimer F. S., 1935, *Animal Life in Palestine*, Jerusalem, pp. 320, 323.
Kripa coelesyriensis —. Bodenheimer F. S., 1935, *Arch. Naturgesch.*, 4 (2): 204.
Metromerus coelesyriensis —. Uvarov B. P., 1943, *Proc. Linn. Soc. Lond.*, 1941–1942: 83, figs. 1c, 2c, 3c.
Metromerus coelesyriensis coelesyriensis —. Bei-Bienko G. Ya. & L. L. Mishchenko, 1951, *Locusts and Grasshoppers of the U. S. S. R. and Adjacent Countries*, I: 259 [in Russian].
Metromerus coelesyriensis —. Ramme W., 1951, *Mitt. zool. Mus. Berl.*, 27: 428.

Medium sized, black or brown. Head of male slightly opisthognathous, in female hypognathous. Frontal ridge punctate, flat, obliterated close to clypeus. Vertex oblong, concave, with indistinct median carina; margins in female rounded, in male semi-parallel. Eyes oblong, slightly longer than subocular groove. Pronotum rugose latero-ventrally, with distinct carinae on dorsal part; lateral carinae slightly curved outwards anteriorly, not extending to posterior margin. Prosternal process cylindrical, slightly pointed at apex, hairy. Mesosternal interspace wider than long; metasternal interspace of female usually square, in male very narrow.
Tegmina short, not reaching, reaching, or extending slightly beyond hind knee; venation at base dense, irregular; median field open or closed. Length of hind femur 2.3–3.2 times its maximum width. Hind tibia shorter than femur, with 8–9 spines on outer line and 9 spines on inner line.
Coloration: Usually uniformly pale brown or black, rarely pale with numerous dark dots and two dorsal longitudinal white lines (Judea population). Tegmina usually of uniform colour. Wings red or reddish; venation dark. The inner surfaces of hind femur and tibia dark reddish in the black forms, sometimes with pale cross-bands;

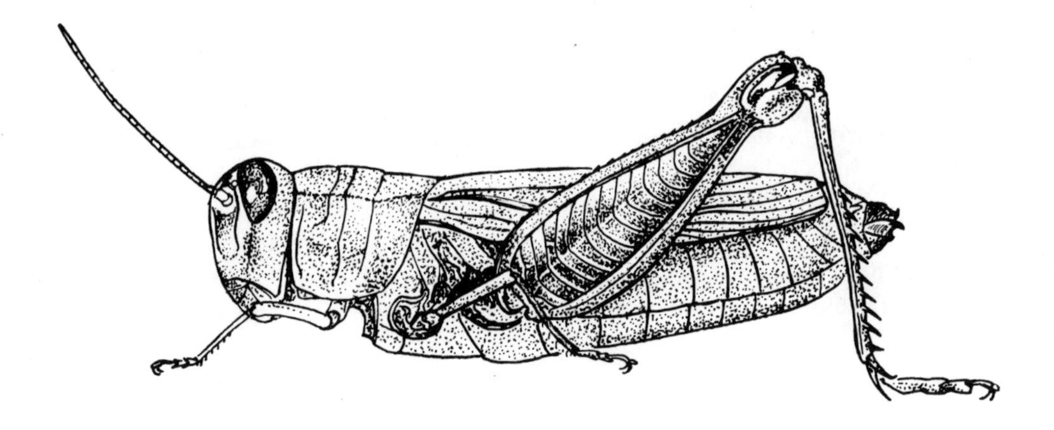

Fig. 52 : *Metromerus coelesyriensis coelesyriensis* (Giglio-Tos, 1893), female

yellowish or reddish in the pale brown forms ; outer side of hind femur with indistinct dark bands and numerous black spots.

Measurements (mm) : Body ♀ 28.2–33.5, ♂ 18.0–32.5 ; tegmina ♀ 17.0–22.5, ♂ 12.5–14.0.

Distribution : Asia Minor, Syria, Lebanon, Israel, Egypt.

Israel : Upper Galilee (1), Jordan Valley (7), Judean Hills (11), Northern Negev (15). Inhabits dry bare mountain slopes in arid regions. Hoppers found from February to May.

Genus CALLIPTAMUS Serville, 1831

Annls Sci. nat. (Zool.), 22 : 284

Caloptenus Burmeister H., 1838, *Handbuch der Entomologie*, Berlin, 2 (2), p. 637.
Calliptenus Stål C., 1873, *Recensio Orthopterorum*, Stockholm, 1 : 38, 72, 73.

Type Species : *Gryllus Locusta italicus* Linnaeus, 1758.

Diagnosis : Small or medium sized. Head hypognathous or slightly opisthognathous. Frontal ridge flat, punctate ; margins diverging near median ocellus, obliterated on clypeus. Vertex depressed or flat ; foveolae absent. Eyes oval. Antennae filiform, shorter than or as long as head and pronotum together. Pronotum almost flat, with three longitudinal carinae extending to posterior margin ; lateral carinae diverging in the middle and usually converging posteriorly (Fig. 57) ; median carina sometimes slightly raised. Three transverse sulci present, the posterior one situated beyond middle of pronotum. Metazona usually more granulate than prozona.

Tegmina fully developed, always covering entire abdomen. Wings transparent, reddish at base. Hind femur short and thick, its length 2.6–3.8 times its maximum width; dorsal carina serrate. Inner surface almost black or with a black vertical band, rarely yellow (Figs. 54, 58). Hind tibia hairy, slightly curved. Male cerci long and slightly curved; two lobes at the apex, the ventral lobe with two black, sharp indentations at the apex (Fig. 55).

Distribution: The genus includes about 15 species distributed in Asia, North Africa and the Mediterranean countries, which are very variable in their general structure, structure of various body parts and their colour. They occur in our region in different ecological habitats, ranging from typical arid ones to areas with dense vegetation.

Key to the Species of Calliptamus in Israel

1. Inner surface of hind femur uniformly yellow, usually with 2–3 black bands which sometimes become fused ventrally, the median one always reaching the ventral margin (Fig. 54). Apical dentations of cerci of male short and usually pressed together. Mesosternal interspace in female usually long and narrow — 2
– Inner surface of hind femur almost entirely black or blackish-violet (Fig. 58). Apical dentations of cerci of male longer, separate. Mesosternal interspace in female wide and short — 3
2. Hind tibia light yellowish or yellowish-grey. — **C. palaestinensis** Ramme
– Hind tibia red or reddish-grey. — **C. palaestinensis erytherocnemis** Ramme
3. Hind tibia yellow. — **C. barbarus pallidipes** Chopard
– Hind tibia red or wine-red. — **C. barbarus deserticola** Vosseler

Calliptamus palaestinensis Ramme, 1930

Figs. 53–55

Type Locality: Ben Shemen, Hadera, Israel.

Gryllus Locusta italicus Linnaeus C., 1758, *Systema Naturae*, 10th ed., 1 : 432.

Calliptamus italicus —. Giglio-Tos. E., 1893, *Boll. Musei Zool. Anat. comp. R. Univ. Torino*, 8 (164) : 8.

Calliptamus italicus —. Krauss H. A., 1909, in: Kneucker A., *Verh. naturw. Ver. Karlsruhe*, 21 : 37.

Calliptamus italicus —. Buxton P. A. & B. P. Uvarov, 1923, *Bull. Soc. R. ent. Égypte*, p. 205.

Calliptamus palaestinensis Ramme W., 1930, *Mitt. zool. Mus. Berl.*, 16 : 395.

Calliptamus palaestinensis —. Bodenheimer F. S., 1930, *Die Schädlingsfauna Palästinas*, Berlin, p. 62.

Calliptamus palaestinensis —. Bodenheimer F. S., 1935, *Animal Life in Palestine*, Jerusalem, pp. 78, 86, 88, 311, 320–323.

Calliptamus palaestinensis —. Bodenheimer F. S., 1935, *Arch. Naturgesch.*, 4 (2) : 202.

Robust, smooth. Head hypognathous. Frontal ridge regularly punctate; margins gradually diverging ventrally, obliterated close to clypeus. Eyes oblong; horizontal diameter 1.5–2.0 times interocular distance. Antennae filiform, shorter than head and pronotum together. Pronotum with a straight, raised median carina, intersected by the first and third transverse sulci. Lateral carinae converging in prozona and metazona, sometimes raised laterally. Ventral margins of lateral lobes rounded. Mesosternal interspace narrow, its margins parallel or slightly diverging posteriorly. Mesosternal interspace in male very narrow, in female square or its length slightly exceeding its width.

Tegmina reaching far or slightly beyond hind knee; median field open; radius-sector with two or three branches. Hind femur thick.

Coloration: Generally brown, light brown, with numerous spots, rarely uniformly brown, and usually with white spots on head and thorax. Two white lines starting from behind the eyes continue on pronotum and on dorsal part of tegmina, sometimes obliterated on head and pronotum, present only on the tegmina. Wings reddish. Inner surface of hind femur usually with three black bands, often fused along the ventral line; basal band on inner-dorsal part of femur often absent (Fig. 54). Hind tibia yellow.

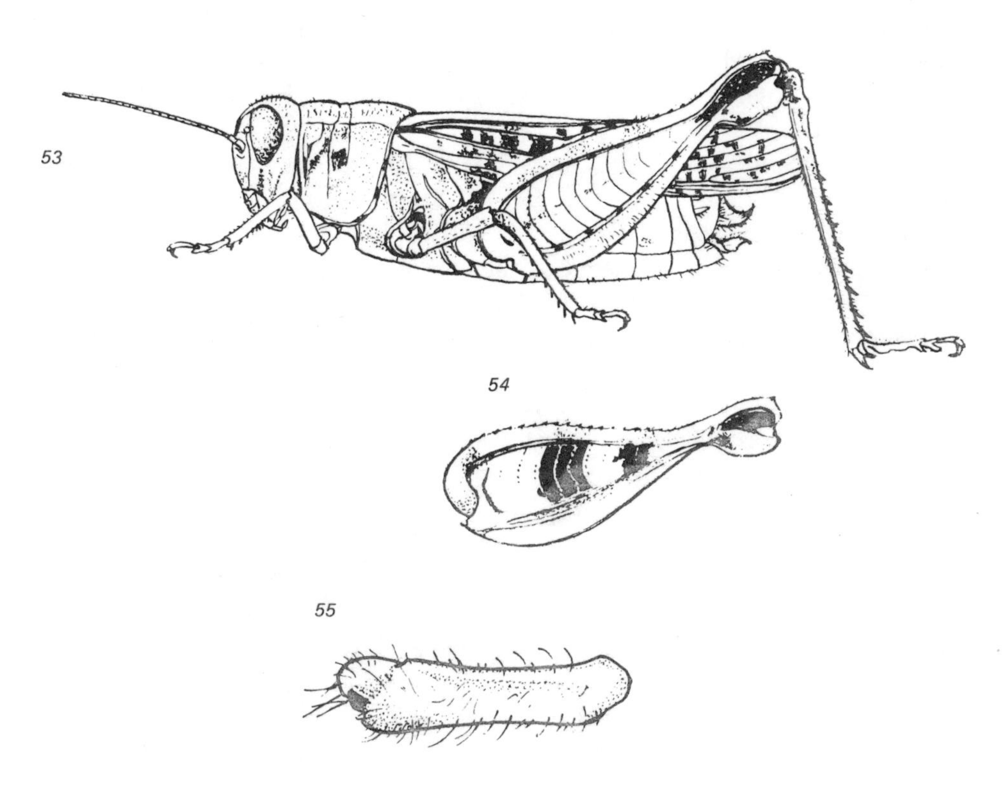

Figs. 53–55: *Calliptamus palaestinensis* Ramme, 1930
53. female; 54. inner side of hind femur; 55. cercus of male

Measurements (mm): Body ♀ 29.5–36.2, ♂ 22.0–26.5; tegmina ♀ 17.5–20.2, ♂ 14.0–17.5.
Distribution, Israel: Upper and Lower Galilee (1, 2), Jordan Valley (7), Coastal Plain (4, 8, 9), Judean Hills (11).
In summer this is one of the most common species in the northern parts of Israel, especially on mountain slopes and in the Coastal Plain, in areas with bushy vegetation.

Calliptamus palaestinensis erytherocnemis Ramme, 1951

Type Locality: 'Syria'.

Calliptamus palaestinensis erytherocnemis Ramme W., 1951, *Mitt. zool. Mus. Berl.*, 27: 313.

More slender than the nominate species, especially males, which have more delicate cerci. Tegmina not reaching hind knee; median field sometimes closed; radius-sector of male with one branch, in female with two branches.
Coloration: Generally as in *C. palaestinensis*, but hind tibia reddish, or reddish-grey and with a pale ring close to knee. Inner surface of hind femur with two large dark bands, one apical and one median; the third black band usually distinct only on the dorsal part or slightly also on the inner part of femur, close to its base. Wings violet-reddish.
Measurements (mm): Body ♀ 28.5–30.5, ♂ 14.5–18.0; tegmina ♀ 17.5–21.0, ♂ 11.5–14.0.
Distribution: Syria, Lebanon, Israel.
Israel: Upper Galilee (1), Northern Coastal Plain (4).
This subspecies is apparently typical of humid mountain regions. In lower regions it occurs together with the nominate subspecies and with subspecies of *C. barbarus.*

Calliptamus barbarus pallidipes Chopard, 1943
Figs. 56–58

Type Locality: 'Morocco' (Paris Museum).

Calliptamus barbarus var. *pallidipes* Chopard L., 1943, *Faune Emp. franç.*, Paris, 1: 404.
Calliptamus babarus pallidipes —. Ramme W., 1951, *Mitt. zool. Mus. Berl.*, 27: 311.

Robust, strong. Head hypognathous. Frontal ridge sparsely punctate; margins parallel or slightly diverging above median ocellus. Vertex depressed, elliptical or elongate; occiput with or without median carina. Lateral carinae of pronotum parallel (Fig. 57) or slightly diverging posteriorly. Median carina linear, in female usually intersected by all three transverse sulci. Ventral margin of lateral lobes straight at least in the middle. Mesosternal interspace square, or long and narrow; metasternal interspace narrow, or lobes contiguous.

Median field of tegmina closed ; radius-sector with two branches. Hind femur wide, its ventral carina slightly higher beyond the middle. Hind tibia with strong spurs ; dorsal inner spur at least 1.5 times as long as dorsal outer spur.

Coloration : Brown, pale brown with a pattern of brown and white blotches and a typical white pattern on head and pronotum. Inner surface of hind femur with a large black area usually extending to ventral margin, merging dorsally with two dark bands which continue on outer surface of femur (Fig. 58). Hind tibia yellow.

Measurements (mm) : Body ♀ 27.0–37.5, ♂ 20.0–33.5 ; tegmina ♀ 26.0–24.5, ♂ 17.5–23.5.

Distribution : North Africa, Israel, Syria, Turkey, Caucasus, Central Asia, Afghanistan.

Israel : Coastal Plain (4, 8, 9), Northern and Central Negev (15, 17).

Field observations show that this subspecies has a greater affinity to areas of vegetation than the following subspecies.

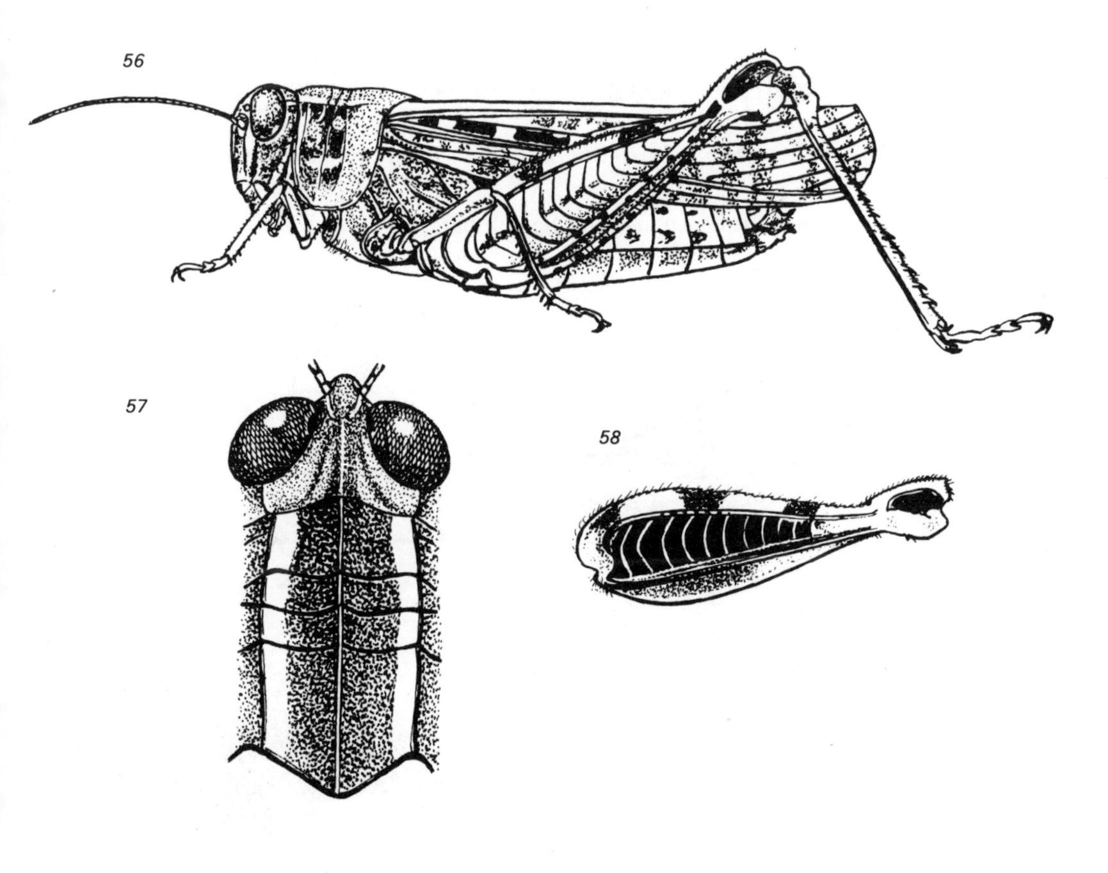

Figs. 56–58 : *Calliptamus barbarus pallidipes* Chopard, 1943
56. female ; 57. head and pronotum ; 58. inner side of hind femur

Calliptamus barbarus deserticola Vosseler, 1902

Figs. 59, 60

Type Locality : Laghouat, Algeria.

Caloptenus italicus var. *deserticola* Vosseler J., 1902, *Zool. Jb.* (Syst.), 16 : 395.
Calliptamus italicus var. *deserticola* —. Werner F., 1908, *ibid.*, 27 : 126.
Caloptenus deserticola —. Kheil N. M., 1915, *Int. ent. Z.*, 9 : 89, 94, 101, 102, fig. 2.
Calliptamus deserticola —. Capra F., 1929, *Annali Mus. civ. Stor. nat. Genova*, 53 : 151.
Calliptamus siculus deserticola —. Werner F., 1932, *Sber. Akad. Wiss. Wien*, (1) 141 : 173.
Calliptamus siculus deserticola —. Bodenheimer F. S., 1935, *Arch. Naturgesch.*, 4 (2) : 204.
Calliptamus barbarus var. *deserticola* —. Ramme W., 1951, *Mitt. zool. Mus. Berl.*, 27 : 428.

Resembling the former subspecies in general, differing in the more robust body, large size and more slender femurs. Ventro-apical parts of hind femur, hind tibia and tarsus red or wine-red. Dark area on inner surface of hind femur black tinged with violet (Fig. 60).
Measurements (mm) : Body ♀ 27.5–39.5, ♂ 19.5–25.5 ; tegmina ♀ 26.8–35.0, ♂ 18.0–24.5.
Distribution : Generally overlapping with *C. b. pallidipes*, but found in more arid habitats.
Israel : Central and Southern Coastal Plain (8, 9), Northern and Central Negev (15, 17).
Copulation observed during August and September, hoppers in March and April.

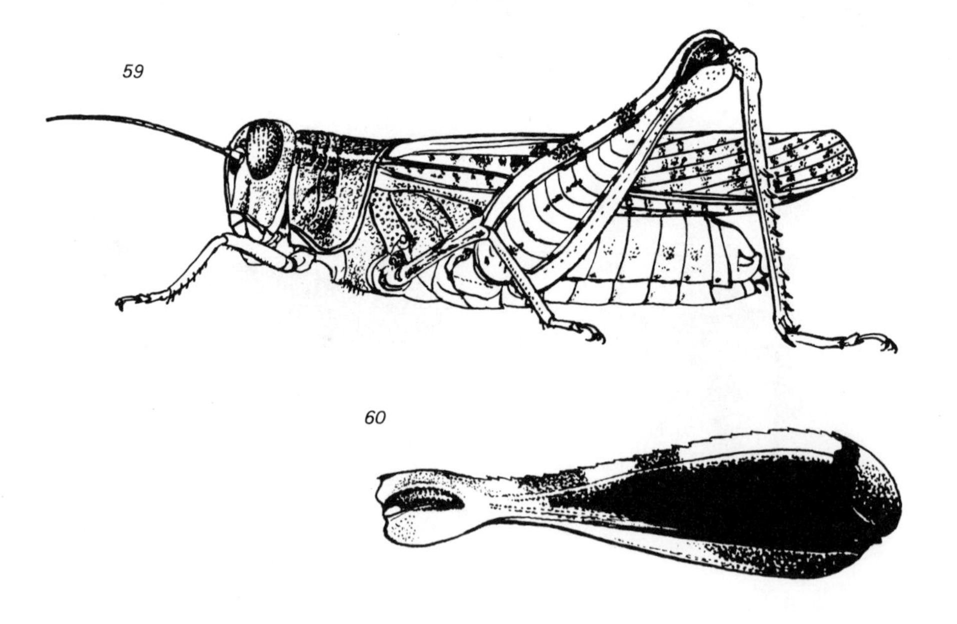

Figs. 59–60 : *Calliptamus barbarus deserticola* Vosseler, 1902
59. female ; 60. inner side of hind femur

Genus SPHODROMERUS Stål, 1873

Recensio Orthopterorum, Stockholm, 1 : 72

Calliptenus (Sphodromerus) Stål C., 1873, *Recensio Orthopterorum*, Stockholm, 1 : 72.
Sphodromerus —. Finot A., 1895, *Annls Soc. ent. Fr.*, 64 : 526, 547.
Kripa Kirby W. F., 1914, *The Fauna of British India, including Ceylon and Burma. Orthoptera (Acrididae)*, London, pp. 195, 257 (partim).

Type Species : *Calliptamus serapis* Serville, 1838.
Diagnosis : Short, strong, hairs sparse. Head hypognathous ; frontal ridge flat, punctate, with margins diverging gradually towards clypeus. Vertex depressed longitudinally ; median carina absent. Eyes large, irregularly oval. Antennae filiform, in males longer than, in females as long as or shorter than head and pronotum together. Pronotum rounded ; lateral carinae indistinct, sometimes marked only in prozona ; median carina distinct along its entire length, slightly raised in metazona, intersected by posterior transverse sulcus situated in the middle of pronotum or beyond it. Posterior angle of metazona obtuse. Hind femur very thick, less than three times as long as wide, its maximum width behind the middle (Fig. 61). Dorsal carina serrate and covered densely with hair. Wings fully developed, extending to or often beyond hind knee ; venation dark. Tegmina wide and short, usually with transverse dark apical stripes.
Distribution : About 17 species are known in North Africa, Somalia, West Asia and North India. Characteristic geophiles, mainly occurring on bare mountains and rocky areas. Two species in our region.

Key to the Species of Sphodromerus in Israel

1. Hind tibia reddish-purple. Posterior transverse sulcus of pronotum situated beyond the middle. In female base of wings light and transparent (Fig. 61). Inner surface of hind femur and inner line of spines reddish-purple. **S. serapis** (Serville)
– Hind tibia and its spines yellow. Posterior transverse sulcus of pronotum situated in the middle, forming a prozona and metazona of equal length. In female basal part of wings slightly purple. Only part of inner surface of femur purple. **S. pilipes** (Janson)

Sphodromerus serapis (Serville, 1838)

Figs. 61, 62 ; Plate II : 2

Type Locality : Dead Sea.

Calliptamus serapis Serville J. G. A., 1838, *Histoire naturelle des Insectes*, in : Roret, *Collection des Suites à Buffon. Orthoptères*, Paris, p. 689.
Caloptenus scriptipennis Walker F., 1870, *Catalogue of the specimens of Dermaptera Saltatoria in the Collection of the British Museum*, London, Part IV, p. 625.

Calliptenus (Sphodromerus) serapis —. Stål C., 1876, *K. svenka VetenskAkad. Handl.*, 4 (57) : 12,
Sphodromerus serapis —. Innes W., 1929, *Mém. Soc. R. ent. Égypte*, 3 (2) : 129, 155.

Robust, medium sized. Head hypognathous ; face smooth, with sparse punctations which form vertical lines on frontal ridge. Vertex depressed ; margins parallel or slightly curved. Occiput swollen, with an indistinct median carina. Pronotum with posterior transverse sulci situated beyond its middle, the first and second curved posteriorly, intersecting the median carina ; sometimes not intersecting the median carina. Anterior part of pronotum with indications of lateral carinae ; metazona densely punctate. Mesosternal interspace with diverging margins, wider than long ; in female as long as mesosternal lobes.
Tegmina wide, rounded apically ; a regular network of diagonal veinlets present on apical part ; radius-sector with four branches. Wings with thick principal veins. Hind femur strong, its length 2.0–2.4 times its maximum width.
Coloration : Brown, pale brown or reddish-brown, males also blackish or blackish-blue. Head and pronotum usually with light blotches. Tegmina of uniform colour or blotched at base ; apical part usually with a pattern of transverse brown bands. Wings of male violet, in female transparent, milky. Hind femur brown dorsally, usually with two dark bands ; inner ventral parts red to dark violet. Inner and dorsal parts of hind tibia reddish-violet to dark violet ; spines usually dark.

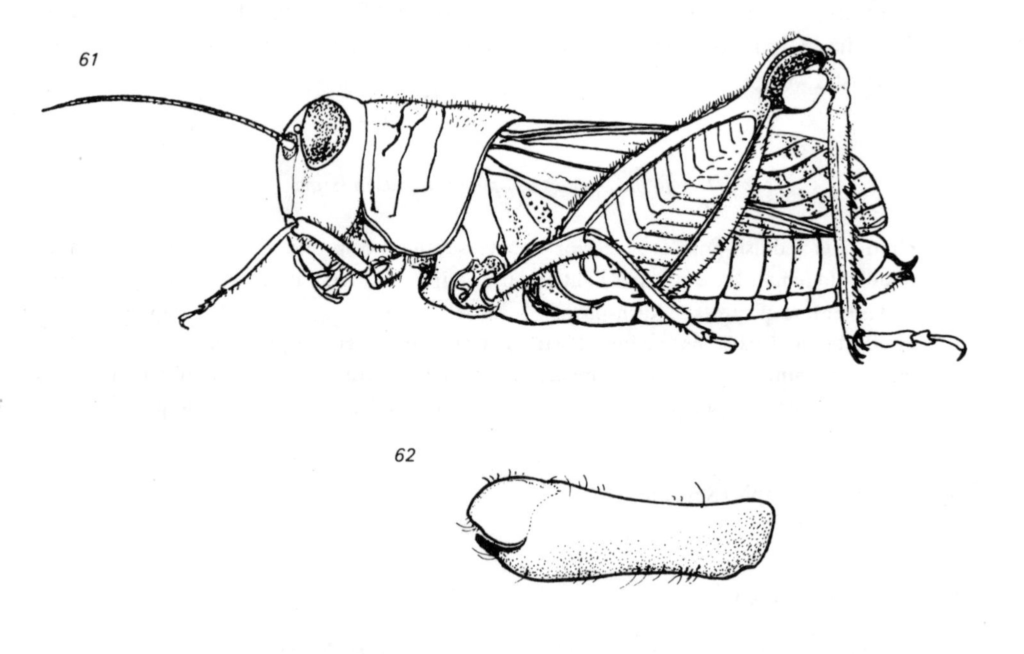

Figs. 61–62 : *Sphodromerus serapis* (Serville, 1838)
61. female ; 62. cercus of male

Measurements (mm) : Body ♀ 31.0–40.5, ♂ 22.5–28.0 ; tegmina ♀ 19.5–25.5, ♂ 12.5–19.0.
Distribution : Egypt, Israel.
Israel : Judean Hills (11), Dead Sea Area (13), Northern and Central Negev (15, 17). A typical inhabitant of rocky mountain slopes, especially in the Southern Judean Hills. Found among bushes of *Zygophyllum* and *Raemuria.* When disturbed, specimens hide in the shadow of stones rather than fly away. Hoppers observed during October and November.

Sphodromerus pilipes (Janson, 1891)

Type Locality : Dead Sea.

Caloptenus pilipes Janson O., 1891, *Orthoptera*, in: Hart H. C., *Fauna and Flora of Sinai, Petra, and Wady Arabah*, London, pp. 183, 185.
Sphodromerus serapis —. Bolivar I., 1892/3, *Rev. biol. nord Fr.*, 5 : 486.
Sphodromerus pilipes —. Buxton P. A. & B. P. Uvarov, 1923, *Bull. Soc. R. ent. Égypte*, p. 204.
Sphodromerus pilipes —. Bodenheimer F. S., 1935, *Arch. Naturgesch.*, 4 (2) : 205.
Sphodromerus pilipes —. Uvarov B. P., 1943, *Proc. Linn. Soc. Lond.*, 1941–1942 : 79.

Robust, short, resembling former species, but rarer. Differs from *S. serapis* in that the hind tibia is almost completely yellow and the inner surface of the hind femur is partly yellow. Pronotum less granulate; posterior transverse sulci in middle of pronotum, forming prozona and metazona of equal length. Mesosternal interspace with parallel margins, less wide than mesosternal lobes. Length of hind femur 2.0–2.1 times its maximum width.
Coloration : Brown ; tegmina of uniform colour at the base, with a pattern of transverse bands at the apex. Wings of male and female slightly violet.
Measurements (mm) : Body ♀ 34.5–36.5, ♂ 23.0–25.0 ; tegmina ♀ 20.5–22.0, ♂ 16.5–17.5.
Distribution : Egypt, Israel.
Israel : Found only in the Judean Hills (11).

Genus CYCLOPTERNACRIS Ramme, 1928
Eos, Madr., 4 : 114

Type Species : *Acridium morbosum* Serville, 1838.
Diagnosis : Medium sized to large, smooth, hairs sparse. Head without distinct carinae ; frontal ridge flat, obliterated ventrally. Eyes large, elliptical ; horizontal diameter more than twice interorbital distance. Antennae longer than head and pronotum together; segments of apical part elongate and densely punctate. Pronotum large,

slightly swollen along median carinae in prozona; lateral carinae slightly marked or absent in prozona.
Tegmina opaque, with dense veins at base and with numerous regular cross-veinlets, not extending posteriorly beyond hind knee. Wings transparent, rounded. Hind femur long, markedly narrowing apically.
The genus includes four or five species, distributed from North Africa to South-West Asia. Most species are typical geophiles and very variable in development of tegmina and wings. Two species in our region.

Key to the Species of Cyclopternacris in Israel and Sinai

1. Inner surface of hind femur almost entirely bluish-black. Hind tibia usually yellow. Mesosternal interspace in female trapezoidal, its margins diverging posteriorly. **C. cincticollis** (Walker)
– Inner surface of hind femur black only near the base and along ventral margin, usually with dark bands. Hind tibia entirely red. Margins of mesosternal interspace almost parallel. **C. morbosa** (Serville)

Cyclopternacris cincticollis (Walker, 1870)
Fig. 63; Plate II: 3

Neotype Locality: Shivta, Negev (Uvarov, 1959) (British Museum).

Caloptenus cincticollis Walker F., 1870, *Catalogue of the specimens of Dermaptera Saltatoria in the Collection of the British Museum*, London, Part IV, pp. 689–690.
Euryphymus cincticollis —. Kirby W. F., 1910, *A synonymic catalogue of the Orthoptera*, Vol. III, Orthoptera Saltatoria, Part II, London, p. 34 (partim).
Calliptamus (Caloptenus) cincticollis —. Innes W., 1929, *Mém. Soç. R. ent. Égypte*, 3 (2): 156.
Cyclopternacris morbosa Serville. Bodenheimer F. S., 1935, *Arch. Naturgesch.*, 4 (2): 209.
Cyclopternacris cincticollis —. Uvarov B. P., 1959, *Ent. Ber. Amst.*, 19: 23.

Medium sized, hairs sparse. Head hypognathous or slightly opisthognathous; frontal ridge wide, flat, sparsely punctate, margins diverging and obliterated towards clypeus. Fastigium slightly concave, nearly pentagonal. Occiput swollen, protruding above eyes, with indications of median carina. Pronotum large; prozona nearly 1.5–1.7 times longer than metazona, divided by first transverse sulcus into two, nearly equal parts. Disc of prozona smooth; metazona and posterior parts of lateral lobes punctate. Median carina prominent; lateral carinae distinct in prozona. Prosternal process round-cylindrical.
Tegmina membranous, more or less shortened, sometimes covering only part of the abdomen or nearly reaching hind knee. Hind femur slender, longer than hind tibia.
Coloration: Generally brown or pale brown. Inner surface of femur almost black, except at upper apex. Hind tibia usually yellow, sometimes with traces of light red; hind tarsus yellow.

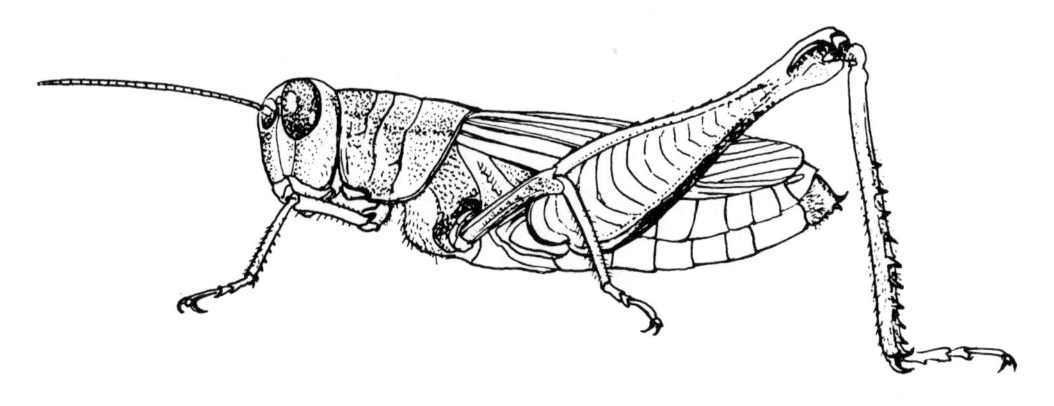

Fig. 63 : *Cyclopternacris cincticollis* (Walker, 1870), female

Measurements (mm) : Body ♀ 33.0–44.5, ♂ 22.0–24.0 ; tegmina ♀ 16.5–29.0, ♂ 11.0–16.0.
Distribution : North Africa, Sinai, Israel, Transjordan.
Israel : Dead Sea Area (13), Northern and Central Negev (15, 17).
Characteristic of mountains in desert regions, with a tendency to form isolated brachypterous populations. Hoppers appear during the winter months (from January to March) ; copulation observed from March to June.

Cyclopternacris morbosa (Serville, 1838)

Type Locality : 'Egypt' (Paris Museum).

Acridium morbosum Serville A., 1838, *Histoire naturelle des Insectes*, In : Roret, *Collection des Suites à Buffon. Orthoptères*, Paris, p. 682.
Cyclopternacris morbosa —. Ramme W., 1928, *Eos, Madr.*, 4 : 114.
Cyclopternacris morbosa —. Uvarov B. P., 1959, *Ent. Ber. Amst.*, 19 : 23–24.

Larger than *C. cincticollis.* Tegmina and wings cover the whole abdomen, usually reaching the hind knee. Head slightly opisthognathous; frontal ridge narrowing apically. Fastigium elongate, pentagonal, margins rounded. Occiput swollen; median carina present. First transverse sulcus situated in middle of pronotum or anterior to it. Lateral carinae distinct in prozona, slightly raised, punctate. Metazona densely punctate, obtuse-angled.
Coloration : Generally brown with a pattern of pale cross-bands on head and pronotum. Tegmina opaque-brown; wings greenish; venation dark, green at apex. Hind femur black ventrally ; inner surface with black and yellow bands and a dark basal

part fused with the black ventral part. Hind tibia red or reddish; dorsal part sometimes yellowish with a black ring close to knee; hind tarsus red or reddish-yellow.
Measurements (mm): Body ♀ 47.0–48.5, ♂ 28.5–30.0; tegmina ♀ 32.5–35.5, ♂ 18.5–19.5.
Distribution, Sinai: Sinai Mountains (22; St. Katharina) and vicinity.
Israel: Central Negev (17).

Genus PAMPHAGULUS Uvarov, 1929

In: *Ergebnisse der Sinai-Expedition*, Leipzig, p. 99

Type Species: *Pamphagulus bodenheimeri* Uvarov, 1929.
Diagnosis: Small, rugose, apterous. Head slightly opisthognathous. Fastigium projecting forwards from between the eyes, with a deeply incised frontal ridge. Vertex deeply depressed; sharp carinulae and median carina present. Occiput rugose. Antennae ensiform, short. Prozona of pronotum long, with a sharp hump (Fig. 64). Metazona very short, granulate, its median carina indistinct. Posterior margin of pronotum nearly straight, slightly curved inwards in middle.
Distribution: North Africa, Sinai, Israel.
One species in Israel.

Pamphagulus bodenheimeri Uvarov, 1929

Fig. 64; Plate I: 3; colour plate: 4

Type Locality: Abu Slima, Suez (British Museum).

Pamphagulus bodenheimeri Uvarov B. P., 1929, in: *Ergebnisse der Sinai-Expedition*, Leipzig, p. 99.

Small, rugose, tuberculate; wings entirely absent. Frontal ridge concave, its sharp margins diverging towards clypeus. Vertex rugose, transversely depressed with a median carina continuing on the occiput. Eyes round, shorter than subocular groove. Antennae short and thick. Pronotum without lateral carinae. Prosternal process collar-like, its apex sharp. Sternum rugose; mesosternal and metasternal interspaces equally wide. Abdominal terga tuberculate; median carina present. Outer surface of hind femur with more or less symmetrical plates. Tibia longer than femur; external apical spine present; 10 spines on inner and outer margins. Spurs long.
Measurements (mm): Body ♀ 20.0–21.5; hind femur ♀ 10.5–11.0; pronotum ♀ 3.6–4.1.
Distribution, Israel: Central Negev (17).
Found on *Artemisia herba-alba* and *Zygophyllum rumosum*.

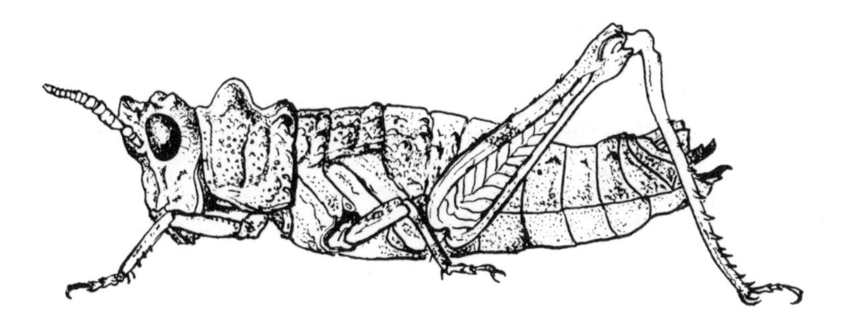

Fig. 64 : *Pamphagulus bodenheimeri* Uvarov, 1929, female

Genus PEZOTETTIX Burmeister, 1840
Z. Ent., Leipzig, 2 : 49–51

Pelecyclus Fieber F. X., 1853, *Lotos*, 3 : 119.
Platyphyma Fischer L. H., 1853, *Orthoptera Europaea*, Leipzig, pp. 298, 373.

Type Species: *Gryllus giornae* Rossi, 1794.
Diagnosis : Apterous, small, smooth, hairs sparse. Head short, slightly opisthognathous. Frontal ridge flat ; margins parallel. Vertex sloping, slightly concave, with a median carina. Fastigium rounded. Foveolae indistinct. Eyes large, oblong, much longer than subocular groove. Occiput with median carina. Pronotum cylindrical, its median carina distinct ; the slightly raised lateral carinae diverging in metazona which is intersected by three transverse sulci. Posterior margin of pronotum rounded (Fig. 66) or slightly curved inwards. Cerci of male conical or widened at apex (Figs. 68, 69). Prosternal process wide, its apex smooth. Vestigial tegmina oblong, oval, narrowing apically, divided into two parts by a median vein (Fig. 67). Hind femur thick, with dark spots on inner side. Arolium larger than claws.
Distribution : Southern Europe, North Africa, West Asia.
Two species in our region.
These species inhabit mountain slopes with dry vegetation, under which they conceal themselves expertly.

Key to the Species of Pezotettix in Israel

1. Cerci of male wide, expanded apically (Fig. 69). Posterior margin of subgenital plate of female with three undulate lobes. **P. judaica** Uvarov
– Cerci of male conical, pointed at the apex (Fig. 68). Subgenital plate of female triangular with a central lobe. **P. curvicerca** Uvarov

Pezotettix curvicerca Uvarov, 1934

Figs 65–68 ; Plate II : 4

Type Locality : Jerusalem (British Museum).

Pezotettix curvicerca Uvarov B. P., 1934, *Eos, Madr.*, 10 : 113, figs. 34, 35C.

Pezotettix curvicerca —. Bodenheimer F. S., 1935, *Animal Life in Palestine*, Jerusalem, pp. 86, 320.

Pezotettix curvicerca —. Bodenheimer F. S., 1935, *Arch. Naturgesch.*, 4 (2) : 201.

Resembles *P. judaica*, but differs in the conical cerci of the male (Fig. 68) and in the triangular margin of the subgenital plate of the female.

Coloration : Brown, dorsal part usually lighter than sides.

Measurements (mm) : Body ♀ 17.0, ♂ 10.5 ; vestigial tegmina ♀ 3.5, ♂ 2.2.

Distribution, Israel : Jordan Valley (7), Judean Hills (11). Rarer than *Pezotettix judaica.*

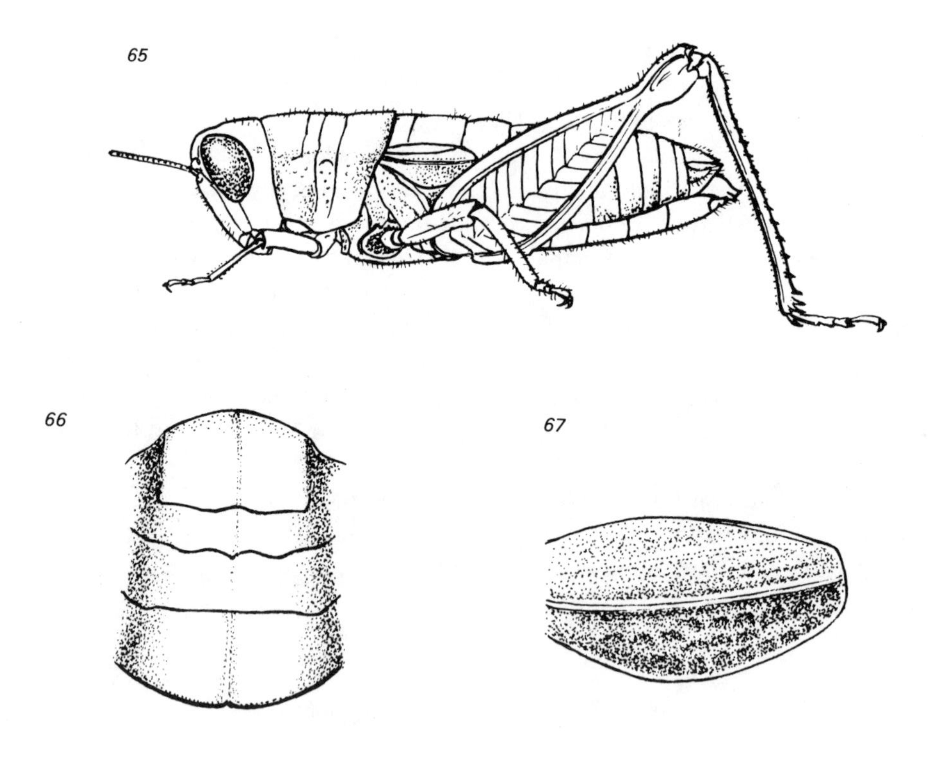

Figs. 65–67 : *Pezotettix curvicerca* Uvarov, 1934
65. female ; 66. pronotum ; 67. tegmen

Pezotettix judaica Uvarov, 1934
Fig. 69

Type Locality : Qiryat 'Anavim (Jerusalem), Israel.

Gryllus giornae Rossi P., 1794, *Mantissa Insectorum*, 2 : 104.
Pezotettix giornae —. Buxton P. A. & N. P. Uvarov, 1923, *Bull. Soc. R. ent. Égypte*, p. 203.
Pezotettix judaica Uvarov B. P., 1934, *Eos, Madr.*, 10 : 113.
Pezotettix judaica —. Bodenheimer F. S., 1935, *Animal Life in Palestine*, Jerusalem, pp. 79, 88, 89, 311, 320, 323.
Pezotettix judaica —. Bodenheimer F. S., 1935, *Arch. Naturgesch.*, 4 (2) : 201.

Small, apterous. Head opisthognathous. Frontal ridge flat, punctate ; margins parallel. Vertex short, as wide as maximum width of frontal ridge, its median carina not very distinct but extending onto occiput. Eyes elliptical, protruding laterally ; vertical diameter 2.5–3.0 times length of subocular groove. Antennae filiform, shorter than head and pronotum together. Pronotum cylindrical ; median carina slightly raised, intersected by three transverse sulci. Lateral carinae distinct or slightly obliterated, diverging posteriorly. Anterior part of prozona before the first sulcus, about the same length as metazona. Prosternal process wide and flattened, usually with inward-curved apex, bent posteriorly towards mesosternum. Cerci of male markedly widened at apex (Fig. 69). Subgenital plate of female trilobed posteriorly. Vestigial tegmina with several longitudinal veins and a network of small veinlets ; their length no more than about 1.5–1.75 times their maximum width.
Coloration : Brown, pale brown or pale, sometimes pronotum and tegmina pale dorsally, brown laterally. Inner side of hind femur usually with two or three black bands. Usually a black line on sides of abdominal terga.
Measurements (mm) : Body ♀ 13.5–17.0, ♂ 12.0–14.0 ; vestigial tegmina ♀ 2.5–3.2, ♂ 2.0–2.3.
Distribution : Syria, Israel.
Israel : Upper and Lower Galilee (1, 2), Coastal Plain (4, 8), Judean Hills (11).
Inhabits areas of dense vegetation, mainly on mountain slopes, hiding among stems and branches. Oviposition in the rainy season ; in captivity during December to January, usually on wet soil. Length of egg-pod 7.0–20.0 mm ; number of eggs 6–60.

68 69

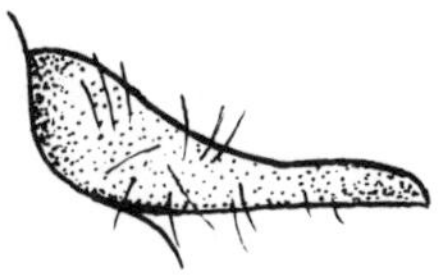

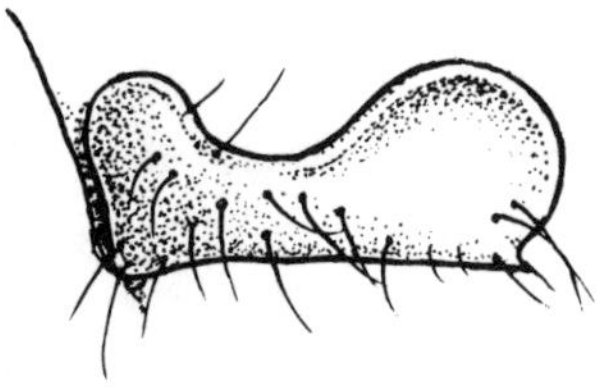

Figs. 68–69 : Cerci of *Pezotettix* males
68. *P. curvicerca* Uvarov, 1934 ; 69. *P. judaica* Uvarov, 1934

Genus TROPIDOPOLA Stål, 1873

Recensio Orthopterorum, Stockholm, 1 : 43, 86

Type Species : *Opsomala fasciculata* Charpentier, 1841.
Diagnosis : Body smooth, cylindrical. Head large, conical, opisthognathous, wider than pronotum at level of eyes. Eyes oval, situated in the longitudinal axis of head. Frontal ridge depressed. Vertex protruding, triangular and flat ; median carina present ; foveolae latero-vertical, triangular. Antennae shorter than head and pronotum together. Pronotum oblong, rounded, with three shallow transverse sulci. Median carina distinct, smooth ; lateral carinae indistinct. Posterior margin of pronotum rounded. Prosternal process triangular, widening apically (Fig. 71). Mesosternal and metasternal lobes contiguous or interspace between them very narrow. Tegmina usually not extending to end of abdomen.
Distribution : The five species of this genus occur in different parts of Asia, Africa and Southern Europe.
One species in our region.
All the species are characteristically phytophilous, inhabiting areas rich in different kinds of Gramineae on which they feed and settle.

Tropidopola longicornis syriaca (Walker, 1871)
Figs. 70–72 ; colour plate : 2

Type Locality : 'Syria'.

Opsomala longicornis Fieber F. X., 1853, *Lotos*, 3 : 98 (partim).
Opsomala syriaca Walker F., 1871, *Catalogue of the specimens of Dermpatera Saltatoria in the Collection of the British Museum*, London, Suppl., Part V, p. 51.
Tropidopola longicornis —. Buxton P. A. & B. P. Uvarov, 1923, *Bull. Soc. R. ent. Égypte*, p. 203.
Tropidopola longicornis longicornis —. Uvarov B. P., 1926, *Eos, Madr.*, 2 : 171.
Tropidopola longicornis —. Bodenheimer F. S., 1935, *Animal Life in Palestine*, Jerusalem, pp. 320, 321.
Tropidopola longicornis longicornis —. Bodenheimer F. S., 1935, *Arch. Naturgesch.*, 4 (2) : 201.
Tropidopola longicornis —. Ramme W., 1951, *Mitt. zool. Mus. Berl.*, 27 : 428.

Medium sized or large. Head markedly opisthognathous. Frontal ridge depressed, hairy, its margins gradually diverging towards clypeus ; foveolae very small, situated laterally ; vertex triangular, its width between the eyes shorter than the distance between its apex and margin of eye. Occiput slightly globular, with or without a median carina. Antennae short ; most segments square or short and wide. Pronotum clearly punctate, especially in metazona ; its posterior margin rounded. Prosternal process widened and flat at the apex (Fig. 71). Mesosternal and metasternal lobes contiguous (Fig. 72). Abdominal sternites 3–6 with hairy areas at the posterior margin. Cerci curved, laterally compressed ; margins of valves of ovipositor dentate. Tegmina narrow, slightly shorter than abdomen ; venation dense. Wings colourless, transparent.

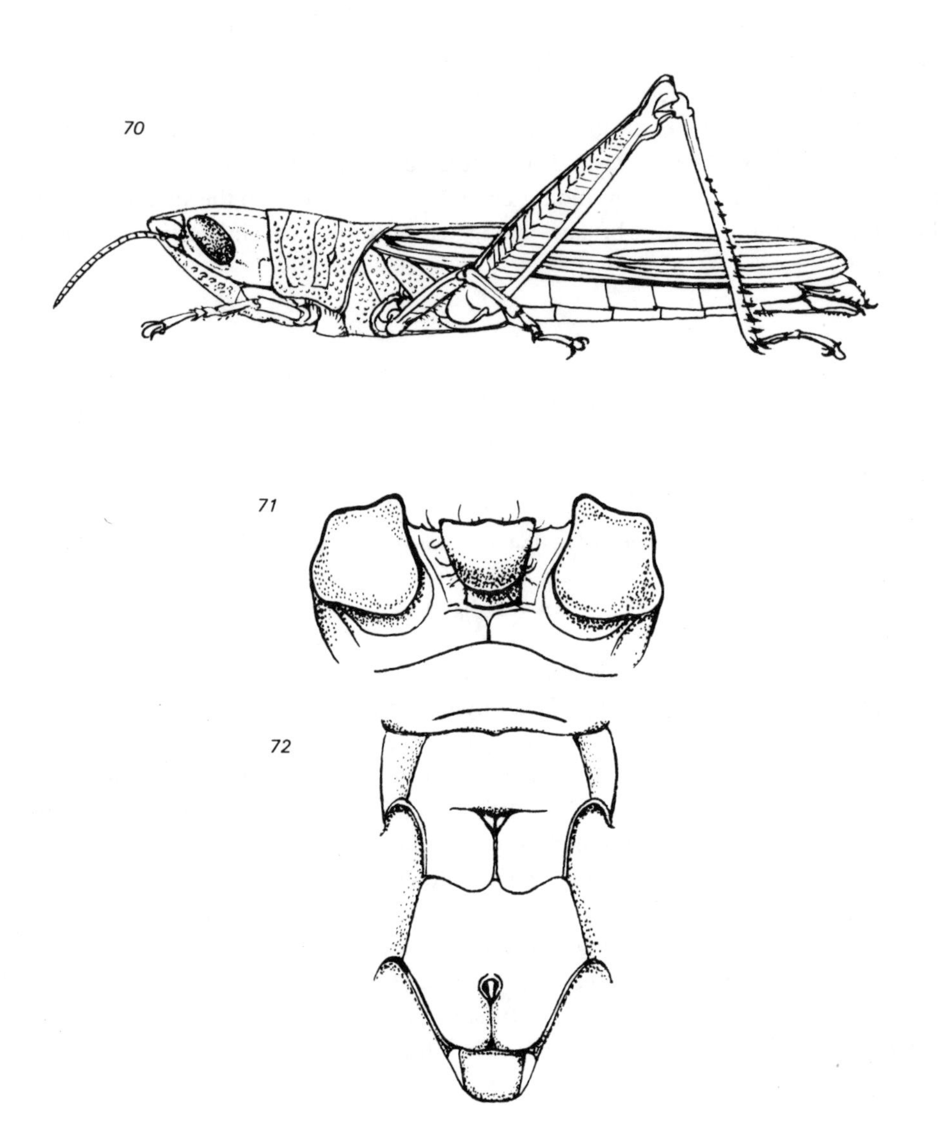

Figs. 70–72 : *Tropidopola longicornis syriaca* (Walker, 1871)
70. female ; 71. prosternal process ; 72. sternum

Coloration : Pale brown, straw-coloured, ochraceous or greenish. Hoppers usually green. Inner surface of hind femur pinkish with dark vertical striation ; hind tibia bluish or bluish-grey.

Measurements (mm) : Body ♀ 33.5–42.5, ♂ 27.5–33.0 ; tegmina ♀ 22.0–27.5, ♂ 18.5–22.0.

Distribution : North Africa, Syria, Asia Minor, Greece and Cyprus.
Israel : Upper and Lower Galilee (1, 2), Coastal Plain (4, 8, 9), Jordan Valley (7), Dead Sea Area (13), Judean Hills (11), Sinai (20, 21).
A typical inhabitant of vegetation near water bodies, usually settling on the stems of *Typha, Phragmites, Ammophila* and other grasses. When disturbed it attaches itself more firmly to the plant's stem, hiding behind it. Attachment to the stem is aided by the large arolium and hairy brushes along the posterior margins of the abdominal plates.

Genus DERICORYS Serville, 1838

Histoire naturelle des Insectes, in : Roret, *Collection des suites à Buffon. Orthoptères*, Paris, pp. 568, 638

Cyphophorus Fischer de Waldheim G., 1846, *Entomographie de la Russie. Orthoptères de la Russie*, pp. 228, 253.
Derocorystes Redtenbacher J., 1889, *Wien. ent. Ztg.*, 8 : 29.

Type Species : *Dericorys albidula* Serville, 1838.
Diagnosis : Medium sized or large. Head hypognathous or slightly opisthognathous. Frontal ridge concave ; margins distinct, converging towards the median ocellus, diverging near the clypeus. Vertex slightly concave, with a distinct median carina which continues on the occiput. Foveolae irregularly shaped, slightly concave. Pronotum in prozona rugose, humped (Fig. 73), flat, densely punctate in metazona ; posterior margin angular. Lateral carinae either distinct or indistinct. Prosternal process short and spine-like (Fig. 74). Mesosternal interspace much wider anteriorly than posteriorly (Fig. 75). Wings fully developed, extending behind hind knees. Distal spine on hind tibia situated close to spurs (Fig. 76).
Distribution : Canary Islands, North Africa to Eastern Mediterranean, Central Asia. The genus includes ten species, two species in our region.

Key to the Species of Dericorys in Israel

1. Wings reddish. Prozonal hump higher than long. Antennae shorter than head and pronotum together (Fig. 73). **D. millierei** Finot & Bonnet
- Wings transparent, yellowish or colourless at base. Prozonal hump less arched, longer than high. Antennae as long as or longer than combined length of head and pronotum. **D. albidula** Serville

Dericorys millierei Finot & Bonnet, 1884
Figs. 73–76

Type Locality : Oran, Algeria (Paris Museum).

Dericorys millierei Finot A. & E. Bonnet, 1884, *Annls Soc. ent. Fr.*, (6) 4, Bull. xxvi–xxvii.
Dericorys millierei —. Buxton P. A. & B. P. Uvarov, 1923, *Bull. Soc. R. ent. Égypte*, p. 203.
Dericorys millierei —. Innes W., 1929, *Mém. Soc. R. ent. Égypte*, 3 (2) : 125.
Dericorys millieri (sic) —. Bodenheimer F. S., 1935, *Animal Life in Palestine*, Jerusalem, pp. 79, 322, 323.
Dericorys millerii (sic) —. Bodenheimer F. S., 1935, *Arch. Naturgesch.*, 4 (2) : 200.

Medium sized, strong, slightly granulate and punctate. Head hypognathous or slightly opisthognathous in male. Margins of frontal ridge raised, slightly diverging between bases of antennae and towards the clypeus ; face concave on both sides of frontal ridge. Fastigium short ; foveolae situated laterally, marginate only posteriorly. Vertex slightly concave, sparsely punctate anteriorly ; margins raised ; median carina extending also on occiput. Eyes oblong, their horizontal diameter slightly longer than the interocular space. Antennae short ; median segments 1.5–2.0 times longer than wide ; basal segments short and wide. Prozona of pronotum raised in a hump along the median carina (Fig. 73) ; posterior transverse sulcus situated beyond the middle. Metazona rugose ; median carina present ; posterior margin obtuse-angled.
Tegmina narrowing apically, with regular sparse cross-venation. Wings elongate ; main veins thick ; in male anterior margin usually opaque. Hind femur serrate dorsally. Hind tibia as long as or longer than hind femur.
Coloration : Generally grey, pale ochraceous or pale brown with a pattern of white lines and bands. Ventral part of body usually whitish. Basal part of wings reddish ; venation dark. Hind femur with or without dark band. Hind tibia yellowish-grey or pale blue. Hoppers usually greenish.
Measurements (mm) : Body ♀ 23.0–38.0, ♂ 15.0–20.0 ; tegmina ♀ 22.0–34.0, ♂ 16.0–17.0.
Distribution : North Africa, Sinai, Israel.
Israel : Jordan Valley (7), Judean Hills (11), Northern and Central Negev (15, 17).
Usually found hiding among branches of *Anabasis, Zygophyllum, Atriplex, Retama* and other bushes dominant in wadis. Hoppers appear from February to May.

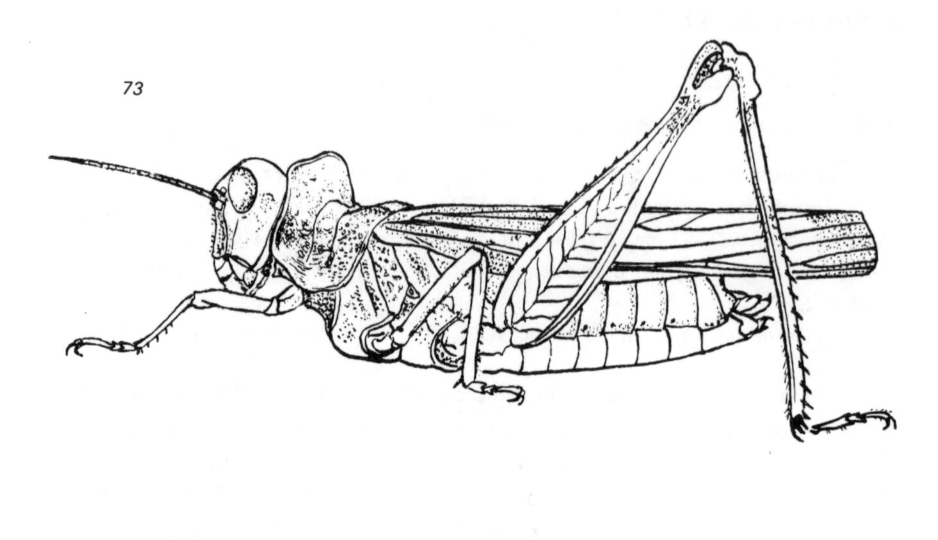

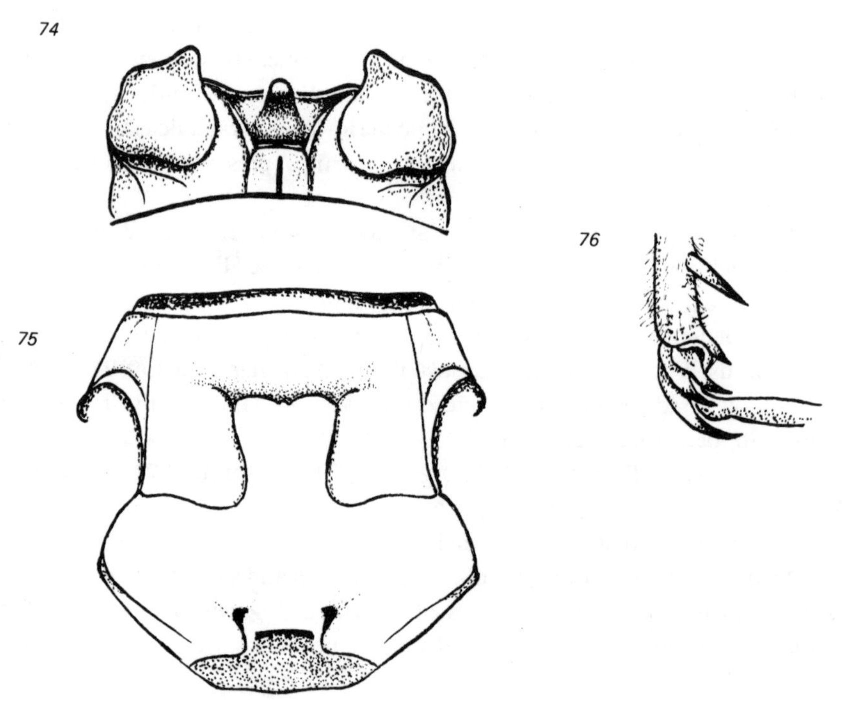

Figs. 73–76 : *Dericorys millierei* Finot & Bonnet, 1884
73. female ; 74. prosternal process ; 75. sternum ; 76. ventral part of hind tibia with external spine

Dericorys

Dericorys albidula Serville, 1838
Plate II : 5

Type Locality : Mount Lebanon, Syria (Paris Museum).

Dericorys albidula Serville J. G. A., 1838, *Histoire naturelle des Insectes*, in : Roret, *Collection des Suites à Buffon. Orthoptères*, Paris, p. 639.
Dericorys acutispina Stål C., 1875, *K. svenska VetenskAkad. Handl.*, 3 (14) : 27.
Derocorystes (Cyphophorus) curvipes Redtenbacher J., 1889, *Wien. ent. Ztg.*, 8 : 29.
Dericorys albidula —. Innes W., 1929, *Mém. Soc. R. ent. Égypte*, 3 (2) : 125.

Larger than *Dericorys millierei*, distinguished by the long antennae, which are longer than head and pronotum together. Wings transparent, colourless or yellowish at base, their apical part opaque, dark. Hind femur shorter than hind tibia, its inner surface whitish or yellowish, with reddish spots. Hind tibia curved, reddish apically, inner surface bluish or bluish-grey. General coloration of female pale grey with white lines.
Measurements (mm) : Body ♀ 51.2, ♂ 42.5–51.2 ; tegmina ♀ 54.6, ♂ 39.4–51.1.
Distribution : North Africa, Israel, Syria, Central Asia, Pakistan.
Israel : Dead Sea Area (13), Central Negev (17).
Hoppers found during April–May.

Subfamily PYRGOMORPHINAE

Diagnosis : Head usually conical, opisthognathous ; frons sloping, forming an acute angle with the vertex, the apex of which projects anteriorly between the eyes (Fig. 84). Only head of genus *Tenuitarsus* almost hypognathous (Fig. 78). Foveolae distinct, completely dorsal, situated near anterior margin of vertex, contiguous anteriorly (Fig. 85), or obliterated. Antennae usually shorter than head and pronotum together, filiform or ensiform. Lateral carinae of pronotum partly or entirely obliterated (Fig. 87). Prosternum with a sharp or collar-like projection between forelegs, covering posterior part of the mouth parts. Wings fully developed or abbreviate ; veins usually not branched. Outer aspect of hind femur without symmetrical plates, often with irregular lines or carinulae, or with a distinct row of oblique lines near the lower margin.
Species of this archaic group occur mainly in warm, subtropical or tropical areas ; many are characteristic phytophiles found in areas with grassy vegetation. Exceptions are genera like *Chrotogonos* and *Tenuitarsus*, which inhabit bare or sandy areas ; morphologically they resemble species of Oedipodinae. Characteristic for all Pyrgomorphinae is a specialized dorsal, dermal gland, which opens between the first and second abdominal terga. This organ secretes a repellent substance, venomous in the genera *Phymateus, Poekilocerus* and *Zonocerus* (Fishelson, 1960 ; Euw et al., 1967)
The Pyrgomorphinae are represented in our region by six genera.

Key to the Genera of Pyrgomorphinae in Israel and Sinai

1. Tegmina vestigial, in form of short lateral plates (Fig. 77). **Pyrgomorphella** Bolivar
- Wings fully developed 2
2. Head hypognathous (Fig. 78). Rows of brown and white tubercles on tegmina. Spurs on hind tibia long, usually longer than first tarsal segment. **Tenuitarsus** Bolivar
- Head opisthognathous. Tegmina without rows of tubercles. Spurs on hind tibia shorter than first tarsal segment. 3
3. Large. Black-blue, with or without pattern of yellow spots and dots (Fig. 80). **Poekilocerus** Serville
- Small. Coloration never black 4
4. Body flat, depressed, extremely rugose. Anterior margin of prosternum raised collar-like. Head, pronotum and tegmina with many sharp tubercles. Wings not extending beyond hind knee (Fig. 82). **Chrotogonus** Serville
- Body smooth, elongate. Anterior margin of prosternum indistinctly raised. Wings extending distinctly beyond hind knee 5
5. Pronotum rugose, with distinct lateral carinae (Fig. 83). **Pyrgomorpha** Serville
- Pronotum smooth, without lateral carinae (Fig. 86). **Macroleptea** Kevan

Genus PYRGOMORPHELLA Bolivar, 1904

Boll. Soc. esp. Hist. nat., 4 : 457

Pyrgomorphella (Pyrgomorphella) Bolivar I., 1909, *Genera Insectorum*, fasc. 90 : 33.
Pyrgomorphella —. Uvarov B. P., 1953, *Publções cult. Co. Diam. Angola*, no. 21, p. 216.

Type Species : *Pyrgomorphella sphenarioides* Bolivar, 1904

Diagnosis : Small, apterous or brachypterous. Head opisthognathous. Frontal ridge narrow. Vertex and occiput with a distinct median carina. Foveolae oval, situated dorsally, slightly arched, contiguous anteriorly. Interocular distance less than diameter of eye. Antennae flat, with wide and short segments ; antennae of male longer than head and pronotum together, shorter in female. Pronotum granulate ; carinae distinct, especially the median carina ; intersected by two transverse sulci, the posterior one situated beyond the middle. Vestigial tegmina situated laterally, not fused dorsally. Closely resembles *Pyrgomorpha*, except in its abbreviate wings.

Distribution : Balkans, Transcaucasus, Northern Iran, South Africa.

One species in our region.

Pyrgomorphella

Pyrgomorphella granosa Stål, 1876
Fig. 77

Type Locality : Beirut.

Pyrgomophella granosa Stål C., 1876, *K. svenska VetenskAkad. Handl.*, 4 (5) : 158.
Pyrgomorphella granosa —. Giglio-Tos, E.,1894, *Boll. Musei Zool. Torino*, 9 (191) : 2.
Pyrgomorphella granosa —. Buxton P. A. & B. P. Uvarov, 1923, *Bull. Soc. R. ent. Égypte*, p. 202.
Pyrgomorphella granosa —. Bodenheimer F. S., 1935, *Arch. Naturgesch.*, 4 (2) : 198.

Small, rugose, with numerous small tubercles. Head opisthognathous, face flat. Frontal ridge narrow ; margins usually straight, diverging ventrally in male, undulate in female. Subocular groove 1.2–1.4 times longer than eye. Fastigium of vertex rounded apically, usually as long as interocular distance. Foveolae elongate, contiguous anteriorly, separated in the middle by the globular vertex. Median carina of occiput visible along its entire length, sometimes slightly obliterated. Antennae short, slightly ensiform in females. Pronotum carinate and granulate ; prozona twice as long as metazona. Mesosternal interspace twice as wide as mesosternal lobe, usually merging posteriorly with the metasternal suture.
Tegmina rounded apically ; margins parallel, reaching or slightly extending beyond anterior margin of first abdominal segment. Dorsal carinae present on abdominal terga. Hind tibia as long as or slightly shorter than hind femur. Arolium longer than claws.
Coloration : Immature specimens yellow or yellowish-grey ; mature ones dark grey or greyish-brown.
Measurements (mm) : Body ♀ 19.0–21.5, ♂ 13.5–17.5.
Distribution : Syria, Israel.
Israel : Upper and Lower Galilee (1, 2), Coastal Plain (4, 8, 9), Judean Hills (11), Jordan Valley (7).

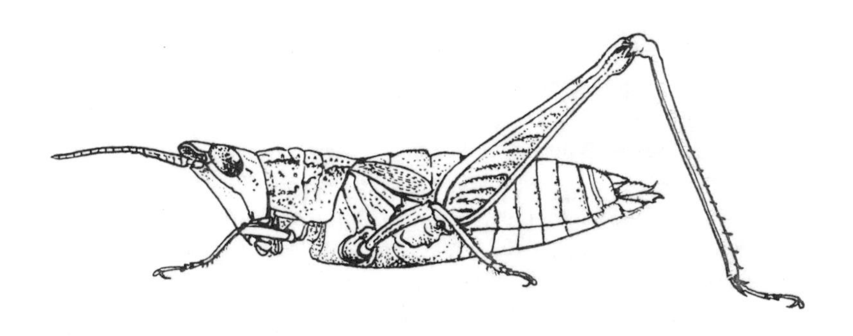

Fig. 77 : *Pyrgomorphella granosa* Stål, 1876, female

Inhabits areas of low, dense and usually dry vegetation, and rocky mountain slopes. Egg-laying females observed from October to January ; hoppers found from April to August. In defense, these grasshoppers activate the dorsal gland and spray a repugnant substance, which smells like cow dung and protects them against predators (Fishelson, 1960).

Genus TENUITARSUS Bolivar, 1904

Bol. Soc. esp. Hist. nat., 4 : 90

Type Species : *Tenuitarsus revoili* Bolivar, 1904.

Diagnosis : Small, body wide, almost smooth. Head hypognathous with slightly projecting vertex. Fastigium sloping ; foveolae situated fronto-dorsally, only partly visible from above. Eyes oval, protruding laterally. Pronotum short, intersected by two transverse sulci ; median carina distinct or obliterated ; lateral carinae absent ; posterior margin obtuse. Prosternum raised collar-like, covering mouth parts.

Tegmina narrow ; wings triangular, slender. Midleg very long, mid femur twice as long as front femur. Spurs of hind tibia as long as first tarsal segment. Tympanum absent on sides of first abdominal segment. Arolium very small, nearly indistinct in the female.

Distribution : Somalia, Egypt, Persia. The two or three known species of this genus resemble the genus *Hyalorrhipis* of the Oedipodinae and also inhabit areas of loose sand.

One species in our region.

Tenuitarsus angustus (Blanchard, 1836)

Figs. 78, 79 ; Plate I : 4

Type Locality : 'Egypt'.

Ommexecha angustum Blanchard E., 1836, *Annls Soc. ent. Fr.*, 5 : 624.
Ommexecha linearis Burmeister H., 1838, *Handbuch der Entomologie*, Berlin, 2 (2) : 657.
Chrotogonus angustatus (sic) —. Bolivar I., 1884, *Ann. Soc. esp. Hist. nat.*, 13 : 38, 41.
Leptoscirtus savignyi Saussure H. de, 1889, *Mitt. schweiz. ent. Ges.*, 8 : 89, 90.
Leptoscirtus linearis —. Krauss H. A., 1890, *Verh. zool.-bot. Ges. Wien*, 40 : 253.
Leptoscirtus angustus —. Jacobson G. G. & V. L. Bianchi, 1902, *Orthoptera and Odonata of the Russian Empire*, St. Petersburg, pp. 191, 271 [in Russian].
Tenuitarsus revoili Bolivar I., 1904, *Boll. Soc. esp. Hist. nat.*, 4 : 90.
Leptoscirtus evansi Uvarov B. P., 1921, *J. Bombay nat. Hist. Soc.*, 27 : 63.
Tenuitarsus angustus —. Uvarov B. P., 1924, *Tech. Bull. Minist. Agric. Egypt*, 41 : 36.
Tenuitarsus angustus —. Bodenheimer F. S., 1935, *Arch. Naturgesch.*, 4 (2) : 198.

Tenuitarsus

Tenuitarsus evansi —. Bei-Bienko G. Ya. & L. L. Mishchenko, 1951, *Locusts and Grasshoppers of the U. S. S. R. and Adjacent Countries*, Moscow, I : 279 [in Russian].
Tenuitarsus evansi —. Kevan D. K. McE., 1953, *Proc. R. ent. Soc. Lond.*, B : 22 (3–4) : 47.

Small, nearly smooth. Head hypognathous, widening dorsally ; vertex slightly protruding. Frontal ridge obliterated below median ocellus, fused dorsally with the foveolae, which are partly visible from above (Fig. 79). Vertex concave, with indistinct traces of median carina. Eyes nearly hemispherical, their horizontal axis longer than the vertical. Antennae short ; apical segment long and undivided (Fig. 79). Pronotum wide and short, usually without longitudinal carinae or only part of the median carina present in metazona. Posterior transverse sulcus situated in middle of pronotum. Prozona rugose and tuberculate (Fig. 79). Prosternum extending like a collar over the mouth parts. Mesosternal interspace wide, fused with the metasternal interspace.

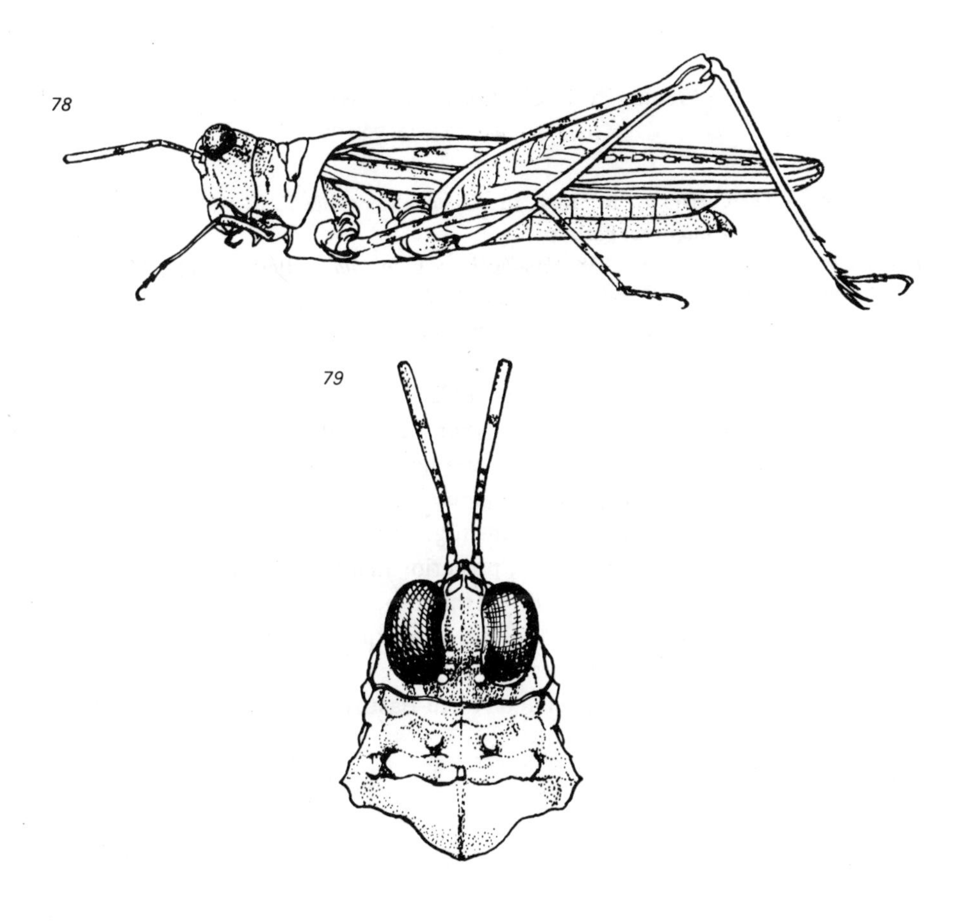

Figs. 78–79 : *Tenuitarsus angustus* (Blanchard, 1836)
78. female ; 79. head and pronotum

Tegmina long and narrow, extending beyond hind knees ; apex pointed ; median vein tuberculate (Fig. 78). Wings slightly longer than tegmina. Mid legs much longer than forelegs. Hind femur narrow, with smooth external surface dorsally and symmetrical carinae ventrally. Hind spurs longer than the first tarsal segment.
Coloration : Sandy, with reddish-brown lines and white spots ; white and dark rings on antennae, sometimes reddish parts on the thorax. Wings colourless or slightly blue.
Measurements (mm) : Body ♀ 14.5–16.5, ♂ 10.5–11.5 ; tegmina ♀ 14.5–15.5, ♂ 9.5–10.5.
Distribution : Egypt, Somalia and Israel.
Israel : Dead Sea Area (13), 'Arava Valley (14), Central and Southern Negev (17, 16). Typical of sandy habitats in wadis. Using hind and mid legs, it digs deep into the sand. Oviposition observed in May ; hoppers found in March and April.

Genus POEKILOCERUS Serville, 1831

Annls Sci. nat. (Zool.), 22 : 275

Poecilocera —. Percheron A. R., 1836, in : Guerin-Meneville F. E. & A. Percheron, *Genera des Insectes, etc. Orthoptera*, Paris, p. 5.
Poecilocera —. Burmeister H., 1838, *Handbuch der Entomologie*, Berlin, 2 (2) : 602, 621.
Poecilocerus —. Stål C., 1873, *Öfvers. VetenskAkad. Forh. Stockh.*, 30 (4) : 51.
Poekilocerus —. Uvarov B. P., 1966, *Grasshoppers and Locusts*, London, 1 : 402.

Type Species : *Gryllus pictus* Fabricius, 1775.
Diagnosis : Medium sized or large, rugose and granulate. head flat dorsally, opisthognathous ; vertex projecting forwards, triangular, deeply intersected by sulcus of frontal ridge. Foveolae obliterated ; occiput with median carina. Antennae filiform, much shorter than head and pronotum together. Pronotum rugose and granulate, intersected by three transverse sulci ; posterior margin rounded. Carinae of pronotum slightly visible or obliterated. Prosternal process between forelegs very distinct. Tegmina fully developed, apically rounded, opaque. Wings well developed, with a network of irregular veinlets.
Distribution : Four species in East and North-East Africa, Sinai, Israel, Transjordan, Arabia, India.
One species in our region.

Poekilocerus bufonius (Klug, 1832)
Figs. 80, 81 ; Plate I : 5 ; Plate V : 6

Type Locality : Alexandria, Egypt.

Decticus bufonius Klug J. C. F., 1832, *Symbolae physicae seu icones et descriptiones insectorum, etc.*, pl. 25, figs. 3, 4, 5.
Poecilocera bufonia —. Burmeister H., 1838, *Handbuch der Entomologie*, Berlin, 2 (2) : 623.
Poekilocerus bufonius —. Serville J. G. A., 1838, *Histoire naturelle des Insectes*, in : Roret, *Collection des Suites à Buffon. Orthoptères*, Paris, p. 599.
Dictyophorus bufonius —. Blanchard E., 1840, in : Castelnau, F. L. Laporte de, *Histoire naturelle des animaux articulés*, Paris, 3 : 39.
Poecilocera bufonia —. Hart H. C., 1891, *Fauna and Flora of Sinai, Petra and Wadi Arabah*, London, p. 183.
Poecilocerus bufonius —. Bolivar I., 1892/93, *Rev. biol. nord Fr.*, 5 : 485.
Poecilocerus vulcanus Serville. Innes W., 1929, *Mém. Soc. R. ent. Égypte*, p. 116.
Poecilocera bufonia —. Bodenheimer F. S., 1932, *Bull. Soc. R. ent. Égypte*, 16 : 30.
Poecilocerus bufonius —. Bodenheimer F. S., 1935, *Arch. Naturgesch.*, 4 (2) : 198.
Poekilocerus bufonius —. Fishelson L., 1960, *Eos, Madr.*, 36 : 41–62.
Poekilocerus bufonius —. Uvarov B. P., 1966, *Grasshoppers and Locusts*, London, 1 : 402.

Medium sized or large, rugose. Head opisthognathous ; face slightly granulate, rugose. Frontal ridge low ; vertex triangular, protruding forwards, its maximum length less than interorbital distance. Foveolae obliterated. Antennae filiform, shorter than head and pronotum together ; number of segments 16–18. Pronotum cylindrical ; longitudinal carinae absent or very indistinct in metazona. Prozona slightly shorter than or as long as metazona. Posterior margin of pronotum rounded ; posterior ventral angle of lateral lobe acute.
Tegmina wide ; apex rounded, shorter than, reaching to, or extending beyond hind knees. Wings shorter than tegmina, with dense dark venation at the apex (Figs. 80, 81).
Coloration : Black-violet, usually with a pattern of yellow-orange dots which form lines on head, pronotum and other parts of the body. A population from the Sinai, without this light pattern, was described as *P. vulcanus.* Collections carried out by the author (during 1969–1974) showed the occurrence of transitional forms between typical *P. bufonius* and *P. vulcanus*, so that the latter should be regarded as a local variety. Hoppers brown, greyish-brown or pale brown, usually with the pattern of yellow dots. The black coloration of adults develops several days after the last moult.
Measurements (mm) : Body ♀ 52.5–68.0, ♂ 35.0–42.5 ; tegmina ♀ 30.0–36.5, ♂ 25.0–28.5.
Distribution : Egypt, Sinai, Israel, Transjordan.
Israel & Sinai : Dead Sea Area (13), Judean Desert (12), ‘Arava Valley (14), Southern Negev (16), Sinai Peninsula (20–23).
This species feeds nearly exclusively on Asclepiadaceae, usually on *Calotropis procera* and *Demia tomentosa* ; the areas of distribution of these plants and of *P. bufonius*

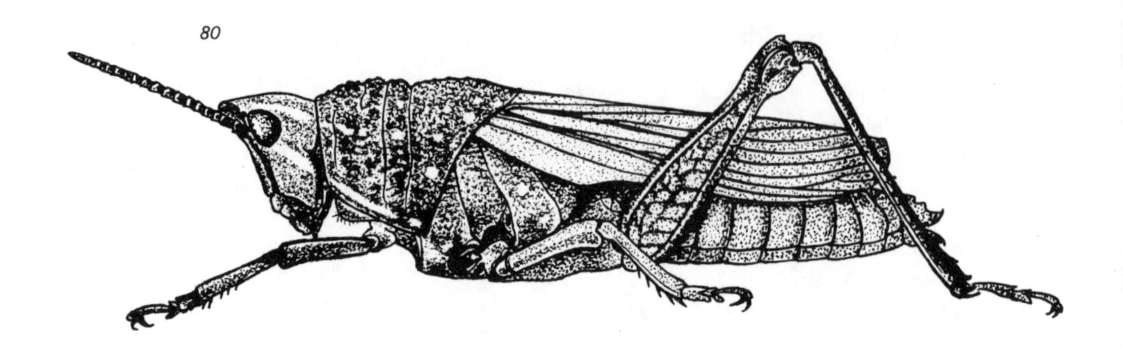

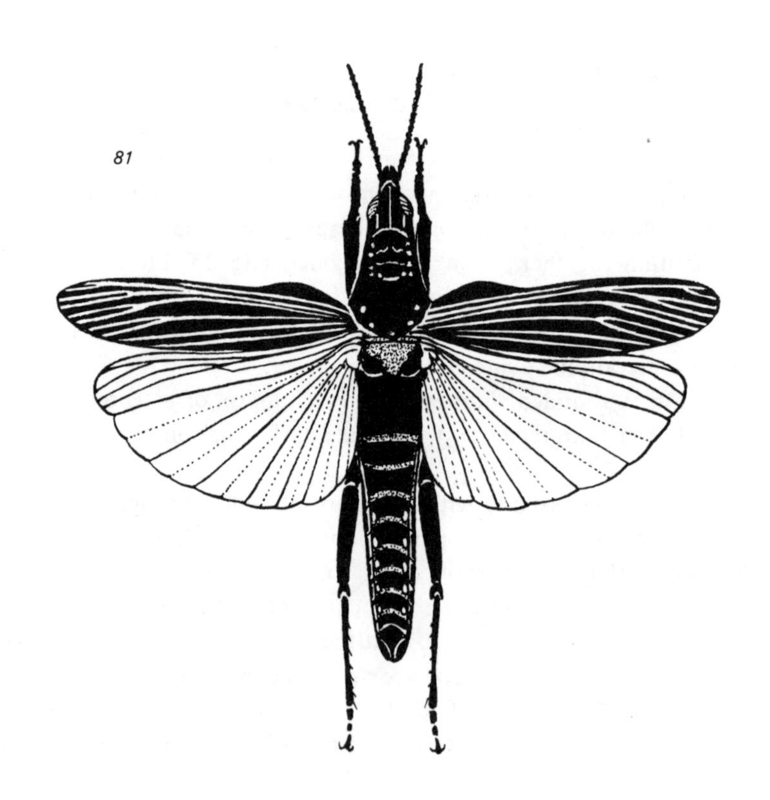

Figs. 80–81 : *Poekilocerus bufonius* (Klug, 1832)
80. female ; 81. male

coincide. This grasshopper usually hides among branches of the plants ; when disturbed, it sprays a repellent secretion from the dorsal gland (Fishelson, 1960). The secretion contains several cardenolides derived from the plants (Euw et al., 1967). Adults are found in winter and early spring (March–April) ; spawning in spring. Egg-pods 10–19 cm long, containing 80–250 eggs. Hoppers observed from December to March.

Genus CHROTOGONUS Serville, 1838

Histoire naturelle des Insectes, in : Roret, *Collection des Suites à Buffon. Orthoptères*, Paris, pp. 569, 698, 702

Ommexecha Blanchard E., 1836, *Annls Soc. ent. Fr.*, 5 : 603 (partim).
Chrotogonus —. Kevan D. K. McE., 1959, *Publções cult. Co. Diam. Angola*, 43 : 113–199.
Chrotogonus —. Kevan D. K. McE., 1968, *Entomologist's mon. Mag.*, 104 : 15.

Type Species : *Ommexecha lugubris* Blanchard, 1836.
Diagnosis : Short, flat, and wide. Head opisthognathous ; vertex projecting, with a median carina and transverse rugulae dorsally. Foveolae situated dorsally, contiguous in front. Eyes rounded, protruding laterally. Pronotum short and wide, distinctly rugose ; posterior angle acute. Posterior angle of lateral lobes raised laterally. Prosternum anteriorly raised collar-like, covering mouth parts ventrally and laterally. Tegmina narrow ; veins with rows of sharp tubercles, not branched.
Distribution : More than 20 species found in Africa and India.
Only one species in our region.

Chrotogonus homalodemus homalodemus (Blanchard, 1836)

Fig. 82 ; Plate I : 6

Type Locality : Senaar, North Africa (Paris Museum).

Ommexecha homalodemum Blanchard E., 1836, *Annls Soc. ent. Fr.*, 5 : 615.
Ommexecha lugubris Blanchard E., 1836, *ibid.*, 5 : 616.
Ommexecha homalodema —. Burmeister H., 1838, *Handbuch der Entomologie*, Berlin, 2 (2) : 656.
Chrotogonus homalodemas (sic) —. Walker F., 1870, *Catalogue of the specimens of Dermaptera Saltatoria in the Collection of the British Museum*, London, Part IV, p. 793.
Chrotogonus lugubris —. Boniteau L., 1904, *Rap. Invas. Criquet Pélérin Egypte*, 19, pl. V.
Chrotogonus homalodemus —. Karny H., 1907, *Sber. Akad. Wiss. Wien*, 57 : 293.
Chrotogonus homalodemus —. Dirsh V. M., 1965, *The African Genera of Acridoidea*, London, p. 111.
Chrotogonus homalodemus homalodemus —. Kevan D. K. McE., 1968, *Entomologist's mon. Mag.*, 104 : 16.

Short and flattened, especially at junction of thorax and abdomen. Head slightly opisthoganathous; vertex strongly projecting forward. Frontal ridge flat; margins undulate, partly obliterated. Foveolae contiguous apically; posterior margins higher than anterior ones. Interocular space depressed, rugose, granulate, with a median carina continuing onto the occiput. A transverse, irregular carina crosses the vertex behind the eyes. Antennae shorter than head and pronotum together; median segments more than twice as long as wide. Pronotum intersected by median transverse sulcus; prozona rugose; metazona tuberculate. Posterior part of lateral lobes raised laterally; lateral carinae crenulate.
Tegmina extending to hind knees or shorter, tapering apically; some veins with one or two rows of curved tubercles. Wings of female much shorter than tegmina, in male as long as or longer than tegmina. Inner spurs of hind tibia 1.5–2.0 times longer than outer spurs.
Coloration: Pale brown, brown or dark brown with a regular pattern of black blotches on the sternum and ventral side of abdomen. Pronotum and tegmina with sparse black shiny tubercles.
Measurements (mm): Body ♀ 20.5–23.0, ♂ 14.0; tegmina ♀ 12.5–14.0, ♂ 10.0; width of thorax ♀ 8.5–9.5, ♂ 5.0.
Distribution: Senegal, Sudan, Egypt, Israel.
Israel: Dead Sea Area (13).
Found on loose soil and sandy areas, usually under dry vegetation. Frequently it digs into the sand, covering itself, leaving only antennae and dorsal part of head visible.

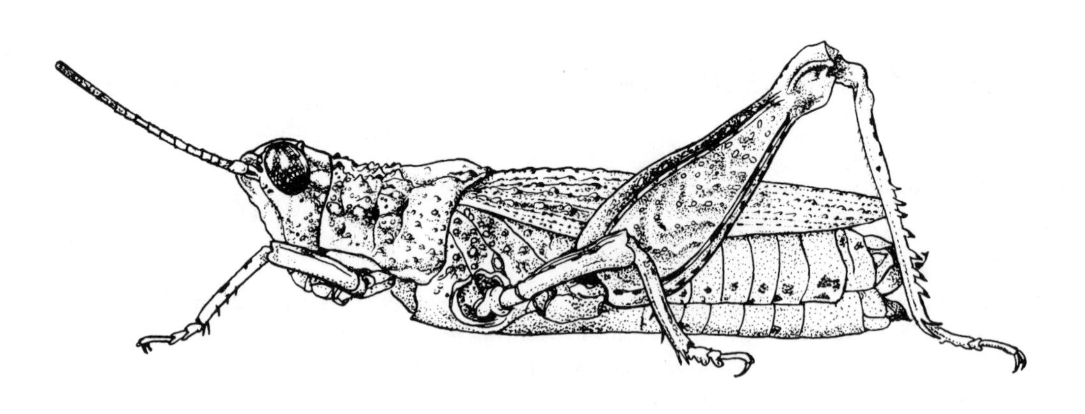

Fig. 82: *Chrotogonus homalodemus homalodemus* (Blanchard, 1836), female

Genus PYRGOMORPHA Serville, 1838

Histoire naturelles des Insectes, in : Roret, *Collection des Suites à Buffon, Orthoptères*, Paris, p. 583.

Truxalis (Pyrgomorpha) Serville J. G. A., 1838, *ibid.*, p. 583.
Pyrgomorpha —. Charpentier T. de, 1841, *Germars Z. Ent.*, 3 : 306.

Type Species : *Acrydium conicum* Olivier, 1791.
Diagnosis : Small, slender, slightly rugose. Head conical, opisthognathous ; eyes nearly round, protruding laterally. Vertex flat dorsally, rounded anteriorly, projecting forward between the eyes, with large contiguous foveolae separated by a shallow groove (Fig. 85). Antennae with 13–17 segments, slightly ensiform, in female narrowing towards apex, shorter than head and pronotum together. Median carina of pronotum distinct along its entire length ; lateral carinae usually partly obliterated. Posterior margin of pronotum rounded ; postero-ventral angle of lateral lobes with sharp processes before the apex (Fig. 83). Prosternum with a tooth-like process. Tegmina delicate ; veins thin, not branched. Wings transparent, sometimes reddish at base.
Coloration : Green or greyish, sometimes reddish.
Distribution : About 22 species, not clearly identified, most of them in Africa. One species in our region.

Pyrgomorpha conica (Olivier, 1791)

Figs. 83–85 ; Plate I : 7

Type Locality : 'Southern France'. Type lost.

Gryllus (Acrida) conicus —. Gmelin J. F., 1790, *Systema Naturae*, revised, 13th ed., Leipzig, 1 (4) : 2056.
Acrydium conicum Olivier G. A., 1791, *Encyclopédie Méthodique, Histoire Naturelle*, 6 : 230.
Truxalis grylloides Latreille P. A., 1804, *Histoire naturelle générale et particulière des Crustacées et des Insectes. Orthoptera, Acrididae*, 12 : 148.
Truxalis rosea Charpentier T. de, 1825, *Horae entomologicae*, Wratislaviae, p. 128.
Truxalis linearis Charpentier T. de, 1825, *ibid.*, p. 129.
Tryxalis rosacea (sic) —. Lucas H., 1849, Insectes, in : *Exploration scientifiques de l'Algérie pendant les années 1846–1849*, Publiée par ordre du Gouvernement, 3 ; 26.
Pyrgomorpha rosea —. Fischer L. H., 1853, *Orthoptera Europaea*, Leipzig, p. 304.
Pyrgomorpha grylloides —. Fieber F. X., 1853, *Lotos*, 3 : 97.
Opomala cingulata Walker F., 1870, *Catalogue of the specimens of Dermaptera Saltatoria in the Collection of the British Museum*, London, Part III, p. 517.
Pyrgomorpha conica —. Jacobson G. & V. L. Bianchi, 1902, *Orthoptera and Odonata of the Russian Empire*, St. Petersburg, pp. 198, 291, fig. 30 [in Russian].
Pyrgomorpha conica —. Buxton P. A. & B. P. Uvarov, 1923, *Bull. Soc. R. ent. Égypte*, p. 202.
Pyrgomorpha conica —. Bodenheimer F. S., 1935, *Arch. Naturgesch.*, 4 (2) : 197.
Pyrgomorpha conica —. Ramme W. 1951, *Mitt. zool. Mus. Berl.*, 27 : 428.

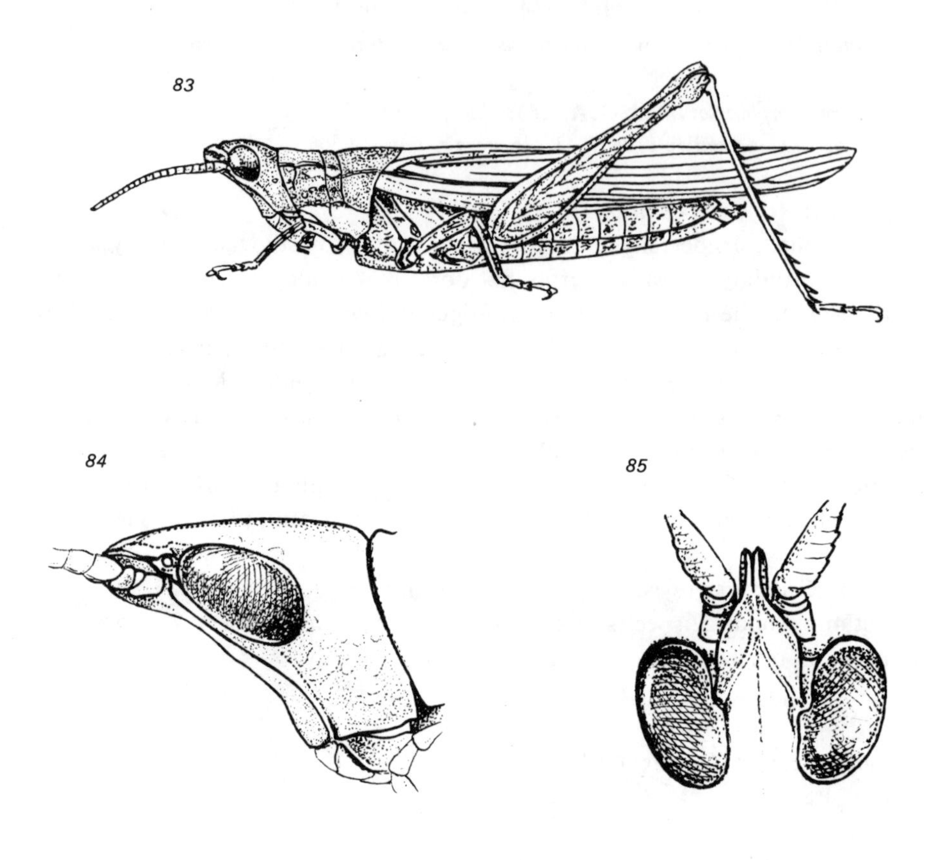

Figs. 83–85 : *Pyrgomorpha conica* (Olivier, 1791)
83. female ; 84. head, lateral ; 85. same, dorsal

Small, males very small. Head hypognathous, rugose ; frons protruding (Fig. 84). Foveolae situated dorsally, large and contiguous (Fig. 85). Antennae short, with 13–17 segments. Lateral carinae of pronotum distinct along their entire length or interrupted or slightly obliterated in metazona. Metazona much shorter than prozona, flat or slightly raised ; median carina linear if present, sometimes absent. Granulation on pronotum very variable. Tooth-like process on postero-ventral angle of prosternum fully developed, sharp, rounded or absent (in a population from the same locality).

Tegmina narrow, pointed apically, extending beyond hind knee.

Coloration : Green, grey or pale brown, sometimes tinged reddish. Hoppers green or grey. Wings transparent, colourless or pinkish at base.

Measurements (mm) : Body ♀ 20.0–30.5, ♂ 12.0–18.0 ; tegmina ♀ 17.0–22.0, ♂ 12.5–16.0.

Distribution: Somalia, Sudan, North Africa, South Europe, Southern U. S. S. R., Central Asia, Israel.
Israel: Upper and Lower Galilee (1, 2), Jordan Valley (7), Coastal Plain (4, 8, 9), Dead Sea Area (13), Northern Negev (15), Sinai (21, 22; St. Katharina).
Typical for areas with low, dense, grass-like vegetation or dense bushes. Hoppers are found throughout the year; they apparently develop without a diapause, or with a very short one, during the hot months of the summer.

Genus MACROLEPTEA Kevan, 1962

Publções cult. Co. Diam. Angola, 60 : 116, 117

Pyrgomorpha (partim) —. Chopard L., 1943, *Faune Emp. franç.*, 1 : 339.

Type Species: *Pyrgomorpha laevigata* Werner, 1914.
Diagnosis: Small to very small, smooth and shiny. Head acute, opisthognathous, with a median carina and rugulae dorsally, especially along the inner eye margins. Frontal ridge narrow, obliterated ventrally. Eyes protruding, rounded. Pronotum cylindrical, smooth. Tegmina extending to middle of hind tibia. Arolium longer than claws.
Distribution: North Africa and Western Asia.
Monotypic.

Macroleptea laevigata (Werner, 1914)
Figs. 86, 87; Plate I: 8

Type Locality: Ain Sefra, Algeria (Vienna Museum).

Pyrgomorpha laevigata Werner F., 1914, *Sber. Acad. Wiss. Wien*, 123 : 399.
Macroleptea laevigata —. Kevan D. K. McE., 1962, *Publções cult. Co. Diam. Angola*, 60 : 118.
Macroleptea laevigata —. Pener M. P., 1966, *Israel J. Ent.*, 1 : 191.

Small, smooth and shiny. Head acute, opisthognathous, with a median carina and numerous rugulae dorsally, especially along inner margins of eye. Face very slightly concave; frontal ridge narrow and low, obliterated ventrally. Fastigium of vertex of female wide and short, in male maximum length like width. Eyes rounded, their diameter equal to or slightly longer than subocular groove. Antennae stout, shorter than head and pronotum together; most of the segments square. Pronotum shiny, cylindrical; prozona much longer than metazona; lateral carinae absent (Fig. 87). Ventral margin of lateral lobe sinuous with a small denticle anterior to the rounded posterior angle. Prosternum only slightly protruding between forelegs. Sternum sparsely punctate; mesosternal lobes narrow, less than half as long as maximum width of mesosternal interspace.

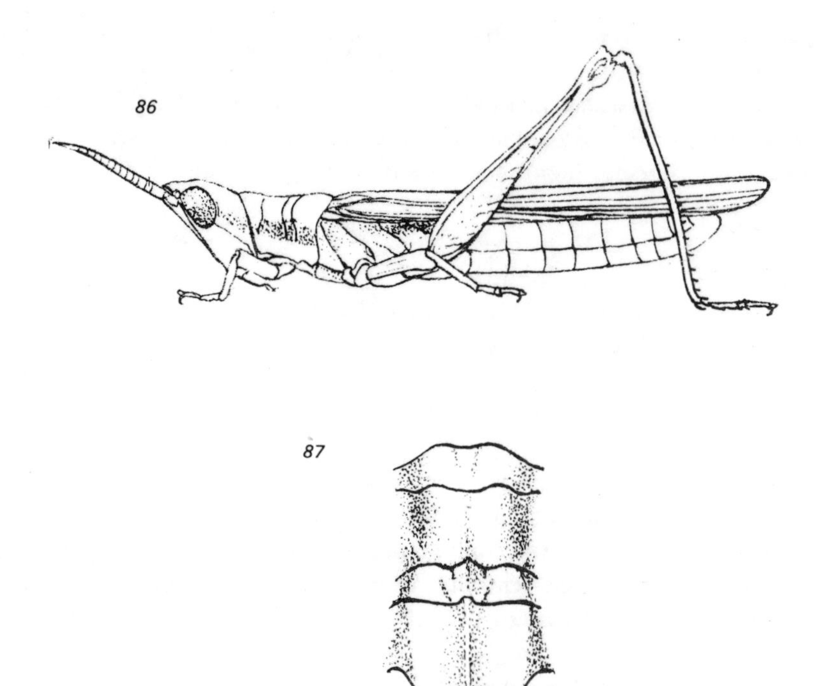

Figs. 86–87 : *Macroleptea laevigata* (Werner, 1914)
86. male ; 87. pronotum

Tegmina narrow, slightly shorter than wings, with very dense, irregular mesh of veinlets in basal part. Hind femur with pointed ventral lobes apically. Hind tibia distinctly widening apically ; inner spurs almost twice as long as outer spurs. Arolium wide, its length equal to or slightly longer than that of claws.

Coloration : Pale brown or greenish, usually with symmetrical brown and white bands extending from eyes, along ventral margins of pronotum, on thorax and partly on tegmina.

Measurements (mm) : Body ♀ 16.0–16.5, ♂ 11.0–13.5 ; tegmina ♀ 14.0–15.0, ♂ 9.0–10.5.

Distribution : North Africa to Israel.

Israel : Northern Negev (15) (south of Be'er-Sheva), Southern Coastal Plain (9) (Nahal Rubin).

Found among bushes on sand dunes, usually occurring together with *Pyrgomorpha conica.* Hoppers observed in August.

Subfamily PAMPHAGINAE

Diagnosis : Medium sized or large, rugose. Head hypognathous, with a typical fastigial furrow, either flat or smooth, or with a frontal ridge projecting forward above the median ocellus (Fig. 93). Vertex wide, concave ; eyes rounded, interocular space exceeding length of eyes. Foveolae absent or indistinct. Antennae long, filiform or ensiform, their segments frequently with distinct carinulae.
Pronotum large, rugose ; lateral carinae usually partly obliterated ; median carina often raised. Anterior margin of pronotum extending slightly over the head. Prosternum forming a raised collar between the forelegs, sometimes with a pointed process (Fig. 91). Mesosternal interspace wide and short. Hind femur wide and strong ; external median field with carinulae and granulation.
Distribution : North Africa to South Europe, Mediterranean and Asia as far as China.

Key to the Genera of the Subfamily Pamphaginae in Israel and Sinai

1. Wings fully developed. Median carina of pronotum sharply incised by the posterior transverse sulcus, usually depressed on metazona (Fig. 90) — 2
- Apterous ; tegmina rudimentary, situated laterally on the thorax (Fig. 93). Median carina of pronotum slightly raised, usually not intersected by the transverse sulcus (Fig. 95) — 4
2. Foveolae absent. Prozona of pronotum laterally compressed, raised along median carina, with a sharp process posteriorly, which bends backwards over the transverse sulcus (Fig. 88). Inner side of hind femur yellowish. — **Eremotmethis** Uvarov
- Foveolae present. Prozona of pronotum without sharp process over the transverse sulcus ; a large dark blue area present on inner side of hind femur — 3
3. Basal parts of wings red. Ventral margin of hind femur undulate (Fig. 90). — **Tmethis** Fieber
- Basal parts of wings yellowish or greenish. Ventral margin of hind femur straight (Fig. 92). — **Utubius** Uvarov
4. Body compressed laterally ; pronotum and abdominal tergites sharply raised and curved posteriorly (Fig. 93). Prosternal spine very narrow. — **Prionosthenus** Bolivar
- Body wide, mostly cylindrical ; pronotum only slightly raised along median carina ; no sharp carinae on the pronotum and abdominal tergites — 5
5. Prosternal spine between the forelegs large, quadrangular. Ventral margin of hind femur smooth without dentations, straight (Figs. 94, 95). Mesosternal interspace longer than wide. — **Acinipe** Rambur
- Prosternal spine low, elongate ; apex rounded, granulate. Ventral margin of hind femur dentate, undulate (Figs. 96, 97, 98). Mesosternal interspace wider than long. — **Ocneropsis** Uvarov

Genus EREMOTMETHIS Uvarov, 1943

Trans. R. ent. Soc. Lond., 93 : 36, 46

Type Species : *Gryllus carinatus* Fabricius, 1775.
Diagnosis : Large ; body more or less smooth, with scattered, sometimes sharp granulations, especially on pronotum. Head hypognathous (Fig. 89), widening ventrally. Facial carinae diverging below ocellus, converging above antennae. Antennae filiform, with oblong segments. Eyes oval. Occiput and genae smooth. Pronotum intersected by transverse sulcus before its middle ; prozona with a roof-shaped structure which ends posteriorly in a tooth-like projection (Fig. 88).
Distribution : From Egypt to Afghanistan and India.
Monotypic.

Eremotmethis carinatus (Fabricius, 1775)

Figs. 88, 89 ; Plate II : 6

Type Locality : Jidda, Arabia.

Gryllus carinatus Fabricius J. C., 1775, *Systema entomologiae*, p. 288.
Acrydium carinatum —. Olivier G. A., 1791, *Encyclopédie Méthodique, Histoire Naturelle*, 6 : 216.
Eremobia carinata —. Serville J. G. A., 1838, *Histoire naturelle des Insectes*, in : Roret, *Collection des Suites à Buffon. Orthoptères*, Paris, p. 706.
Eremobia continuata Serville J. G. A., 1838, *ibid.*, p. 707.
Thrinchus continuatus —. Walker F., 1870, *Catalogue of the specimens of Dermaptera Saltatoria in the Collection of the British Museum*, London, Part IV, p. 794.
Eremobia carinata —. Giglio-Tos E., 1893, *Boll. Musei Zool. Anat. comp. R. Univ. Torino*, 8 (164) : 7.
Tmethis carinatus —. Jacobson G. G. & V. L. Bianchi, 1902, *Orthoptera and Odonata of the Russian Empire*, St. Petersburg, pp. 197, 285 [in Russian].
Tmethis continuatus —. Jacobson G. G. & V. L. Bianchi, 1902, *ibid.*, pp. 197, 286.
Tmethis aegyptius Uvarov B. P., 1924, *Tech. Bull. Minist. Agric. Egypt*, 41 : 34, pl. 3, figs. 41, 42.
Tmethis carinatus —. Uvarov B. P., 1934, *Eos, Madr.*, 10 : 100.
Tmethis carinatus continuatus —. Bodenheimer F. S., 1935, *Arch. Naturgesch.*, 4 (2) : 197.
Eremotmethis carinatus —. Uvarov B. P., 1943, *Trans. R. ent. Soc. Lond.*, 93 : 46.

Large, robust, wide and depressed anteriorly, Head comparatively small, hypognathous. Frontal ridge only slightly concave, sometimes flat with margins converging below the median ocellus and obliterated near clypeus. Subocular groove 1.5 times as long as diameter of eye ; interocular space flat or slightly concave, wider than diameter of eye. Foveolae absent or only traces present. Vertex usually with a median carina and numerous lateral carinulae. Pronotum granulate with a regular row of granules or a swelling on posterior margin and lateral lobes, rarely without granulation. Pronotum projecting over occiput anteriorly (Fig. 89) ; raised prozonal median carina intersected by two transverse grooves ; with spinous projection posteriorly.

Tegmina wide, extending posteriorly beyond middle of the hind tibia. Wings transparent, bluish-yellowish with a complete or interrupted dark band. Inner aspect of hind femur yellowish ; dorsal carina serrate.

Coloration : Coloration and granulation very variable, changing from uniform brown to varying patterns of light and brown bands. Under wings, first abdominal tergites bluish, either rugose-granulate or smooth. The colour patterns of the hoppers are very variable.

Measurements (mm) : Body ♀ 48.0–65.0, ♂ 37.0–45.0 ; tegmina ♀ 45.0–50.0, ♂ 33.0–38.0.

Distribution : Egypt, Israel, Arabia, Persia, Afghanistan.

Israel : Judean Desert (12), Dead Sea Area (13), Negev (15, 16, 17).

The adults are very heavy, and when sensing danger they remain on the ground hidden by their cryptic coloration, instead of taking flight.

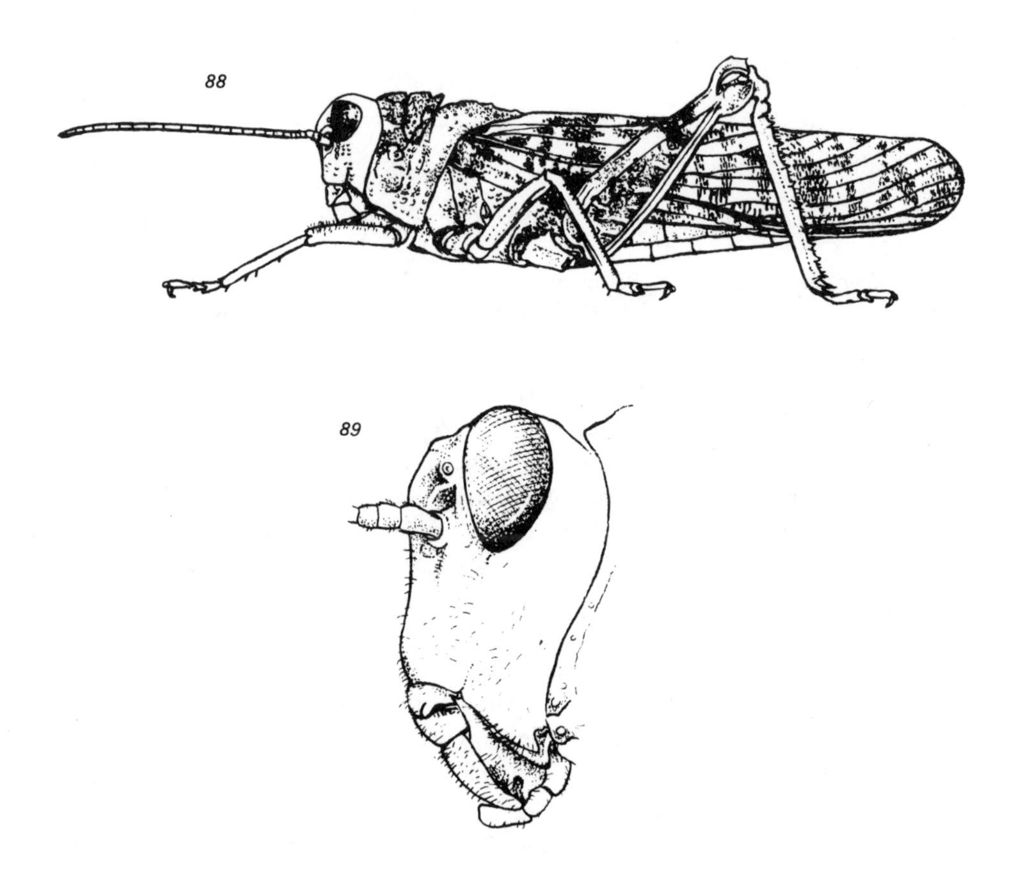

Figs. 88–89 : *Eremotmethis carinatus* (Fabricius, 1775)
88. female ; 89. head, lateral

Genus TMETHIS Fieber, 1853

Lotos, 3 : 128

Eremobia (nom. preoc.) Serville J. G. A., 1838, *Histoire naturelle des Insectes*, in : Roret, *Collection des Suites à Buffon. Orthoptères*, Paris, pp. 596, 704.
Tmethis (nom. substit.) Fieber F. X., 1853, *Lotos*, 3 : 128.

Type Species : *Gryllus cisti* Fabricius, 1787.
Diagnosis : Head hypognathous, narrowing dorsally. Frontal ridge narrow, with a deep groove. Vertex narrow, not more than one-fifth of the width of the eye. Metazona of pronotum convex, with or without a raised median carina. Tegmina fully developed. Wings red or reddish. Tympanic lobe covering half of the tympanic organ.
Distribution : West and North Africa, Eastern Mediterranean.
One species in the Asiatic part of distribution range.

Tmethis pulchripennis asiaticus Uvarov, 1943

Figs. 90, 91 ; Plate II : 7 ; colour plate : 7

Type Locality : Khedeira [Hadera], Israel.

Tmethis pulchripennis asiaticus Uvarov B. P., 1943, *Trans. R. ent. Soc. Lond.*, 93 : 66, figs. 14, 21, 36, 69, 70, 73.
Tmethis pulchripennis asiaticus —. Bei-Bienko G. Ya. and L. L. Mishchenko, 1951, *Locusts and Grasshoppers of the U. S. S. R. and Adjacent Countries*, Moscow, I : 318 [in Russian].
Tmethis pulchripennis asiaticus —. Ramme W., 1951, *Mitt. zool. Mus. Berl.*, 27 : 427.

Medium sized to large, flattened, wide, rugose and granulate. Head hypognathous. Frontal ridge either extremely concave or flat. Vertex wider than interocular space. Eyes round. Antennae in female as long as head and pronotum together, in male longer. Pronotum intersected deeply by posterior transverse groove, strongly raised in prozona along median carina (Fig. 90). The median carina in the metazona distinctly raised (f. *typica*), slightly raised (f. *incristata*) or smooth (f. *laeviuscula*). Dorsum of pronotum rugose, usually serrate close to posterior margin, or with a raised rib. Prosternum swollen, usually smooth (Fig. 91).
Tegmina fully developed ; median field open. Basal part of wings red with complete or incomplete black band sending a short radial branch along anterior margin towards the base. Hind femur short, wide, compressed ; ventral and dorsal carinae sharp. Inner aspect blue, becoming yellow or orange apically. Hind tibia yellow or yellowish-orange.
Coloration : Coloration of body varying according to the structure and colour of the ground.
Measurements (mm) : Body ♀ 36.0–37.0, ♂ 27.0–31.0 ; tegmina ♀ 27.0–38.0, ♂ 22.0–24.0.

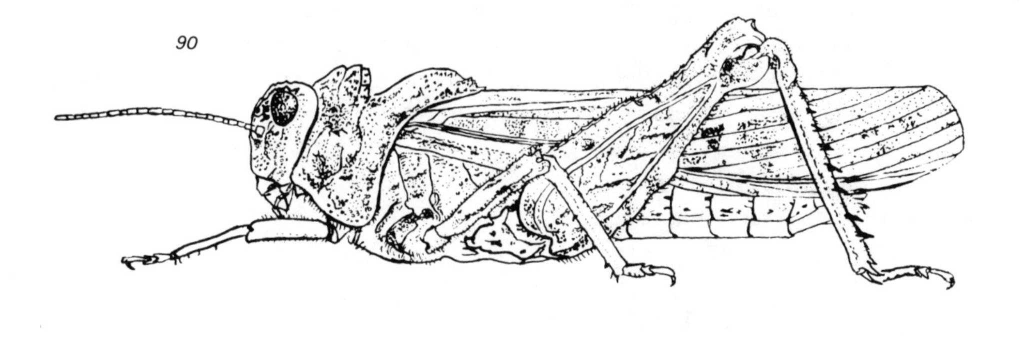

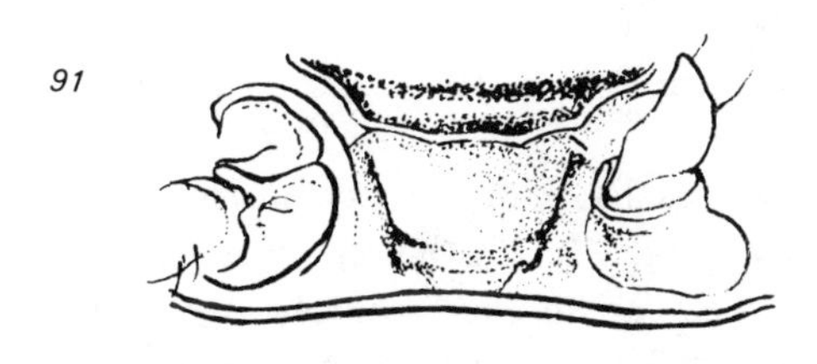

Figs. 90–91 : *Tmethis pulchripennis asiaticus* Uvarov, 1943
90. female ; 91. prosternum

Distribution : Western Iran to Sinai.
Israel : Upper and Lower Galilee (1, 2), Carmel (3), Coastal Plain (4, 8, 9), Judean Hills (11), Jordan Valley (7), Dead Sea Area (13), Northern and Central Negev (15, 17), Northern Sinai (20).
This species is the most euryecous amongst the Pamphaginae in our region. It occurs in typical arid habitats, as well as on hills and in areas with dense vegetation.
All three forms (*typica, incristata* and *laeviuscula*) were found in the same areas, f. *laeviuscula* being the most common.

Genus UTUBIUS Uvarov, 1936
J. Linn. Soc. (Zool.), 39 (268) : 544

Type Species : *Utubius zahrae* Uvarov, 1936.
Diagnosis : Body slightly compressed. Head hairy laterally, in male slightly opisthognathous or hypognathous. Frontal ridge obliterated below median ocellus. Eyes round ; interocular distance less than diameter of eye. Vertex slightly concave, usually with

numerous small transverse grooves. Antennae longer than head and pronotum together ; basal segments flat ; apical segments oblong-oval. Pronotum large and triangular ; median carina very distinct, slightly raised. Prozona short, narrow, rugose and raised along median carina. Metazona twice as long as prozona, its posterior angle rounded.
Distribution : Syria, Israel, Arabia.
Monotypic with several subspecies.

Utubius syriacus syriacus (Bolivar, 1893)
Fig. 92 ; Plate II : 8

Eremocharis syriaca Bolivar I., 1893, *Rev. biol. nord Fr.*, 5 : 483.
Utubius syriacus syriacus —. Uvarov B. P., 1936, *J. Linn. Soc.* (Zool.), 39 (268) : 544.

Large, strong, flattened dorsally, hairy ventrally. Head hypognathous or slightly opisthognathous in male (Fig. 92). Fastigium short, flat ; fastigial foveolae either situated before lateral ocelli or absent. Frontal ridge obliterated or only slightly concave. Eyes round, their long axis shorter than the subocular groove. Pronotum compressed in prozona, raised anteriorly along median carina, rugose and granulate in metazona. Prosternum between legs smooth.
Wings long, extending beyond hind knees for nearly half their length ; yellowish-greenish at base, with a dark band in the middle, extending from anterior margin to inner margin. Basal half of tegmina brown, not transparent, with dense venation ; apical part transparent with square, regular venation. Inner side of hind femur blue, outer side divided into irregular squares ; dorsal and ventral carinae haïry and granulate. Inner side of hind tibia yellowish. Arolium small, less than half the length of the claws.

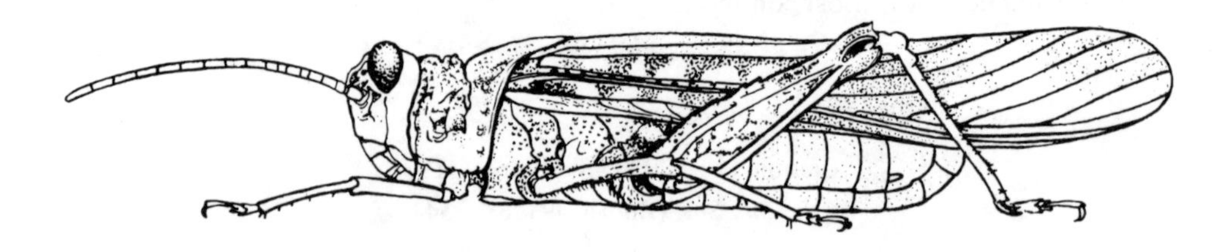

Fig. 92 : *Utubius syriacus syriacus* (Bolivar, 1893), male

Coloration : Mainly pale brown, with bluish lines and dark and light spots, resembling colour of the ground.
Measurements (mm) : Body ♀ 45.0–49.0, ♂ 31.0–35.0 ; tegmina ♀ 50.0–51.0, ♂ 35.5–38.0.
Distribution, Israel : Southern Negev (16), 'Arava Valley (14).
Typical of the hammadas of the 'Arava and wadis. Usually cryptic coloured, and varying according to surroundings. An excellent flier ; when disturbed, it can cover 50–80 m without landing.

Genus PRIONOSTHENUS Bolivar, 1878

An. Soc. esp. Hist. nat., 7 : 423–452

Type Species : *Pamphagus galericulatus* Stål, 1876.
Diagnosis : Small to large, compressed laterally. Head hypognathous or slightly opisthognathous ; frons strongly projecting between bases of antennae. Pronotum and abdominal terga sharply raised, curving posteriorly (Fig. 93). Prosternal process narrow, with tubercles at the apex.
Distributed in mountainous areas of Lebanon, Syria and Israel. Until a revision of all the apterous Pamphaginae is made, there is apparently only one species in the genus *Prionosthenus*, i.e., *Prionosthenus galericulatus* Stål.

Prionosthenus galericulatus (Stål, 1876)

Fig. 93

Type Locality : Tiberias, Israel.

Pamphagus galericulatus Stål C., 1876, *K. svenska VetenskAkad. Handl.*, 4 (5) : 28.
Pamphagus galericulatus —. Bolivar I., 1892/93, *Rev. biol. nord Fr.*, 5 : 484.
Pamphagus galericulatus —. Giglio-Tos E., 1893, *Boll. Musei Zool. Anat. comp. R. Univ. Torino*, 8 (164) : 7.
Pamphagus galericulatus —. Navas L., 1911, *Revista Montserratina*, Barcelona, p. 2.
Prionosthenus galericulatus —. Buxton P. A. & B. P. Uvarov, 1923, *Bull. Soc. R. ent. Égypte*, p. 202.
Prionosthenus galericulatus —. Bodenheimer F. S., 1935, *Arch. Naturgesch.*, 4 (2) : 199.

Medium to large, compressed laterally. Head rugose ; frons projecting forward from between the eyes, straightening below median ocellus. Frontal ridge sharp, narrow ; its marginal carinae diverge slightly above the clypeus, meet between the antennae and continue dorsally with the lateral carinulae of the fastigium. Vertex concave between eyes, with a median carina and numerous transverse carinulae ; vertex in female wider than length of longitudinal axis of eye, in male narrower. Antennae short, widened at the base, narrowing apically. Pronotum rugose; median carina raised, arc-like. Posterior transverse sulcus distinct, situated close to posterior margin, forming a short metazona which is distinctly raised. Prosternal process sharply raised, with numerous granules at the apex.

Vestigial tegmina narrow at the base, widening apically, extending to first abdominal segment, and sometimes covering the tympanal opening. Abdominal tergites compressed laterally, especially in males, forming a row of sharp protrusions along the dorsum; apex of abdomen usually curved upwards, especially in males. Outer median field of hind femur covered by a network of granulate lines. Arolium distinct, its length more than half that of claws. Inner pair of spurs on hind tibia much longer than outer spurs.

Coloration: Ochraceous, greenish-brown or brown, with or without pale parts or patterns. Dark blotch present on inner side of hind femur; inner side of posterior tibia dark violet.

Measurements (mm): Body ♀ 39.0–48.5, ♂ 20–26; tegmina ♀ 5.5–8.5, ♂ 4.0–5.0; hind femur ♀ 10.5–12.5, ♂ 8.5–9.5.

Distribution: Syria, Lebanon, Israel.

Israel: Upper and Lower Galilee (1, 2), Carmel (3, 4), Jordan Valley (7), Judean Hills & Desert (11, 12).

Inhabits rocky mountain slopes, usually in areas of dense vegetation. Oviposition during April–June. In captivity the eggs are deposited at a depth of 8–10 cm; length of egg-pod 4 cm; number of eggs in pod 7–60. Hoppers appear at the beginning of winter, after the first rains, and are found until May; they are able to survive at low temperatures for two months or longer.

Fig. 93: *Prionosthenus galericulatus* (Stål, 1876), female

Genus ACINIPE Rambur, 1838

Faune entomologique de l'Andalousie. Orthoptera, Paris, 2 : 68

Orchamus Stål C., 1876, *K. svenska VetenskAkad. Handl.*, 4 (5) : 30.

Type Species : *Acinipe hesperica* Rambur, 1838.

Diagnosis : Body cylindrical, rugose and granulate. Head short, hypognathous ; fastigium projecting anteriorly between bases of antennae. Vertex flat, intersected by groove of frontal ridge, with a longitudinal carina extending towards the clypeus. Antennae flat, ensiform ; segments oblong ; articulations narrowed. Eyes oblong, shorter than subocular sulcus ; pronotum arc-like, intersected in the posterior fourth by a transverse sulcus ; posterior margin slightly curved inwards in the middle. Prosternal process nearly quadrangular, granulate dorsally with a collar-like extension behind mouth parts.

Vestigial tegmina covering the tympanum and the posterior margin of the first abdominal tergite. Hind femur strong, thick ; carinae straight. Hind tibia long and hairy ; inner spurs longer than outer spurs. Studies by Dirsh (1958) showed that the species formerly included in the genus *Orchamus* should be ascribed to *Acinipe.*

Distribution : Transcaucasus, Turkey, Eastern Mediterranean.

Five to six species which vary markedly in granulation and rugosity.

Key to the Species of Acinipe in Israel and Sinai

1. Body white, with longitudinal grey bands. Concavity on vertex continuing on occiput, on both sides of the median line. Numerous tuberculate longitudinal carinae on both sides of median carina of pronotum and on the abdominal terga (Fig. 94). **A. zebratus** (Brunner-Wattenwyl)

– Body brown, greyish-brown. Longitudinal carinae absent on pronotum and abdominal terga 2

2. Prosternal process with numerous dense tubercles (Fig. 95). Frontal ridge usually reaching clypeus. Hind tibia bluish. **A. hebraeus** (Uvarov)

– Prosternal process with a few scattered tubercles. Frontal ridge obliterated below median ocellus. Hind tibia violet-reddish. **A. davisi** (Uvarov)

Acinipe zebratus (Brunner-Wattenwyl, 1882)

Fig. 94 ; Plate III : 1

Type Locality : 'Syria' (Vienna Museum).

Pamphagus zebratus Brunner von Wattenwyl C., 1882, *Prodromus der europäischen Orthopteren*, Leipzig, pp. 197, 199.

Orchamus zebratus —. Kirby W. F., 1910, *A synonymic catalogue of the Orthoptera*, Vol. III, Orthoptera Saltatoria, Part II, London, p. 349.

Pamphagus zebratus —. Boyd A. W., 1918, *Bull. Soc. ent. Égypte*, 5 : 104.
Orchamus zebratus —. Innes W., 1929, *Mém. Soc. R. ent. Égypte*, 3 (2) : 122.
Orchamus zebratus —. Bodenheimer F. S., 1935, *Arch. Naturgesch.*, 4 (2) : 199.

Long, rugose. Head short. Face granulate, rugose, with distinct carinae, especially in male. Vertex and occiput more or less smooth, concave along the longitudinal carina. Antennae shorter than head and pronotum together ; segments rounded. Pronotum rugose, with a row of black tubercles at the anterior and posterior margin. Dorsum rugose and with rows of tubercles lateral to median carina. Anterior abdominal terga with longitudinal carinulae ; posterior margins with dark zones.

Vestigial tegmina densely granulate, usually veinless. Inner side of hind femur with scattered specks and black tubercles. Hind tibia purple or purple-grey. Arolium of female very small, in male less than half as long as claws.

Coloration : Generally whitish with a pattern of dark lines.

Measurements (mm) : Body ♀ 46.5–59.5, ♂ 27.5–30.0 ; vestigial tegmina ♀ 7.0–8.0, ♂ 5.0–6.0 ; hind femur ♀ 17.0–19.5, ♂ 12.0–13.0.

Distribution : Israel, Sinai, Egypt.

Israel : Northern and Central Negev (15, 17), Sinai (20, 21).

Usually found among the branches of *Retama, Polygonum, Artemisia, Haloxylon.* Typical phytophiles, they hide among the lower branches of bushes and crawl to other side of stem on sensing danger. Oviposition from March to May, in dry soil at depth of about 7–8 cm. Wall of egg-pod hard and strong ; number of eggs 22–90. Diapause can be interrupted by wetting. Hoppers appear after the first rains during January–February.

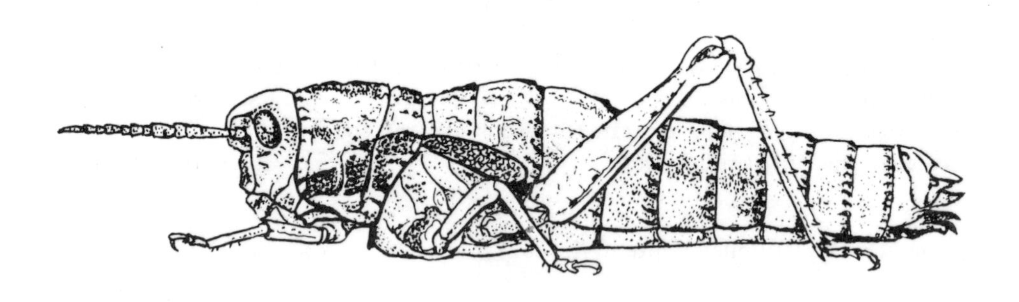

Fig. 94 : *Acinipe zebratus* (Brunner-Wattenwyl, 1882), female

Acinipe hebraeus (Uvarov, 1942)
Figs. 95 ; Plate III : 2

Type Locality : Mount Carmel, Israel (British Museum).

Pamphagus foreli (Pictet & Saussure). Giglio-Tos E., 1893, *Boll. Musei. Zool. Anat. comp. R. Univ. Torino*, 8 (164) : 7.
Orchamus yersini (Brunner-Wattenwyl). Bodenheimer F. S., 1935, *Arch. Naturgesch.*, 4 (2) : 199.
Orchamus hebraeus Uvarov B. P., 1942, *Trans. Am. ent. Soc.*, 67 : 348.

Large. Head hypognathous, swollen. Occiput with slightly marked longitudinal carinae. Raised margins of frontal ridge gradually diverging from median ocellus towards clypeus. Fastigium of vertex sloping. Antennae ensiform ; articulation between basal segments distinct ; apical segments narrower, more separate, chain-like. Pronotum with light and dark tubercles ; median carina arc-like along its entire length, intersected posteriorly by only one transverse sulcus. Prosternal process square with numerous tiny tubercles at the apex. Metasternal lobes nearly contiguous.
Coloration : Brown or greyish-brown. Hind tibia very hairy ; outer surface grey, inner surface bluish-grey or bluish. Arolium nearly as long as claws.
Measurements (mm) : Body ♀ 62.0–76.5, ♂ 40.0 ; residual tegmina ♀ 9.0–11.0, ♂ 6.0 ; hind femur ♀ 25.5–29.5, ♂ 17.5.
Distribution : Endemic to the Carmel Ridge and vicinity (apparently closely related to *O. yersini* from Lebanon and Syria).
Israel : Carmel Ridge (3), Northern Coastal Plain (4), Upper Galilee (1).
Inhabits areas with dense and rich vegetation.

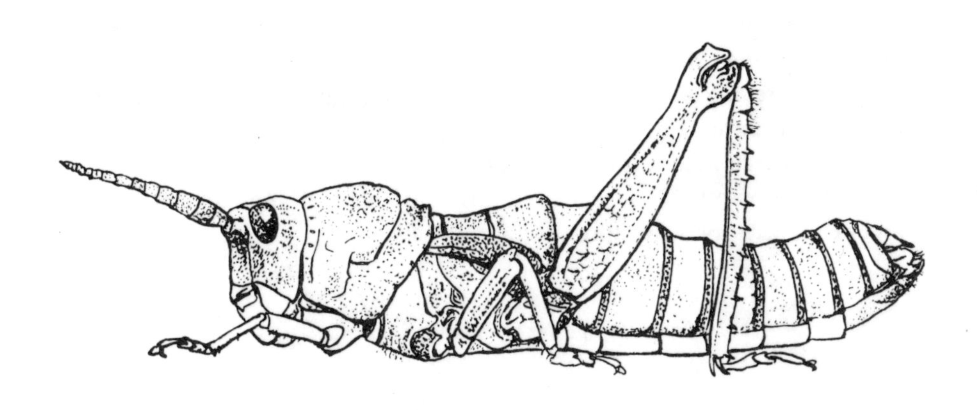

Fig. 95 : *Acinipe hebraeus* (Uvarov, 1942), female

Acinipe davisi (Uvarov, 1949)

Type Locality : South-West Anatolia.

Orchamus davisi Uvarov B. P., 1949, *Bull. Soc. Fouad Ier Ent.*, 33 : 8.

Large, very rugose ; integument with numerous black shiny tubercles. Frontal ridge entirely obliterated below median ocellus. Punctations on occiput often forming continuous lines. Pronotum arc-like, very rugose ; posterior margin nearly straight in female. Prosternal process with a few scattered tubercles at the apex. Sternal plate very hairy.
Coloration : Brown or greyish-brown ; lighter areas on face, pronotum and outer side of posterior femur. Inner surface of hind tibia reddish-violet ; spines strong.
Measurements (mm) : Body ♀ 65.5–80.0 ; tegmina ♀ 9.0–11.0 hind femur ♀ 25.5–27.5.
Distribution : Turkey, Syria, Northern Israel.
Israel : Upper Galilee (1), Northern Coastal Plain (4).

Genus OCNEROPSIS Uvarov, 1942
Trans. Am. ent. Soc., 67 : 348

Type Species : *Pamphagus bethlemita* Bolivar, 1893.
Diagnosis : Large or medium sized, slightly depressed, rounded laterally. Head hypognathous. Eyes round, their diameter much shorter than subocular groove. Occiput globose, with a median carina usually accompanied by a network of carinulae. Face nearly square ; fastigium strongly sloping, triangular. Antennae slightly ensiform. Pronotum thick ; median carina raised, slightly intersected posteriorly by a transverse sulcus ; posterior margin straight or curved inwards. Mesosternal interspace wider than long. Metasternal interspace of male nearly four times wider than long, and twice as long as wide in female. Ventral margin of lateral lobe undulate, sometimes also dentate.
Distribution : Eastern Mediterranean countries.
Found usually on basalt soil on bare or nearly bare areas of mountain slopes. Five closely related species, of which three occur in Israel.

Key to the Species of Ocneropsis in Israel

1. Occiput usually smooth, only with median carina (Fig. 96). Pronotum smooth or slightly granulate, either intersected by posterior transverse sulcus or not ; posterior margin straight. Inner surface of hind femur bluish. **O. kneuckeri** (Krauss)
– Occiput with a network of carinulae (Fig. 99). Pronotum strongly granulate or rugose, distinctly intersected by transverse sulcus ; posterior margin curved inwards 2

2. Eyes almost round. Pronotum sharply slopin behind transverse sulcus (Fig. 97). Pronotum and abdominal terga very rugose, with strong elongate tubercles. Inner surface of hind femur and tibia red or reddish. **O. bethlemita** (Bolivar)
– Eyes oval, elongate. Pronotum continuous on same level behind transverse sulcus (Fig. 98). Inner surface of hind femur and tibia blackish-blue. **O. lividipes** n. sp.

Ocneropsis kneuckeri (Krauss, 1909)
Fig. 96

Type Locality : 'Lebanon' (2,500 m).

Pamphagus kneuckeri Krauss H. A., 1909, in : Kneucker A., *Verh. naturw. Ver. Karlsruhe*, 21 : 99, 120.

Large, granulate and rugose, but less so than the other species. Head hypognathous. Margins of frontal ridge reaching clypeus. Vertex flat or slightly concave, usually with numerous tubercles, posteriorly present also on the partly obliterated median carina. Eyes elliptical ; as long as subocular groove in male, slightly shorter in female. Antennae shorter than head and pronotum together ; 14 segments, the basal segments partly fused, the 7th almost twice as long as the 6th. Pronotum rugose, especially in male ; median carina arc-like or almost straight ; posterior transverse sulcus usually obliterated or only slightly marked.
Vestigial tegmina slightly curved, in females only partly reaching or covering the tympanal organ ; network of veinlets present. First three abdominal terga rugose, with low crests along the median line ; in males continuing over most of the dorsal part. Hind femur with serrate dorsal and ventral marginal carinae, the ventral carina undulate. Hind tibia shorter than femur, with nine spines on the inner and outer margin.
Coloration : Ochraceous-grey ; margins of abdominal terga brown. Inner surface of femur and tibia bluish ; tarsus pale orange.
Measurements (mm) : Body ♀ 50.0–60.2, ♂ 29.5–36.5 ; vestigial tegmina ♀ 6.0–8.2, ♂ 4.5–5.0 ; pronotum ♀ 12.0–14.0, ♂ 7.0–8.5.
Distribution : Lebanon, Syria, Israel.
Israel : Mount Hermon (19).
Found on Mount Hermon from 1,600 to 2,100 m altitude, on bare areas covered with gravel, sometimes near snow. Adults and copulating pairs found during July–August. Hoppers prominent on account of orange colouring of inner surfaces of hind femur and tibia. Younger stages are found mainly in May–June, but as premature stages are also found during October–November. The population on Mount Hermon apparently hibernates as juveniles and maturates only during the following year.

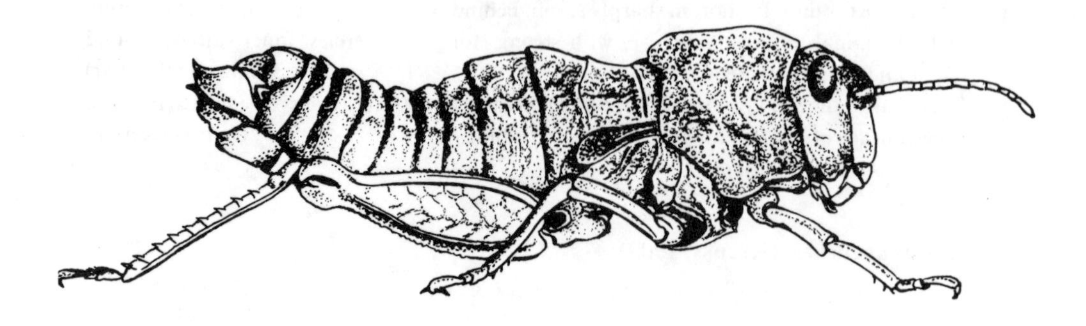

Fig. 96 : *Ocneropsis kneuckeri* (Krauss, 1909), female

Ocneropsis bethlemita (Bolivar, 1893)
Fig. 97 ; Plate V : 7

Type Locality : 'Palestine'.

Pamphagus bethlemita Bolivar I., 1892/93, *Rev. biol. nord Fr.*, 5 : 484.
Prionosthenus bethlemita —. Bolivar I., 1916, *Genera Insectorum*, fasc. 170 : 15.
Prionosthenus bethlehemita —. Bodenheimer F. S., 1935, *Arch. Naturgesch.*, 4 (2) : 200.
Ocneropsis bethlemita —. Uvarov B. P., 1942, *Trans. Am. ent. Soc.*, 67 : 348.

Strong, very rugose, carinate, with dorsal protrusions on all abdominal terga. Frontal ridge projecting ventrally, reaching the clypeus. Vertex flat, its margins sharp, with a median carina posteriorly with smooth areas on both sides in the female. Eyes almost round, their diameter much shorter than the subocular groove. Pronotum abruptly sloping behind transverse sulcus ; metazona usually with raised shoulders and inwards-curved posterior margin.
Vestigial tegmina with a very dense, irregular network of veinlets. Hind femur slightly flattened ; dorsal and ventral margins undulate. Hind tibia as long as femur, in male usually with 10, rarely with nine, spines on the inner and outer margin.
Coloration : Usually dark grey, with a pattern of light spots. Inner surface of hind femur and hind tibia pinkish-violet or pinkish.
Measurements (mm) : Body ♀ 53.0–61.0, ♂ 28.5–32.0 ; vestigial tegmina ♀ 8.0–10.0, ♂ 6.0–6.5 ; hind femur (length) ♀ 20.0–22.0, ♂ 13.5–17.0 ; hind femur (width) ♀ 5.0.
Distribution : Syria, Israel.
Israel : Eastern Upper Galilee (1), Golan Heights (18).
Usually found on basalt soil, in areas with dense vegetation.

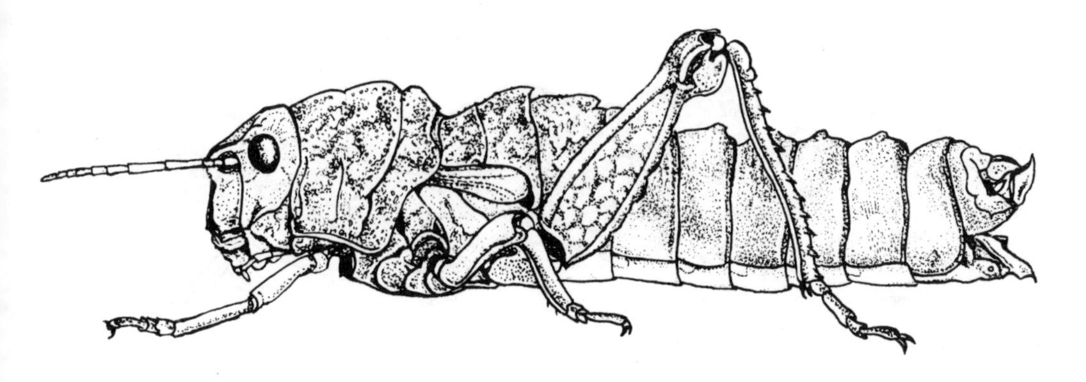

Fig. 97 : *Ocneropsis bethlemita* (Bolivar, 1893), female

Ocneropsis lividipes n. sp.
Figs. 98–102

Type Locality : Mount Hermon.

Body round in cross-section or slightly higher than wide, markedly rugose posteriorly, sparsely rugose anteriorly. Head more or less smooth, especially in female ; frontal ridge slightly concave, slightly protruding anteriorly, intersected below median ocellus by a transverse groove (Fig. 101). Vertex flat or slightly concave, with sharp margins. Occiput with a network of carinulae on both sides of median carina (Figs. 99, 100). Eyes oval, their vertical diameter in males as long as subocular groove, in females shorter. Antennae short, with 11 or 12 segments (Fig. 102). Median carina of pronotum granulate ; posterior margin straight or slightly curved inwards ; transverse sulcus shallow. Parallel ridges formed by rugose surface of pronotum on both sides of median carina.
Vestigial tegmina very narrow at the base, slightly widening apically and covering the tympanum, their surface with dense irregular veins. First three abdominal terga with ridges, granulate and with a central crest on the posterior margin ; crest and ridges also present on the other terga in the male. Dorsal margin of hind femur serrate, ventral margin undulate and with serrations on the outside parts. Hind tibia with nine spines on inner margin, 9–10 on outer margin.
Coloration : Brown with a pattern of pale lines ; a pattern of black and white lines present on the posterior margins of abdominal terga. Inner surface of hind femur and tibia dark blue or blue ; tarsi pale yellowish.
Measurements (mm) : Body ♀ 58.0–64.0, ♂ 35.0–42.0 ; vestigial tegmina ♀ 8.5–9.0, ♂ 6.5–7.0 ; pronotum ♀ 11.5, ♂ 7.5.
Distribution, Israel : Mount Hermon, 2,200 m altitude (18).

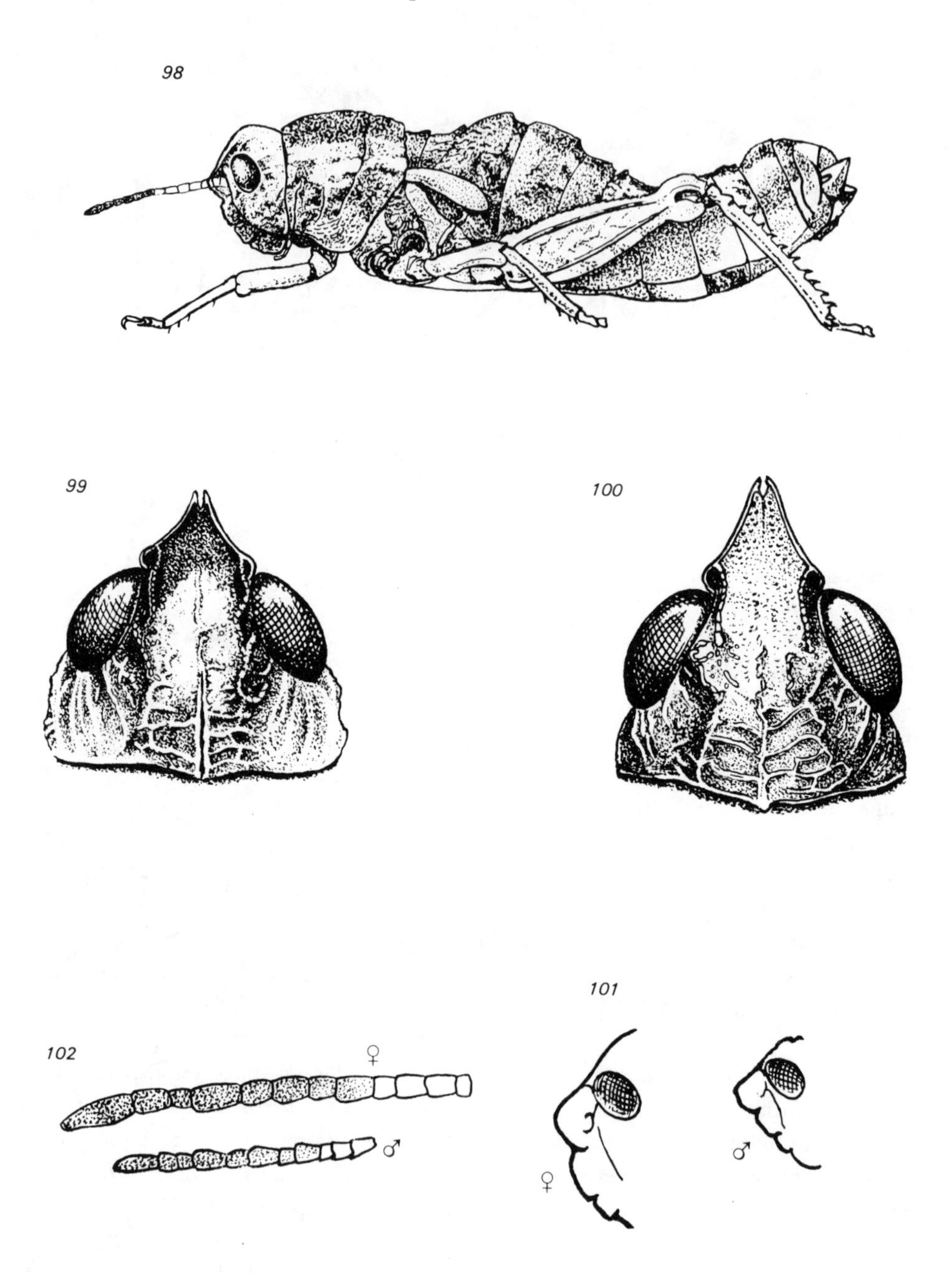

Figs. 98–102 : *Ocneropsis lividipes* n. sp.
98. female ; 99. head of female, dorsal ; 100. head of male, dorsal ; 101. head profiles of female and male ; 102. antennae of female and male

The name *O. lividipes* refers to the blue colour of the inner surface of the hind femur. Differs from other species of the genus *Ocneropsis* by the oval eyes, flat pronotum behind the transverse sulcus and blackish-blue inner surfaces of hind femurs. The holotype (male) and six paratypes (two males and four females) are deposited in the Entomological Collection of the Department of Zoology, Tel Aviv University.

Subfamily EGNATIINAE

Diagnosis : Small. Head slightly opisthognathous ; antennae filiform ; eyes rounded, protruding laterally. Foveolae triangular, usually distinct, margins sometimes obliterated. Pronotum rounded ; carinae partly obliterated ; posterior transverse sulcus situated in middle of pronotum or slightly in front of it. Prosternum slightly convex ; mesosternal suture undulate, curved backwards, bending between mesosternal lobes (Fig. 104).

Tegmina narrow ; intercalary vein distinct in median field, usually short ; field open. Arolium small, often rudimentary. Males of many species with rugose posterior abdominal segments.

Distribution : North Africa and South-West Asia.

Two genera in our region.

Key to the Genera of Egnatiinae in Israel and Sinai

1. Head smooth dorsally. Prozona of pronotum without median carina. **Leptoscirtus** Saussure
– Head with ridges and rugulae dorsally. Prozona of pronotum with distinct median carina. **Egnatioides** Vosseler

Genus LEPTOSCIRTUS Saussure, 1888

Mém. Soc. Phys. Genève, 30 (1) : 23, 24, 72

Leptoscirtus —. Innes W., 1929, *Mém. Soc. R. ent. Égypte*, 3 (2) : 40, 50.
Leptoscirtus —. Uvarov B. P., 1929, in : *Ergebnisse der Sinai-Expedition*, Leipzig, p. 97.
Leptoscirtus —. Bei-Bienko G. Ya. & L. L. Mishchenko, 1951, *Locusts and Grasshoppers of the U. S. S. R. and Adjacent Countries*, Moscow, I : 373 [in Russian].
Leptoscirtus —. Dirsh V. M., 1961, *Bull. Br. Mus. nat. Hist. (Ent.)*, 10 (9) : 412.

Type Species : *Leptoscirtus aviculus* Saussure, 1888.

Diagnosis : Small, smooth. Head without rugulae dorsally, slightly opisthognathous.

Vertex sloping; foveolae triangular, sometimes indistinct. Prozona of pronotum without carinae; metazona with only a median carina. Pronotum intersected by three distinct transverse sulci, the posterior one situated in front of middle of pronotum. Tegmina developed; intercalary vein spurious.
Distribution: North Africa, Arabia, Iran.
One unidentified species in our region.

Leptoscirtus sp.

Small, smooth, resembling *Sphingonotus.* Head of female slightly opisthognathous; frontal ridge slightly concave, protruding between bases of antennae, and with diverging margins below median ocellus. Fastigium sloped; foveolae only partly visible from above, their margins raised. Vertex concave; lateral margins raised, continuing beyond the eyes as a row of low tubercles. Occiput smooth; median carina present. Eyes protruding laterally, their vertical diameter longer than the subocular groove. Pronotum slightly saddle-shaped, intersected by three transverse sulci, the posterior one situated in front of its middle. Prozona smooth; metazona densely punctate, its posterior margin rounded. Prosternum with an oblong process; mesosternal and metasternal interspaces wide and short.
Tegmina extending beyond hind knees, membranous apically, with a sparse mesh of oblong cells. Ventral outer spur of hind tibia twice as long as inner spur. Arolium very small.
Coloration: Pale brown; face and lateral lobes of pronotum white. Brown bands and blotches on tegmina; wings colourless. Inner surface of hind femur dark, with a light apical band; on dorsal part, two dark bands merging with the dark grey colour of the outer surface.
Measurements (mm): Body ♀ 13.0; tegmina ♀ 11.5; hind femur ♀ 7.5.
Distribution, Israel: One female collected south of Be'er-Sheva (1954) and a second one in Sinai (August 1968).
As the genus has not been revised taxonomically, the identification of this species is not possible.

Genus EGNATIOIDES Vosseler, 1902

Zool. Jb. (Syst.), 16: 361

Egnatius Stål C., 1876, *K. svenska VetenskAkad. Handl.*, 4 (5): 25.

Type Species: *Egnatioides striatus* Stål, 1876.
Diagnosis: Small, smooth, hairs sparse. Head raised above pronotum, with numerous small ridges and rugulae towards occiput. Frontal ridge protruding between

antennae, flat, with diverging margins below median ocellus. Vertex sloping, narrow; foveolae not visible from above. Lateral and median carinae of pronotum present at least in prozona. Prosternum globular, smooth. Tegmina with a spurious vein in median field.
Distribution: Sahara, North Africa, Central Asia, Iran.
One species in our region.

Egnatioides coerulans (Krauss, 1893)
Figs. 103, 104

Type Locality: Mecheria, Algeria (Stuttgart Museum).

Egnatius coerulans Krauss H. A., 1893, *Jh. Ver. vaterl. Naturk. Württemb.*, 49: xcv.
Egnatioides coerulans —. Uvarov B. P., 1926, *Eos, Madr.*, 2: 357.

Small, smooth. Head slightly opisthognathous, raised above pronotum, especially in male. Frontal ridge protruding between antennae; margins parallel to the median ocellus, then slightly separating and diverging towards clypeus. Facial carinae extending straight towards lateral ocelli. Fastigium sloping; foveolae small, their ventral margin obliterated. Vertex narrow, its lateral margins converging between the eyes and connected by a small transverse ridge. Occiput rugose and swollen in parts. Pronotum short, rounded posteriorly, intersected by three transverse sulci, the posterior one situated in middle of pronotum (male) or before it (female). Metazona with a median carina; anterior part of prozona with median and lateral carinae. Prozonal lateral lobe rugose, its posterior margin rounded. Antennae in female 1.5 times as long as head and pronotum together, in male nearly twice as long. Prosternum globular, oblong; mesosternal suture deeply arched between mesosternal lobes (Fig. 104).
Tegmina narrow, extending beyond hind knees; median field open; intercalary vein spurious. Hind femur with sharply raised dorsal and ventral margins; hind tibia with seven spines on inner line and 8–10 on outer line. Arolia absent.
Coloration: Sandy-brown with numerous brown blotches. Brown and light rings on the antennae; two triangular brown spots on upper surface of hind femur. Hind tibia greyish-blue.
Measurements (mm): Body ♀ 13.5–14.5, ♂ 11.0; tegmina ♀ 13.0–14.0, ♂ 10.5.
Distribution: North and East Africa, Israel.
Israel: Northern Negev (15).
Inhabits areas of loose sand, usually found together with *Dociostaurus genei.*

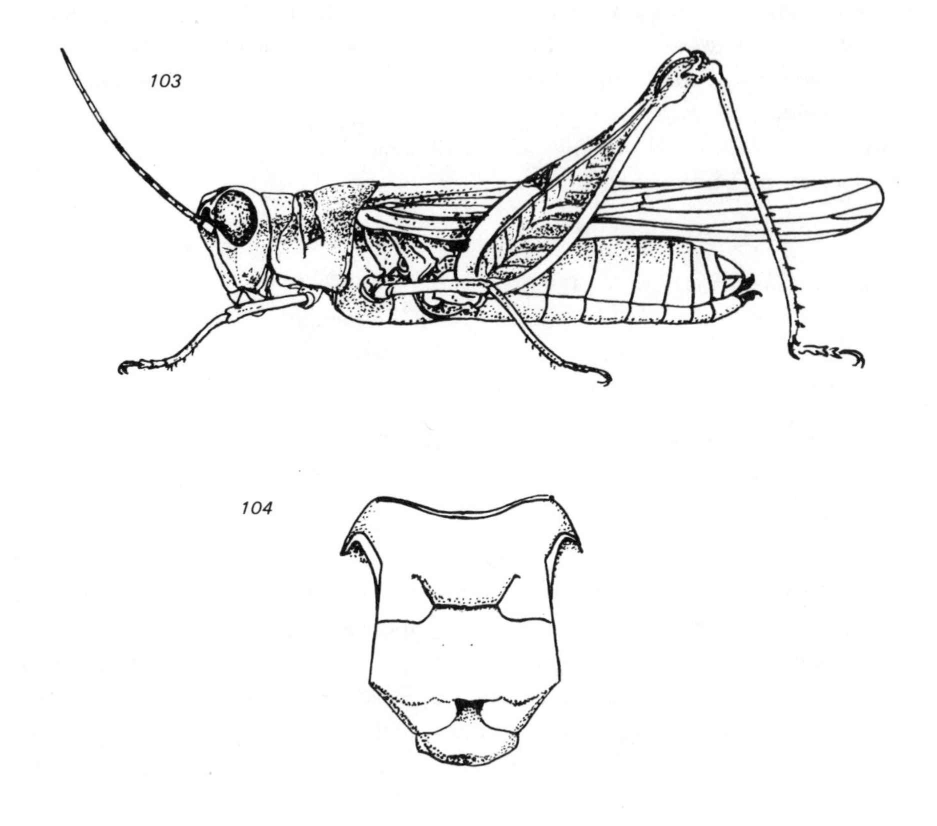

Figs. 103–104 : *Egnatioides coerulans* (Krauss, 1893)
103. female ; 104. sternum

Subfamily ACRIDINAE

Diagnosis : Head opisthognathous, except in *Eremogryllus* Krauss, where it is hypognathous. Antennae widened at the base, ensiform or filiform, with flattened segments (Fig. 108). Foveolae distinct, situated laterally, square or oblong (Fig. 136), sometimes absent ; head smooth in *Duroniella* Bolivar (Fig. 117).
Pronotum usually with three distinct carinae. Prosternum smooth ; projecting process absent. In most species body height exceeds width of thorax. Antennae long, usually widened at the base. Outer median field of femur with two rows of symmetrical plates ; inner surface usually with a row of minute serrations forming the characteristic sound-producing apparatus. Wings usually transparent.

Acridinae

Coloration : Body green or pale grey (ochraceous) ; often a blending of both colours in the same individual. Most species have large arolia and strong claws.
The Acridinae form a typically phytophilous group inhabiting areas with dense vegetation, except *Bodenheimerella* and *Eremogryllus*, which inhabit bare and dry desert habitats. Most species spend their life on plants, which serve as both environment and food source. Usually found sitting along stems, or jumping from one plant to another, without descending to the ground. Propagation in the spring, egg-laying at end of spring and development of hoppers in early summer.

Key to the Genera of Acridinae in Israel and Sinai

1. Posterior knee lobes of femur with sharp apical projections (Fig. 109). Head large, conical, larger than pronotum (Fig. 107). — 2
– Knee lobes rounded (Fig. 119). Head shorter than or as long as pronotum — 3
2. Front femur longer than pronotum. Arolium small, generally shorter than claws, often very small. Lateral carinae of pronotum usually diverging from the median one, especially in the metazona (Figs. 110, 111). — **Truxalis** Fabricius
– Front femur shorter than or as long as pronotum. Arolium larger, at least as long as claws. Lateral carinae of pronotum parallel to the median one or only slightly diverging (Fig. 114). — **Acrida** Linnaeus
3. Foveolae present, often only partly visible from above or only from a frontal view (Fig. 157) — 4
– Foveolae absent. Vertex smooth (Fig. 122). — **Duroniella** Bolivar
4. Foveolae square or nearly square (Figs. 136, 138). All antennal segments of equal length — 5
– Foveolae narrow, at least twice as long as wide (Fig. 146). Antennal segments of different forms — 10
5. Posterior spurs very long, usually as long as or longer than first segment of hind tarsus (Fig. 124). Head nearly hypognathous (Fig. 123). — **Eremogryllus** Krauss
– Posterior spurs shorter than above. Head opisthognathous — 6
6. Metasternal space reduced in female ; metasternal lobes contiguous — 7
– Metasternal space distinct ; metasternal lobes not contiguous — 8
7. Lateral carinae of pronotum distinct along their entire length, intersected by one transverse sulcus (Fig. 126). Foveolae only partly visible from above. Ventral spur on inner side of hind tibia twice as long as dorsal spur. — **Stenohippus** Uvarov
– Lateral carinae of pronotum obliterated in the middle, intersected by three transverse sulci. Foveolae visible from above. Inner spurs of hind tibia nearly equally long. Outer apex of wings with a large dark spot (Figs. 127, 128). — **Ramburiella** Bolivar
8. Light X-shaped marking on pronotum (Fig. 133). First tarsal segment as long as the other two segments together. Antennae longer than head and pronotum together — 9
– No marking on pronotum. First tarsal segment shorter than the other two together. Antennae shorter than head and pronotum together (Fig. 129). Small species that inhabits southern deserts. — **Bodenheimerella** Uvarov
9. Occiput with a distinct longitudinal carina, irregularly granulate on both sides (Fig. 131). — **Notostaurus** Bei-Bienko

- Occiput smooth, lacking longitudinal carina (Fig. 135). **Dociostaurus** Fieber

10. Antennae ensiform, widened at the base, narrowing apically (Fig. 160) 11

- Antennae filiform ; most segments of equal length (Fig. 143). **Chorthippus** Fieber

11. Head sharp, opisthognathous, conical. Lateral carinae of pronotum almost straight, gradually diverging posteriorly (Fig. 152). Foveolae not visible from above. **Ochrilidia** Stål

- Head almost hypognathous or slightly opisthognathous. Lateral carinae of pronotum not straight, usually interrupted by transverse sulci 12

12. Smooth ; lateral carinae on prozona of pronotum converging towards the median carinae (Fig. 163). Hind femur slender ; arolium large. **Xerohippus** Uvarov

- Rugose-granulate ; lateral carinae on prozona straight, on metazona diverging (Fig. 164). Hind femur short and hairy (Fig. 165). Arolium very small. **Notopleura** Krauss

Genus TRUXALIS Fabricius, 1775

Systema entomologiae, p. 279

Tryxalis —. Brullé G. A., 1835, in : Audouin J. V. & Brullé, G. A., *Histoire naturelle des insectes*, 9 : 216.

Acrida L. Bolivar I., 1876, *An. Soc. esp. Hist. nat.*, 5 : 102.

Acridella Bollivar. Kirby W. F., 1910, *A synonymic catalogue of the Orthoptera*, Vol. III Orthoptera Saltatoria, Part II, London, p. 95.

Truxalis —. Dirsh V. M., 1951, *Eos, Madr.*, Tomo estraord. (1950), pp. 135, 151.

Truxalis —. Dirsh V. M., 1961, *Bull. Br. Mus. nat. Hist. (Ent.)*, 10 (9) : 414.

Type Species : *Truxalis nasutus* Fabricius, 1775 = *Gryllus (Acrida) nasuta* Linnaeus, 1758.

Diagnosis : Large species. Head conical, opisthognathous, longer than pronotum ; antennae filiform (Fig. 108). Eyes situated anteriorly, elongate. Pronotum usually saddle-shaped, with single transverse sulcus and lateral carinae diverging posteriorly from the median carina (Figs. 110, 111).

Hind femur longer than pronotum ; posterior knee with sharp spines on dorsal and ventral lobes (Fig. 109). Arolium very short, less than half as long as the claws.

Coloration : Usually green and light brown. Wings red or purple, often transparent.

Distribution : Southern Europe, Asia, Africa.

Several species are found in the Negev, only one in the north of Israel. Most species inhabit areas with low, often sparse vegetation.

Key to the Species of Truxalis in Israel

1. Antennae very long, narrow, filiform, longer by a third than head and pronotum together, with very long segments. Tegmina with rounded apex. Basal part of wings dark blue in female, reddish in male. **T. longicornis** (Krauss)

- Antennae shorter, as long as head and pronotum together. Coloration of basal part of wings not as in *T. longicornis* 2

2. Transverse sulcus divides the pronotum into a prozona and a metazona of the same length (Fig. 105). Mesosternal interspace wider than high. Posterior angle of lateral lobe rounded (Fig. 106). **T. eximia** (Eichwald)
- Transverse sulcus posterior to middle of pronotum, forming a prozona which is longer than the metazona. Mesosternal interspace less wide than high. Posterior angle of lateral lobe acute or with a posteriorly curved process 3
3. Pronotum saddle-shaped ; prozona cylindrical (Fig. 110). Median carina indistinct ; lateral carina undulate, obliterated at posterior and anterior margins. Lateral lobe with right-angled posterior apex. Arolium small, narrow at apex. **T. procera** Klug
- Pronotum not saddle-shaped ; median carina distinct along its entire length. Posterior angle of lateral lobe with a sharp posteriorly curved process (Fig. 112). **T. grandis** Klug

Truxalis eximia (Eichwald, 1830)

Figs. 105, 106

Type Locality : Transcaucasus.

Truxalis eximius Eichwald E., 1830, *Zoologia specialis*, Vilnae, 2 : 239.

Acridella nasuta L. Kirby W. F., 1910, *A synonymic catalogue of the Orthoptera*, Vol. III, Orthoptera Saltatoria, Part II, London, p. 95 (partim).

Truxalis eximia eximia —. Dirsh V. M., 1951, *Eos, Madr.*, Tomo estraord. (1950), p. 213.

Antennae filiform, shorter than head and pronotum together ; basal segment not completely divided. Vertex rounded apically, concave, with a carinula extending behind the eyes. Pronotum saddle-shaped with a posterior transverse sulcus along the middle. Prozona smooth ; metazona with a pattern of linear punctations (Fig. 105). Lateral carinae of pronotum straight, diverging distinctly in metazona and converging towards the hind margin (Fig. 106). Posterior angle of the lateral lobe rounded. Mesosternal sutures curved inwards, the interspace with wide apex ; mesosternal interspace square.

Tegmina extending beyond hind knees ; apex obtuse ; anterior margin curved. Hind femur long, extending beyond abdomen ; knees with sharp, pointed lobes. Arolium small.

Coloration : Green, ochraceous or brown with a white line along the middle of the tegmina, often interrupted by the marginal brown colour. Metanotum and base of abdomen bluish or bluish-purple. Wings of female purple at base, other parts yellowish, veins brown.

Measurements (mm) : Body ♀ 61.0–72.0, ♂ 42.0–50.0 ; tegmina ♀ 55.0–61.0, ♂ 37.0–44.0 ; front femur ♀ 11.5–14.0, ♂ 8.5–11.0 ; antennae ♀ 21.5–27.5, ♂ 16.0–19.0.

Distribution : Caucasus, Central Asia, Arabia, Persia, India, Israel.

Israel : Jordan Valley (7), Coastal Plain (8, 9), Judean Hills (11), Negev (15, 16, 17). It mainly inhabits southern arid areas ; found also in sheltered wadis with bush vegetation.

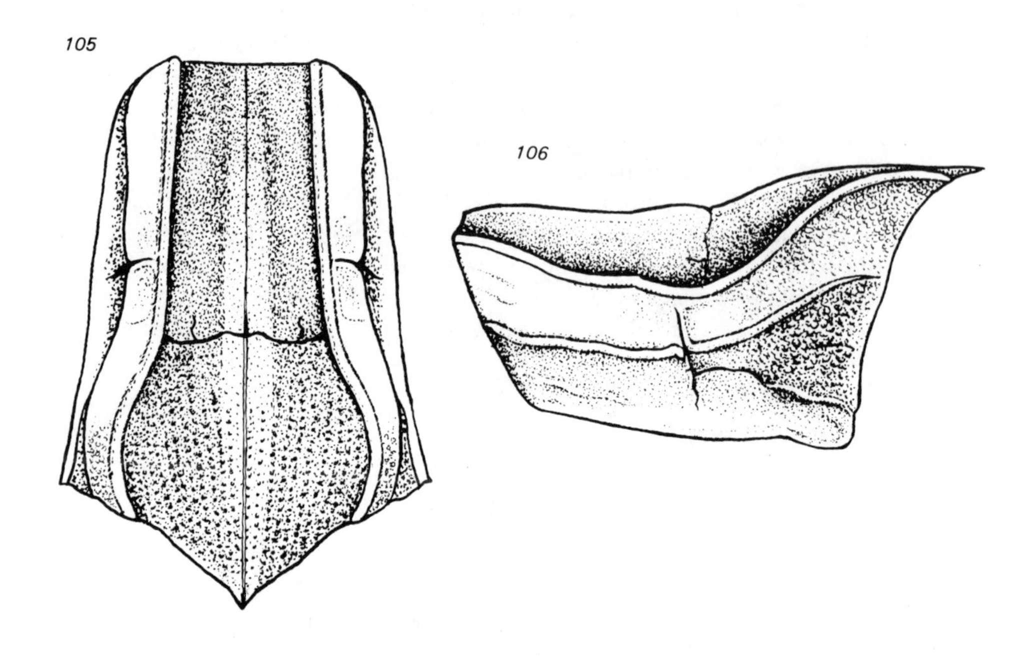

Figs. 105–106 : *Truxalis eximia* (Eichwald, 1830)
105. pronotum, dorsal ; 106. same, lateral

Truxalis procera Klug, 1830
Figs. 107–110 ; Plate III : 3

Type Locality : 'Arabia felix' = South Arabia (Berlin Museum).

Truxalis procera Klug J. C. F., 1830, *Symbolae physicae*, Vol. 4, pl. 16, fig. 2.
Tryxalis klugii Fieber F. X., 1853, *Lotos*, 3 : 97 (partim).
Acridella nasuta L. Kirby W. F., 1910, *A synonymic catalogue of the Orthoptera*, III, Orthoptera Saltatoria, Part II, London, p. 95 (partim).
Acridella procera —. Shaw W. B. K., 1933, *Entomologist*, 66 : 177.
Truxalis procera —. Dirsh V. M., 1951, *Eos, Madr.*, Tomo estraord. (1950), pp. 154, 183.

Slender, cylindrical and long, especially head and pronotum (Figs. 107, 110). Antennae as long as or shorter than head and pronotum together, gradually narrowing in the female, nearly ensiform in the male ; distal segments punctate. Frons deeply concave, swollen around median ocellus ; frontal ridges prominent. Fastigium obtuse ; carinula raised. Pronotum cylindrical (Fig. 110) ; transverse sulcus situated posterior to the middle. Lateral carinae parallel, sinuate in prozona, slightly raised and diverging in metazona. Posterior margin of metazona obtuse- or acute-angled, with sides curving slightly inwards.

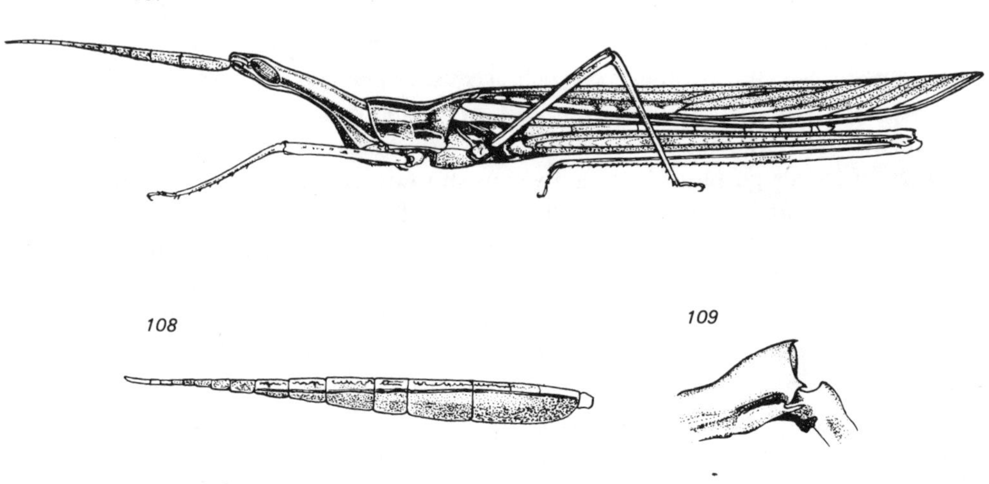

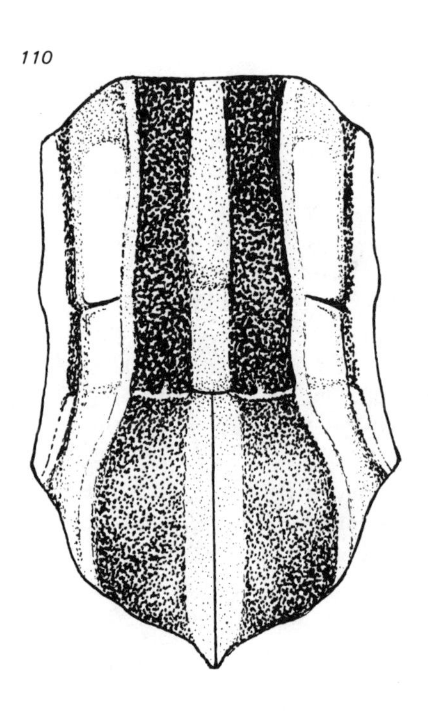

Figs. 107–110 : *Truxalis procera* Klug, 1830
107. female ; 108. antenna; 109. hind knee ; 110. pronotum, dorsal

Mesosternal interspace narrow, twice as long as wide; curved inwards. Metasternal interspace square. Tegmina narrow, at least 10 times longer than wide, with narrow subacute apex extending beyond hind knee. Hind femur long; knee with short lateral lobe on the dorsal side and a long spine on both ventral sides (Fig. 109).
Coloration: Brown, brownish-grey or ochraceous. Antennae pale apically. Prominent pattern of white lines present on head, pronotum and the ventral side. Distal abdominal terga dark violet; bluish-grey and pale rings on legs. Wings of male yellowish, of female purple at base and yellowish towards apex.
Measurements (mm): Body ♀ 72.0–86.0, ♂ 53.0–60.0; femur ♀ 43.0–49.5, ♂ 31.5–35.5; pronotum ♀ 1.20–13.0. ♂ 7.5–8.5; tegmina ♀ 59.0–68.0, ♂ 43.0–45.0; antennae ♀ 22.0–31.0, ♂ 21.0–23.0.
Distribution: North and East Africa, Israel, Arabia, Persia.
Israel: Central and Southern Coastal Plain (8, 9), Northern Negev (15).
Mainly inhabits sandy areas with very low and sparse vegetation, merging well in its surroundings, though no green phase has been observed. One of the most common species of *Truxalis* in the Coastal Plain. Its feeds on Gramineae and other ephemeral plants. Courtship accompanied by long chirping sounds. Hoppers observed in March–April and September–October.

Truxalis grandis Klug, 1830
Figs. 111, 112

Type Locality: 'Upper Egypt' (Berlin Museum).

Truxalis grandis Klug J. C. F., 1830, *Symbolae physicae*, Vol. 2, pl. 15, fig. 1 (♀).
Truxalis miniata Klug J. C. F., 1830, *ibid.*, pl. 18, fig. 1 (♂).
Tryxalis grandis —. Lucas H., 1857, *Annls Soc. ent. Fr.* (3) 5: Bull. xl.
Tryxalis miniata —. Gerstaecker A., 1869, *Arch. Naturgesch.*, 35 (1): 215.
Tryxalis unguiculata var. *miniata* —. Krauss H. A., 1892, *Wien. ent. Ztg.*, 11 (5): 148.
Acrida miniata —. Jacobson G. G. & V. L. Bianchi, 1902, *Orthoptera and Odonata of the Russian Empire*, St. Petersburg, pp. 176, 214 [in Russian].
Acridella grandis —. Kirby W. F., 1910, *A synonymic catalogue of the Orthoptera*, Vol. III, Orthoptera Saltatoria, Part II, London, p. 95.
Acridella miniata —. Buxton P. A. & B. P. Uvarov, 1923, *Bull. Soc. R. ent. Égypte*, p. 184.
Acridella nasuta var. *miniata* —. Innes W., 1929, *Mém. Soc. R. ent. Égypte*, 3 (2): 17.
Acridella nasuta var. *grandis* —. Innes W., 1929, *ibid.*, p. 16.
Acridella grandis —. Bodenheimer F. S., 1935, *Arch. Naturgesch.*, 4 (2): 178.
Acridella miniata —. Bodenheimer F. S., 1935, *Animal Life in Palestine*, Jerusalem, pp. 86, 311, 320.
Truxalis grandis —. Dirsh V. M., 1951, *Eos, Madr.*, Tomo estraord. (1950), pp. 169, 170.

Antennae of females markedly filiform, their apex sharply narrowing; moderately ensiform and narrow in males. Fastigium of vertex broad and rounded. Facial frontal ridges low; additional longitudinal carinae on both their sides. Pronotum flat (Fig.

111) ; lateral carinae almost straight, slightly diverging posteriorly. Metazona granulate, either obtuse- or acute-angled. Mesosternal interspace three times as long as wide ; apex wide. Metasternal interspace twice as long as wide.
Tegmina long, extending beyond abdomen and posterior knees; outer margins curved ; apex pointed. Lateral lobes of hind knee short, acute-angled. Arolium large, one-half or two-thirds the length of the claws.
Coloration : Pale brown, pale green or green, with white lines on pronotum and thorax. Tegmina with or without a pale longitudinal line, sometimes framed dorsally by a brown field. Base of wing of female blue or bluish-violet, of male pinkish-red.
Measurements (mm) : Body ♀ 70.0–88.0, ♂ 44.0–47.5 ; tegmina ♀ 64.0–74.0, ♂ 36.0–40.0 ; antennae ♀ 22.0–27.0, ♂ 16.0–19.0 hind femur ♀ 13.0–17.0, ♂ 8.0–9.0.
Distribution : North and East Africa, Arabia, Israel, Syria.
Israel : Upper Galilee (1), Coastal Plain (4, 8, 9), Jordan Valley (7), Judean Hills (11), ‘Arava Valley (14), Northern Negev (15).
The most common species of *Truxalis* ; found throughout the year. Inhabits areas with dense green vegetation.

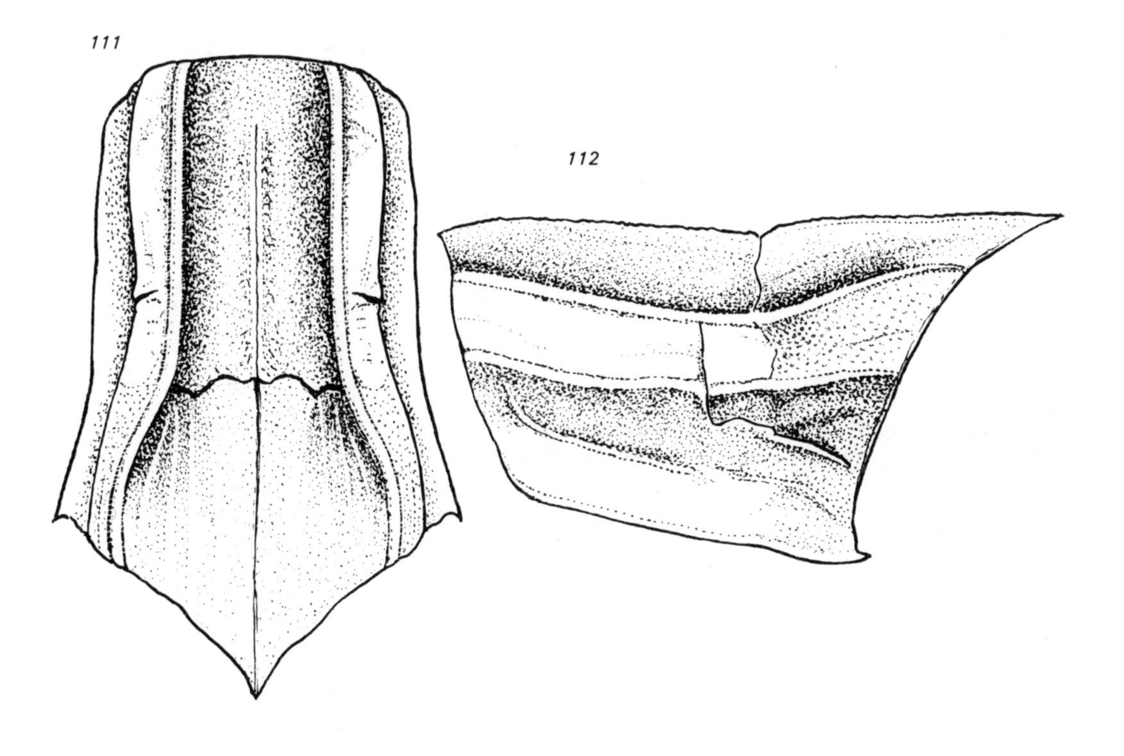

Figs. 111–112 : *Truxalis grandis* Klug, 1830
111. pronotum, dorsal ; 112. same, lateral

Truxalis longicornis (Krauss, 1902)

Type Locality : 'South Arabia' (Vienna Museum).

Acrida (Acridella) longicornis Krauss H. A., 1902, *Sitz. Akad. Wiss. Wien*, 7 : 2.
Acrida longicornis —. Krauss H. A., 1907, *Denkschr. Akad. Wiss. Wien*, 71 (2) : 5, pl. 1, fig. 3.
Acridella longicornis —. Bodenheimer F. S., 1935, *Arch. Naturgesch.*, 4 (2) : 179.
Truxalis longicornis —. Dirsh V. M., 1951, *Eos, Madr.*, Tomo estraord. (1950), p. 163.

Antennae narrow, filiform, very long with long segments, the 3–5 basal segments incompletely divided. Fastigium of vertex rounded apically, narrower than interocular space, with lateral fields raised and median carinula prominent. Pronotum with raised granulate median carina. Lateral carinae parallel to the median carina in prozona, diverging and elevated in metazona. Prozona roof-shaped ; metazona densely punctate, flat, with angular apex and inwards-curving shoulders.
Tegmina broad, extending only slightly beyond hind knees ; apex rounded.
Coloration : Pale brown, sometimes greenish, with a pattern of white lines on the head, pronotum and outer parts of the tegmina. Base of wings of female bluish-black ; the colour sometimes spreads over the entire wing. Base of wings of male red. Most of the veins and veinlets dark.
Measurements (mm) : Body ♀ 78.0 ; tegmina ♀ 63.0 ; antennae ♀ 34.0.
Distribution : From Somalia to East Africa, Sahara, Arabia, Israel.
Israel : Central Negev (17 ; one ♀ collected).
This is a typical desert form and is the rarest species of *Truxalis* in our area.

Genus ACRIDA Linnaeus, 1758

Systema Naturae, 10th ed., 1 : 427

Gryllus Acrida Linnaeus C., 1758, *Systema Naturae*, 10th ed., 1 : 427.
Acrida —. Stål C., 1873, *Recensio Orthopterorum*, Stockholm, 1 : 88, 95, 104.
Acrida —. Dirsh V. M., 1949, *Eos, Madr.*, 25 : 15–47.

Type Species : *Gryllus (Acrida) turritus* Linnaeus, 1758.
Diagnosis : Large, smooth. Head markedly opisthognathous, longer than pronotum (Fig. 113). Eyes large, elliptical, their longitudinal axis parallel to the longitudinal axis of the head. Antennae filiform, shorter than head and pronotum together. Pronotum cylindrical ; posterior margin obtuse- or acute-angled. Median carina distinct along its entire length ; carinae on lateral lobe parallel to each other and to median carina (Fig. 114).
Tegmina narrow, extending beyond abdomen. Arolium as long as or longer than claws.
Distribution : Africa, several species in Europe, Asia and Australia.
One species in Israel.

Acrida

Acrida bicolor (Thunberg, 1815)
F gs. 113–115 ; Plate III : 4

Type Locality : 'South Africa' (Uppsala Museum).

Truxalis bicolor Thunberg C. P., 1815, *Mém. Acad. Sci. St.-Pétersb.*, 5 : 266.
Truxalis pellucida Klug J. C. F., 1830, *Symbolae Physicae*, pl. 18, figs. 5, 6, 7, 9.
Acrida bicolor —. Stål C., 1873, *Recensio Orthopterorum*, Stockholm, 1 : 97.
Tryxalis nasuta L. Brunner von Wattenwyl C., 1882, *Prodromus der europäischen Orthopteren*, Leipzig, p. 88.
Tryxalis unguiculata Rambur, Giglio-Tos E., 1893, *Boll. Musei Zool. Anat. comp. R. Univ. Torino*, 8 (164) : 1.
Tryxalis stålii Bolivar, I., 1893, *Feuill. jeun. Nat.*, 23 (275) : 162, 164.
Acrida ståli —. Burr M., 1902, *Trans. R. ent. Soc. Lond.*, 1902 : 147–169.
Acrida turrita L. Kirby A. H., 1910, *A synonymic catalogue of the Orthoptera*, Vol. III, Orthoptera Saltatoria, Part II, London, p. 94.
Acrida orientalis Bolivar I., 1919, *Bull. Soc. ent. Fr.*, 1919 : 242.
Acrida deminuta Bolivar I., 1922, *Orthoptères*, in : *Voyage de M. le Baron Maurice de Rothschild en Ethiopie et en Afrique Orientale Anglaise* (1904–5), Paris, p. 183.
Acrida turrita —. Buxton P. A. & B. P. Uvarov, 1923, *Bull. Soc. R. ent. Égypte*, p. 184.
Acrida turrita var. *pellucida* —. Innes W., 1929, *Mém. Soc. R. ent. Égypte*, 3 (2) : 14.
Acrida tessellata Sjöstedt Y., 1931, *Ark. Zool.*, 23A. (17) : 5, pl. 3, fig. 2.
Acrida leopoldi Sjöstedt Y., 1934, *Bull. Mus. r. Hist. nat. Belg.*, 10 (4) : 1.
Acrida turrita —. Bodenheimer F. S., 1935, *Animal Life in Palestine*, Jerusalem, pp. 86, 311, 320.
Acrida turrita —. Bodenheimer F. S., 1935, *Arch. Naturgesch.*, 4 (2) : 177.
Acrida anatolica Dirsh V. M., 1949, *Eos, Madr.*, 25 : 34.
Acrida pellucida —. Dirsh V. M., 1949, *ibid.*, 21, 30.
Acrida pellucida palaestina Dirsh V.M., 1949, in : Dirsh V. M. & B. P. Uvarov, 1953, *Tijdschr. Ent.*, 96 (3) : 231.
Acrida bicolor —. Uvarov B. P., 1953, *Publções cult. Co. Daim. Angola*, no. 21 : 165.
Acrida bicolor —. Dirsh V. M., 1954, *Bull. Soc. Fouad Ier Ent.*, 38 : 124, 140.

Large, laterally compressed. Head conical, markedly opisthognathous (Fig. 113). Vertex long, obtuse ; fastigium granulate, its carinula partly visible on occiput. Pronotum usually roof-shaped, intersected by the posterior transverse sulcus beyond the middle. Lateral carinae parallel to the median one, all granulate (Fig. 114). Lateral lobes with protruding longitudinal carinulae (Fig. 115) and acute posterior angles. Mesosternal interspace longer than wide ; metasternal interspace square.
Tegmina narrow, slightly curving towards apex, acute-angled, with a projecting false intercalary vein. Wings triangular, oblong, extending beyond abdomen and hind knees. Front femur shorter than pronotum. Sharp spine present on dorsal lobes of hind femur. Arolium large ; as long as or longer than claws.
Coloration : Generally brown, pale brown, brownish-green or green ; rarely with

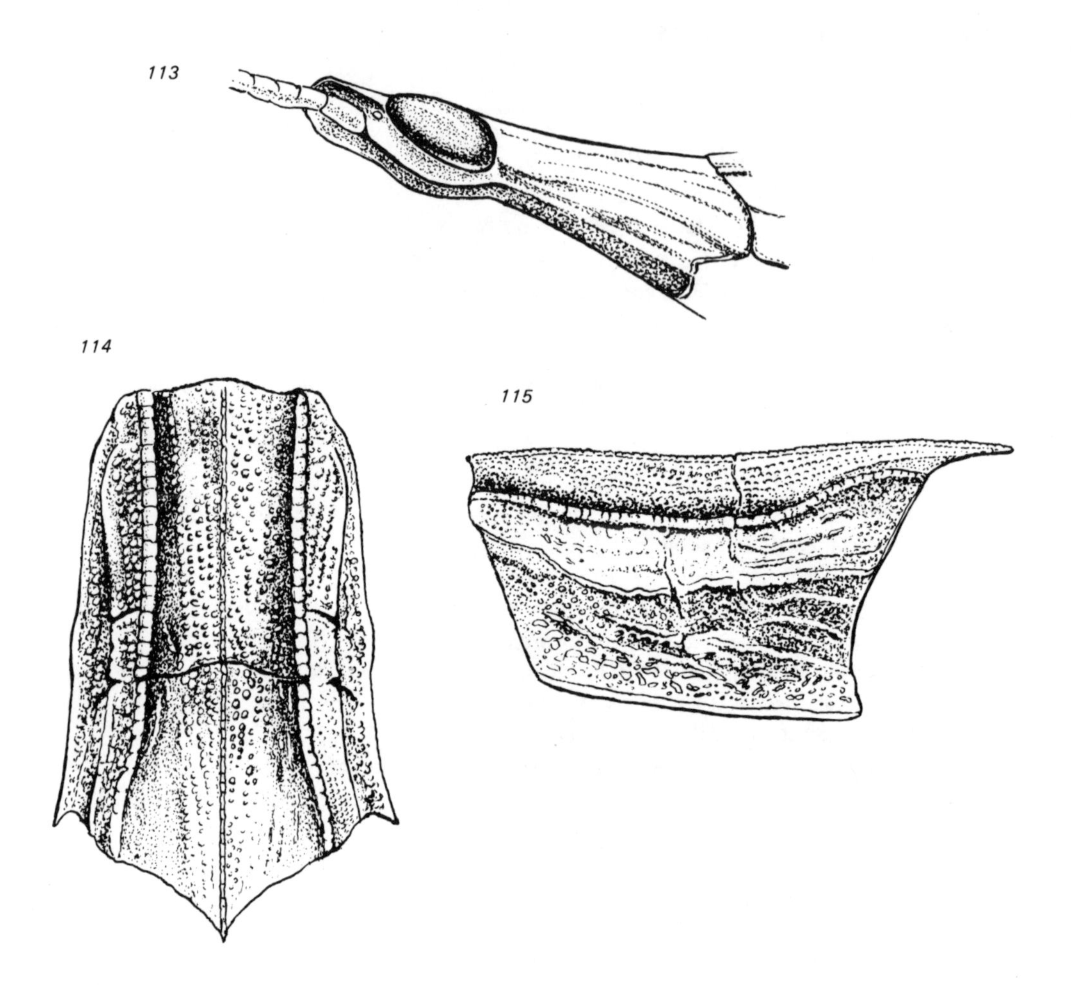

Figs. 113–115 : *Acrida bicolor* (Thunberg, 1815)
113. head ; 114. pronotum, dorsal ; 115. same, lateral

white lines on thorax. Tegmina with a pattern of white and brown blotches. Wings transparent, greenish or yellowish.

Measurements (mm) : Body ♀ 55.0–71.0, ♂ 37.0–43.0 ; tegmina ♀ 46.0–58.5, ♂ 31.0–35.0 ; front femur ♀ 7.5–10.0, ♂ 5.0–5.7 ; hind femur ♀ 15.0–22.0, ♂ 12.5–15.0.

Distribution : East Africa, Eastern Mediterranean.

Israel : Upper Galilee (1), Jordan Valley (7), Coastal Plain (4, 8, 9), Northern Negev (15).

Occurs in the northern parts of the country, in areas with dense, green vegetation, in cultivated and uncultivated fields, along rivers, lakes and fishponds. Usually found clinging to stalks, blending with surroundings. When taking off, it emits a high chirping-like sound. During reproduction, chirping males attract females. Oviposition observed in March, November and December ; hoppers and adults occur during most of the year. This grasshopper feeds on various Gramineae and other ephemeral grasses. So far only one species of the genus *Acrida* found in our region. Several attempts have been made to divide the species, but recent research has shown that the various forms are variations of a single species.

Genus DURONIELLA Bolivar, 1908

Mém. Soc. r. ent. Belg., 16 : 100

Type Species : *Duronia fracta* Krauss, 1890.
Diagnosis : Small, smooth, laterally compressed (Fig. 116). Head short, opisthognathous. Vertex slightly protruding, smooth ; foveolae absent (Fig. 122). Antennae ensiform (Fig. 118), shorter in female, longer in male than head and pronotum together. Frontal ridge with carinulae gradually diverging towards clypeus.
Pronotum roof-shaped, intersected by one posterior transverse sulcus. Lateral carinae of pronotum parallel to the median one, diverging in metazona (Fig. 117). Arolium small, less than half as long as claws.
Coloration : Dark or pale brown, sometimes green on sides of head, pronotum, thorax, tegmina and hind femur. Males usually darker. Tegmina dark grey, sometimes with white lines. Distal part of wings mainly greyish.
Distribution : South-East and North Africa, Caucasus, Asia.
Two species in our region.

Key to the Species of Duroniella in Israel

1. Greater part of wings dark. Antennae, especially of female, short with 12–14 segments, markedly widened at the base ; all the segments flat (Fig. 118). **D. laticornis** (Krauss)
– Wings transparent, often with a dark apical spot. Antennae narrow, slightly widened at the base, with 18–20 segments, of which the apical ones are cylindrical (Fig. 121). **D. lucasi** (Bolivar)

Duroniella laticornis (Krauss, 1909)

Figs. 116–119 ; Plate III : 5

Type Locality : Jerusalem.

Duronia laticornis Krauss H. A., 1909, in : Kneucker A., *Verh. naturw. Ver. Karlsruhe*, 21 : 118, 119, figs. 11, 12.
Duroniella laticornis —. Buxton P. A. & B. P. Uvarov, 1923, *Bull. Soc. R. ent. Égypte*, p. 185.
Duroniella laticornis —. Bodenheimer F. S., 1935, *Animal Life in Palestine*, Jerusalem, pp. 28b, 86, 87, 310, 320, 323.

Duroniella laticornis —. Bodenheimer F. S., 1935, *Arch. Naturgesch.*, 4 (2): 179.

Small to medium sized, smooth. Head slightly opisthognathous, with concave frontal ridges which gradually diverge towards the clypeus. Eyes of male longer than subocular groove, those of female shorter. Fastigium of vertex short, that of male as long as or longer than the minimal distance between the eyes, shorter in the female; usually with a prominent carinula. Antennae ensiform, flat. Carinae of pronotum slightly raised, especially the median one. Lateral carinae in prozona parallel to median one, diverging in metazona (Fig. 117). Lateral lobes of pronotum as high as wide.
Tegmina rounded apically, extending only slightly beyond abdomen and hind knees; intercalary vein present. Posterior arolia narrow, small or half the length of claws.
Coloration: Brown, dark brown, often partly green or with light bands on head and sides of thorax, and on margin of tegmina and hind femur. Males usually darker. Tegmina with dark, dense network of veins. Hind legs dark.
Measurements (mm): Body ♀ 22.5–26.0, ♂ 16.0–16.5; tegmina ♀ 18.0–20.5, ♂ 12.5–13.0.
Distribution: Near East.

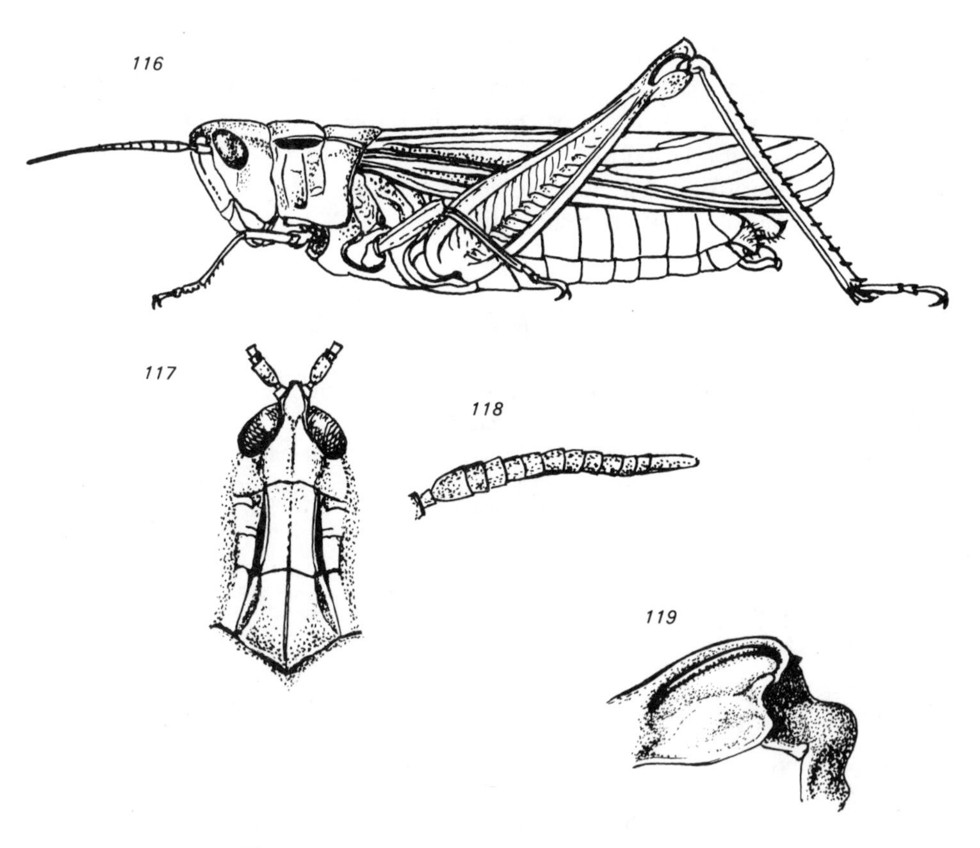

Figs. 116–119: *Duroniella laticornis* (Krauss, 1909)
116. female; 117. head and pronotum; 118. antenna; 119. hind knee

Israel : Upper and Lower Galilee (1, 2), Coastal Plain (4, 8, 9), Jordan Valley (7), Judean Hills (11).
Most common in wet and partly dry habitats, in low grasses. In autumn, after the first rains, with appearance of the hoppers, this species is the most common in the northern parts of the country. Adults first observed in January, forming the densest population from March to May, which is also the period of copulation and egg-laying.
Feeds on grasses.

Duroniella lucasi (Bolivar, 1881)
Figs. 120–122

Type Locality : Oran, Algeria (Madrid Museum).

Phleoba (Duronia) lucasii Bolivar I., 1881, *An. Soc. esp. Hist. nat.*, 10 : 502.
Duronia lucasi —. Krauss H. A., 1890, *Verh. zool.-bot. Ges. Wien*, 40 : 260.
Duronia fracta Krauss H. A., 1890, *ibid.*, p. 260.
Duronia (Phlaeoba) lucasi —. Vosseler J., 1902, *Zool. Jb. (Syst.)*, 17 : 354.
Duronia lucasi —. Krauss H. A., 1902, *Verh. zool.-bot. Ges. Wien*, 52 (4) : 231, 238.
Duroniella lucasii —. Bolivar I., 1914, *Mém R. Soc. esp. Hist. nat.*, 8 (5) : 183.
Duroniella fracta —. Buxton P. A. & B. P. Uvarov, 1923, *Bull. Soc. R. ent. Égypte*, p. 184.
Duroniella lucasi —. Innes W., 1929, *Mém. Soc. R. ent. Égypte*, 3 (2) : 26.
Duroniella lucasi —. Bodenheimer F. S., 1935, *Animal Life in Palestine*, Jerusalem, p. 323.
Duroniella lucasi —. Bodenheimer F. S., 1935, *Arch. Naturgesch.*, 4 (2) : 179.

Smooth, small to medium sized. Head hypognathous. Eyes as long as subocular groove. Fastigium of vertex short, its maximal length before the eyes as long as or shorter than the shortest line between the eyes. A prominent carinula usually extends posteriorly over the vertex, accompanied laterally by numerous carinulae (Fig. 122). Antennae narrow, ensiform, shorter than head and pronotum together ; apical segments cylindrical (Fig. 121). Pronotum rugose or smooth ; transverse sulcus situated beyond the middle ; sometimes with traces of a second sulcus dividing the prozona into two equal parts. Lateral carinae of disc parallel, slightly converging in prozona and diverging in metazona.
Tegmina rounded apically, extending beyond hind knees ; typical intercalary vein present. Arolium of hind legs small, less than half as long as claws.
Coloration : Pale, pale brown and only rarely green in parts. Wings transparent or with a slightly dark reticulation. White line present on the tegmina.
Measurements (mm) : Body ♀ 22.0–24.0, ♂ 14.0–14.5 ; tegmina ♀ 17.5–20.0, ♂ 12.5–13.0.
Distribution : North Africa, Near East.
Israel : Jordan Valley (7).
Found on low mountain slopes and areas with dry vegetation. This is apparently a more thermophilic species than *D. laticornis* and occurs also in the summer.

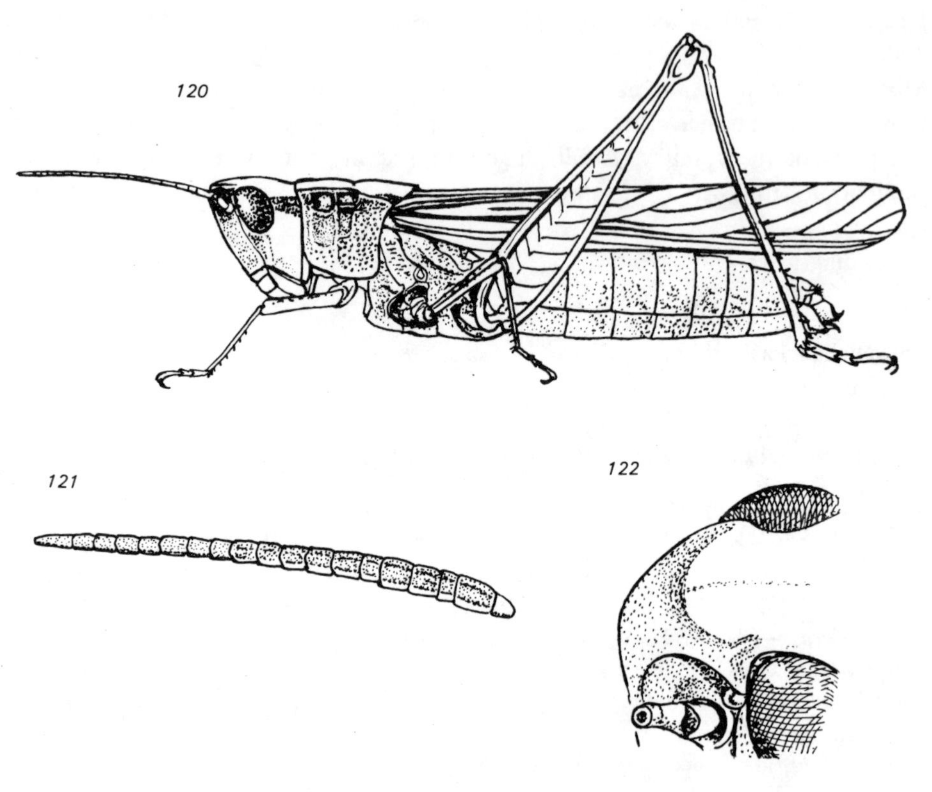

Figs. 120–122 : *Duroniella lucasi* (Bolivar, 1881)
120. female ; 121. antenna ; 122. fastigium

Genus EREMOGRYLLUS Krauss, 1902

Verh. zool.-bot. Ges. Wien, 52 : 238

Sphingonotina Chopard L., 1943, *Faune Emp. franç.*, Paris, 1 : 323.
Eremogryllus —. Dirsh V. M., 1961, *Bull. Br. Mus. nat.Hist. (Ent).*, 10 (9) : 413.

Type Species : *Eremogryllus hammadae* Krauss, 1902.

Diagnosis : Small, hairy. Head only slightly opisthognathous ; antennae as long as or shorter than head and pronotum together. Pronotum with lateral and median carinae interrupted by the posterior transverse sulcus. Tegmina with projecting radial vein. Cerci of male with inwards-curved and acute apex. Spurs of hind legs very long.
Distribution : North Africa.
One species in our region.

Eremogryllus hammadae Krauss, 1902

Figs. 123, 124

Type Locality : Ouargla to Ghardaja, Sahara (Berlin Museum).

Eremogryllus hammadae Krauss H. A., 1902, *Verh. zool.-bot. Ges. Wien*, 52 (4) : 231, 239, figs. 4–6.
Leptopternis quadriocellata Werner F., 1932, *Sber. Akad. Wiss. Wien*, (1) 141 : 146, fig. 10.
Eremogryllus quadriocellata —. Uvarov B. P., 1934, *Ann. Mag. nat. Hist.*, (10) 14 : 473.
Eremogryllus hammadae —. Chopard L., 1938, *Mém. Soc. Biogéogr.*, 6 : 222.

Small, robust. Head slightly opisthognathous. Eyes round, their diameter shorter than the subocular groove. Facial ridges protrude between the eyes, diverging strongly towards clypeus. Foveolae quadrangular, narrowing towards inner margin. Vertex concave, half as wide as long. Antennae as long as or shorter than head and pronotum together. Lateral lobes of pronotum rounded. Lateral carinae on disc obliterated in the middle, distinct only on the anterior part of the prozona and partly on the metazona, forming oblique shoulders near the transverse sulcus. Wings transparent, extending beyond abdomen and hind knees ; radial vein distinct. Hind tibia with very long spurs ; upper internal spur at least twice as long as uper external spur (Fig. 124). Arolium absent.
Coloration : Sandy-yellow, with white and brown spots.
Measurements (mm) : Body ♀ 14.0–16.0, ♂ 11.0–13.0.

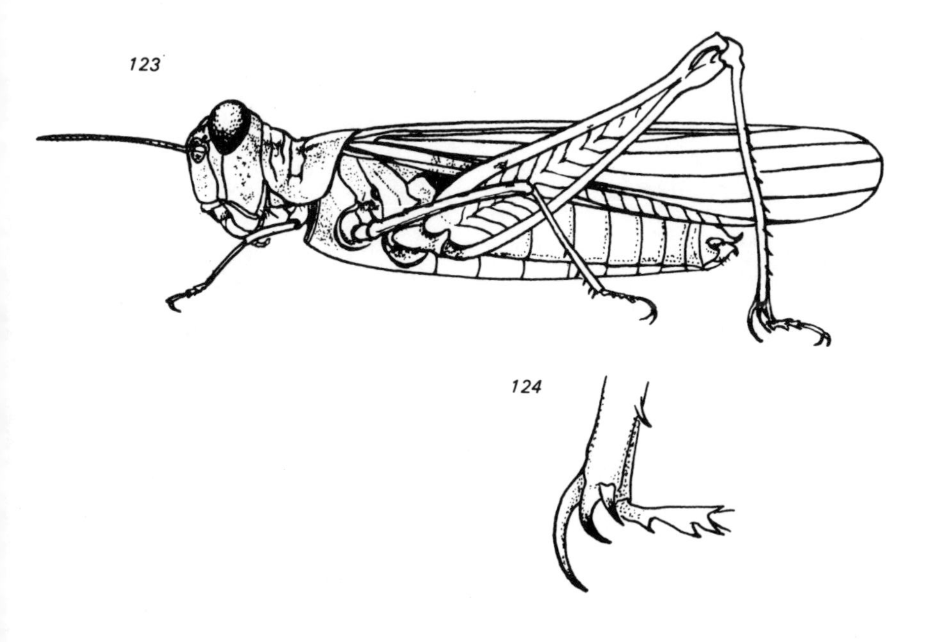

Figs. 123–124 : *Eremogryllus hammadae* Krauss, 1902
123. female ; 124. hind spurs

Distribution : North Africa, Israel.
Israel & Sinai : Northern Negev (15), Northern Sinai (20).
Inhabits areas of loose sand. The colour pattern blends perfectly with the environment. This species digs into the sand by movements of body and legs, thus concealing itself so that only the antennae and slightly raised eyes remain above the surface. It is easily located by the loud cricket-like sound emitted during courtship.

Genus STENOHIPPUS Uvarov, 1926

Trans. ent. Soc. Lond., 1925 : 423

Eremippus Uvarov B. P., 1926, *Eos, Madr.*, 2 (4).

Type Species : *Chorthippus (Stauroderus) xanthus* Karny, 1907.
Diagnosis : Small. Head short, opisthognathous. Vertex concave, shorter than the interocular space. Foveolae quadrangular, slightly longer than wide, partly visible from above. Antennae filiform. Pronotum roof-shaped, intersected in the middle by one transverse groove ; lateral carinae distinct along their entire length. Spurs long.
Distribution : Canary Islands, East Africa.
One species known, *Stenohippus bonneti* (Bolivar, 1885), of which one subspecies occurs in Israel.

Stenohippus bonneti orientalis Uvarov, 1933

Figs. 125, 126 ; Plate III : 6

Type Locality : Jericho.

Stenohippus bonneti orientalis Uvarov B. P., 1933, *Ann. Mag. nat. Hist.*, Ser. 10, 11 : 666.
Stenohippus bonneti orientalis —. Bodenheimer F. S., 1935, *Arch. Naturgesch.*, 4 (2) : 183.

Small, laterally compressed. Head opisthognathous, especially in the male. Eyes elongate, their longitudinal axis longer than subocular groove. Frontal ridge flat, punctate, its margins straight or gradually diverging towards clypeus. Vertex shorter than the interocular space, concave, without sharp margins. Foveolae only in female partly visible from above, 1.5 times as long as wide, their frontal part rounded. Pronotum slightly roof-shaped, the single transverse sulcus intersecting it in the middle (Fig. 126). Lateral carinae distinct along the entire length of the pronotum, converging towards the middle. Tegmina rounded apically, extending for a third of their length beyond hind knees ; costal field membranous with numerous oblique venules ; median field divided into regular squares by cross-venules which are much denser in female. Radius-sector with two branches. Inner ventral spine of tibia twice as long as the inner dorsal one. Arolium well developed, only slightly shorter than claws. Mesosternal interspace as long as wide ; the sutures diverge slightly posteriorly in the female. Metasternal lobes partly contiguous.

Coloration : Pale brown or ochraceous ; lateral parts of thorax usually lighter than dorsal parts. Base of tegmina dark ; margins transparent. Usually two brown bands on upper surface of hind femur.

Meaurements (mm) : Body ♀ 19.0–20.0, ♂ 18.0–20.0 ; tegmina ♀ 14.0–16.0, ♂ 13.0–16.0.

A typical grass-dwelling species, found usually in areas with dense vegetation, hiding along the stems. It feeds on soft Gramineae.

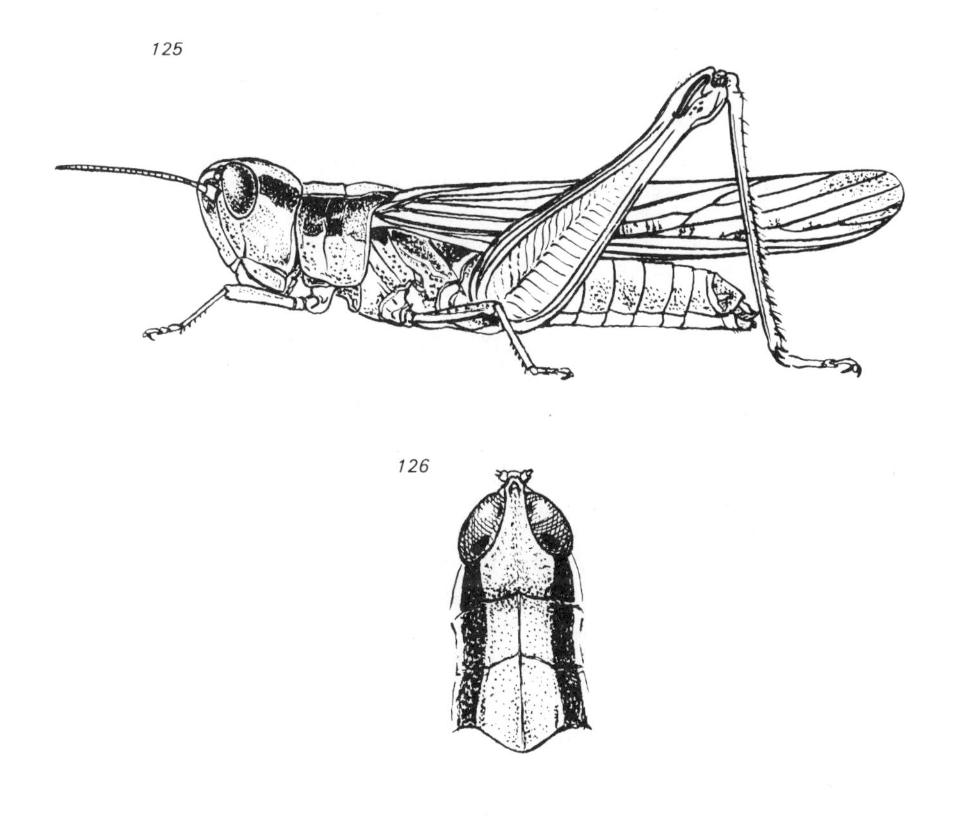

Figs. 125–126 : *Stenohippus bonneti orientalis* Uvarov, 1933
125. female ; 126. head and pronotum

Genus RAMBURIELLA Bolivar, 1906

Bol. Soc. esp. Hist. nat., 6 : 397

Ramburia (nom. preoc.) Bolivar I., 1897, *Actas Soc. esp. Hist. nat.*, 26 : 168.
Pallasiella Kirby W. F., 1910, *A synonymic catalogue of the Orthoptera*, Vol. III, Orthoptera Saltatoria, Part II, London, p. 168.
Ramburiella —. Uvarov B. P., 1923, *Trans. R. ent. Soc. Lond.*, 1928 : 158.

Type Species : *Gryllus hispanicus* Rambur, 1838.
Diagnosis : Head swollen, short, slightly hypognathous ; foveolae visible from above, square. Eyes oval, elongate. Antennae filiform, longer than head and pronotum together. Lateral carinae of pronotum obliterated in the middle, intersected by three transverse sulci ; median carinae distinct ; posterior margin of dorsum rounded. Wings well developed, extending beyond hind knees.
Distribution : The genus includes four species in North Africa, South-West and South-East Europe, Central Asia, Caucasus to Pakistan.
One species in Israel.

Ramburiella turcomana (Fischer-Waldheim, 1833)
Figs. 127, 128

Type Locality : 'Turcmania'.

Oedipoda turcomana Fischer de Waldheim G., 1833, *Bull. Soc. Imp. Nat. Moscou*, 6 : 384.
Oedipoda truchmana Fischer de Waldheim G., 1848, *Orthoptera Imperii Russici*, Moscow, p. 313.
Stethophyma turcomana —. Brunner von Wattenwyl C., 1882, *Prodromus der europäischen Orthopteren*, Leipzig, p. 140.
Pallasiella turcomana —. Kirby W. F., 1910, *A synonymic catalogue of the Orthoptera*, Vol. III, Orthoptera Saltatoria, Part II, London, p. 168.
Ramburiella truchmana —. Buxton P. A. & B. P. Uvarov, 1923, *Bull. Soc. R. ent. Égypte*, p. 190.
Ramburiella turcomana —. Uvarov B. P., 1930, *Eos, Madr.*, 6 (4) : 365.
Ramburiella truchmana —. Bodenheimer F. S., 1935, *Animal Life in Palestine*, Jerusalem, pp. 320, 323.
Ramburiella truchmana —. Bodenheimer F. S., 1935, *Arch. Naturgesch.*, 4 (2) : 185.
Ramburiella truchmana —. Ramme W., 1951, *Mitt. zool. Mus. Berl.*, 27 : 397, 425.

Head short, opisthognathous ; frontal ridge smooth, slightly convex around median ocellus, obliterated towards clypeus ; fastigial foveolae square, their anterior part obliterated. Vertex before eyes twice as wide as long, with a slightly visible carinula continuous posteriorly with the indistinct carina on the fastigium. Eyes elongate, longer than subocular groove. Pronotum punctate ; lateral carinae convergent anteriorly, obliterated in middle and intersected by three transverse sulci. Median carina slightly raised, intersected by the posterior sulcus which divides it into a prozona and metazona of equal length.

Ramburiellla

Tegmina extending beyond hind femur ; 6–10 dark blotches appearing along their middle longitudinal axis (Fig. 128). Wings transparent, with a dark apical spot, sometimes found also on apex of tegmina. Hind femur slender with rounded apical lobes ; three dark bands on external and internal median areas. Outer spurs of hind tibia only half as long as inner spurs. Arolium of hind legs large, as long as claws.
Coloration : Brown with a pattern of white lines, pronounced mainly on the pronotum, especially along the median carina and the converging lateral ones ; lines extending also along margins of tegmina. The lines extending anteriorly on the head marginate by dark brown fields. Hind knees black ; hind tibia pale grey.
Measurements (mm) : Body ♀ 31.0–43.0, ♂ 25.0–33.0 ; tegmina ♀ 20.0–32.0, ♂ 18.0–24.0.
Distribution : South-West Europe, Middle East, Central Asia.
Israel : Lower Galilee (2), Coastal Plain (4, 8, 9), foothills of Judea (10), Judean Hills (11), Jordan Valley (7).
It inhabits areas with dense vegetation, hiding on branches. On sensing danger it seeks refuge in the grass.

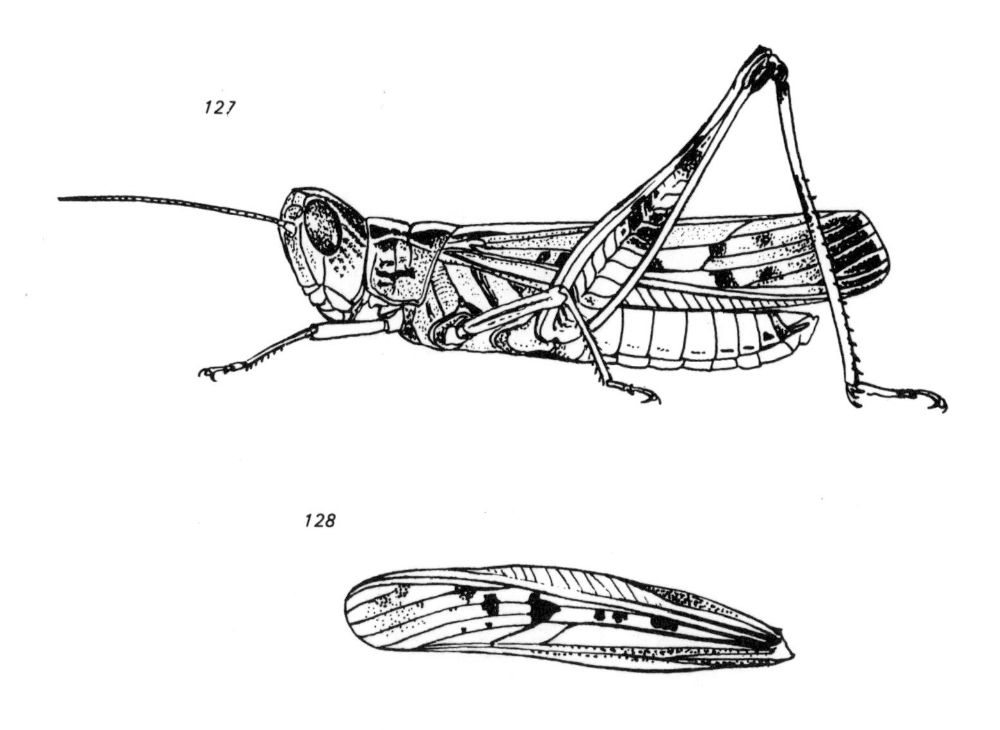

Figs. 127–128 : *Ramburiella turcomana* (Fischer-Waldheim, 1833)
127. male; 128. tegmen

Genus BODENHEIMERELLA Uvarov, 1933
Ann. Mag. nat. Hist., Ser. 10, 11 : 668, Figs. 3, 4

Type Species : *Bodenheimerella jordanica* Uvarov, 1933.
Diagnosis : Small. Head slightly swollen, hypognathous. Eyes round, protruding laterally ; antennae filiform, slightly shorter than head and pronotum together ; foveolae square (Fig. 129). Pronotum intersected only by posterior transverse sulcus ; prozona longer than metazona ; wings rounded, short, extending to end of abdomen. The genus contains one species found in Transjordan and Israel.

Bodenheimerella jordanica Uvarov, 1933
Fig. 129

Type Locality : Jericho.

Bodenheimerella jordanica Uvarov B. P., 1933, *Ann. Mag. nat. Hist.*, Ser. 10, 11 : 668, Figs. 3, 4.
Bodenheimerella jordanica —. Bodenheimer F. S., 1935, *Arch. Naturgesch.*, 4 (2) : 184.

Small. Head large, slightly opisthognathous ; eyes round, protruding laterally (Fig. 129) ; frontal ridge flat, its margins parallel in greater part, diverging only near clypeus. Vertex deeply concave, with a distinctly raised carinula, longer than wide. Foveolae concave, oblong, slightly trapezoidal, contiguous in front and only partly visible from above. Antennae filiform, shorter than head and pronotum together. Median carina on dorsum distinct, intersected by posterior transverse sulcus beyond middle of pronotum. Lateral carinae obliterated in middle, distinct only on anterior parts of prozona and forming shoulders on metazona. Lateral lobes of pronotum

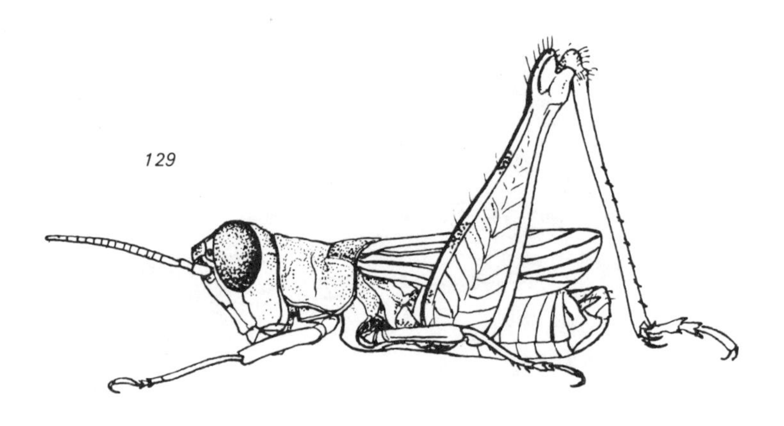

Fig. 129 : *Bodenheimerella jordanica* Uvarov, 1933, male

rounded, narrow, with a deep vertical sulcus in the middle. Wings short, plate-like, not reaching end of abdomen. Hind legs thick, hairy, with strong spurs and minute arolia.
Coloration : Ochraceous, blending with the ground.
Measurements (mm) : Body ♀ 14.0–15.0, ♂ 9.0 ; tegmina ♀ 6.0–7.5, ♂ 5.5 ; hind femur ♀ 8.0–9.0, ♂ 8.0.
Distribution : Jordan Valley to Saudi Arabia.
Israel : Dead Sea Area (13), 'Arava Valley (14; from Jericho to 'Ein Hazeva).
A typical geophilous species found on hammadas with very sparse vegetation. Its small size and colour afford it good concealment in its habitat. Adults found from August to October. Nymphs collected at end of April and in early May.

Genus NOTOSTAURUS Bei-Bienko, 1933

Bol. Soc. esp. Hist. nat., 33 : 337, 338

Stauronotus Fischer. Brunner von Wattenwyl C., 1882, *Prodromus der europäischen Orthopteren*, Leipzig, pp. 84, 135 (partim).
Dociostaurus Fieber. Uvarov B. P., 1921, *Bull. ent. Res.*, 11 (4) : 397 (partim).
Notostaurus Bei-Bienko G. Ya., 1933, *Bol. Soc. esp. Hist. nat.*, 33 : 337, 338 (Generic revision).

Type Species : *Stauronotus anatolicus* Krauss, 1896.
Diagnosis : Small or medium sized. Head large, opisthognathous or almost hypognathous. Foveolae slightly elongate. Occiput with a median carina (Fig. 131), rugose, with distinct X-shaped marking on pronotum ; prozona longer than metazona. Wings either abbreviate or fully developed ; median field with dense, irregular network of veinlets.
Distribution : Four species known from the U. S. S. R., Caucasus and Asia.
Two species in our region.

Key to the Species of Notostaurus in Israel

1. Tegmina and wings fully developed, reaching posteriorly to or extending beyond hind knees (Fig. 130). **N. anatolicus** (Krauss)
– Tegmina and wings abbreviate, posteriorly not covering end of abdomen (Fig. 132). **N. cephalotes** (Uvarov)

Notostaurus anatolicus (Krauss, 1896)
Figs. 130, 131 ; Plate III : 7

Type Locality : 'Turkey'.

Stauronotus anatolicus Krauss H. A., 1896, *Zool. Jb.* (Syst.), 9 : 560.
Dociostaurus anatolicus —. Kirby W. F., 1910, *A synonymic catalogue of the Orthoptera,* Vol. III, Orthoptera Saltatoria, Part II, London, p. 153.
Dociostaurus anatolicus —. Ebner R., 1919, *Arch. Naturgesch.*, 85 : 165.
Dociostaurus anatolicus —. Uvarov B. P., 1923, in : Buxton P. A. & B. P. Uvarov, *Bull. Soc. R. ent. Égypte*, p. 187.
Notostaurus anatolicus —. Bodenheimer F. S., 1935, *Animal Life in Palestine*, Jerusalem, pp. 86, 311, 323.
Notostaurus anatolicus —. Bodenheimer F. S., 1935, *Arch. Naturgesch.*, 4 (2) : 183.
Notostaurus anatolicus —. Ramme W., 1951, *Mitt. zool. Mus. Berl.*, 27 : 424.

Medium sized and robust. Head opisthognathous, especially in males ; raised above pronotum. Frontal ridge of females flat, its margins almost parallel ; of males concave, margins diverging towards clypeus. Vertex short, concave ; occiput with median carina and numerous carinulae, swollen. Foveolae quadrangular, slightly longer than wide. Antennae much longer than head and pronotum together. Pronotum slightly swollen in prozona, intersected by the posterior transverse sulcus situated beyond middle of pronotum (Fig. 131). Lateral carinae converging in prozona, diverging in metazona, intersected by all three sulci, obliterated between first and third sulci. Prosternum swollen, pointed apically. Maximum width of mesosternal interspace as wide as or wider than mesosternal lobe.
Tegmina extending to hind knees, their anterior margin curved ; median field wide ; venation dense and irregular ; radius-sector with single branch. Wings transparent. Hind femur thick, its length 3.2–3.5 times its maximum width ; dorsal carina ending at the knee in a short spine. Outer ventral spur of hind tibia slightly longer than dorsal one and more than twice as long as inner ventral spur.
Coloration : Dorsum usually pale, ochraceous with a light line extending from head to pronotum. Lateral lobes of pronotum brown or pale brown, also bordering the white X-shaped marking, the brown coloration continuing over the eyes and the base of the antennae. Tegmina with several brown blotches in the middle, usually transparent apically and with a pale line on the costal field. Wings transparent, their bases reddish. Three oblique, sometimes indistinct brown bands present on inner surface of hind femur, usually not reaching ventral margin, extending onto dorsal part and sometimes onto outer surface. Hind tibia pale or pale grey ; hind knee of male black.
Measurements (mm) : Body ♀ 24.5–32.5, ♂ 18.0–24.0 ; tegmina ♀ 19.0–22.5, ♂ 13.5–18.5.
Distribution : Southern Russia, Iran, Caucasus, Asia Minor, eastern Mediterranean. Israel : Upper and Lower Galilee (1, 2), Samaria (6), Coastal Plain (4, 8, 9), foothills of Judea (10), Judean Hills (11).

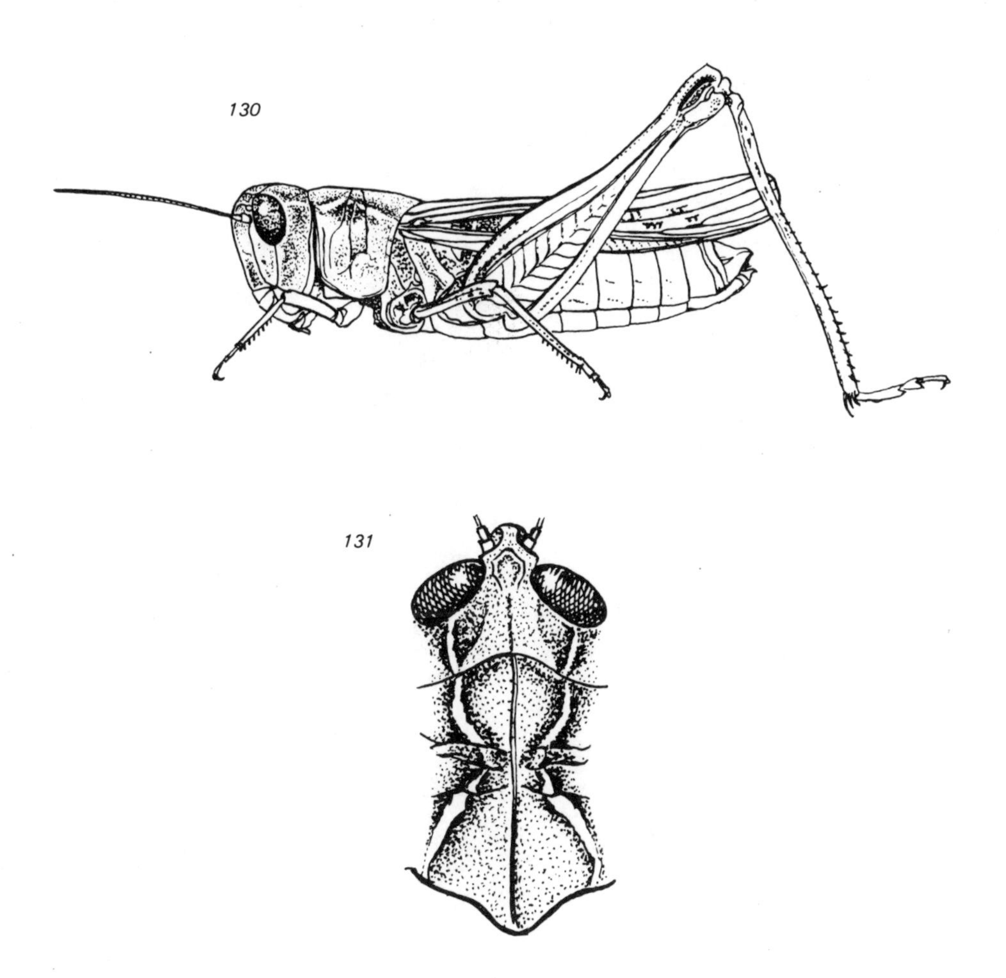

Figs. 130–131 : *Notostaurus anatolicus* (Krauss, 1896)
130. female ; 131. head and pronotum

It inhabits mountain slopes with dense, grassy vegetation ; found also in ploughed and on harvested fields. Feeds mostly on Gramineae. On sensing danger, it seeks refuge among dry plant material. In summer, on the western slopes of the Jordan Valley, this species occurs in dense populations together with *Calliptamus palaestinensis* Ramme.

Notostaurus cephalotes (Uvarov, 1923)
Figs. 132, 133; Plate IV: 1

Type Locality: 'Palestine'.

Dociostaurus cephalotes Uvarov B. P., 1923, in: Buxton P. A. & B. P. Uvarov, *Bull. Soc. R. ent. Égypte*, p. 187, fig. 1.
Dociostaurus cephalotes —. Bodenheimer F. S., 1935, *Animal Life in Palestine*, Jerusalem, pp. 321, 323.
Dociostaurus cephalotes —. Bodenheimer F. S., 1935, *Arch. Naturgesch.*, 4 (2): 182.

Small, brachypterous. Head large, swollen, opisthognathous. Frontal ridge densely punctate, slightly concave below vertex, then becoming flat and widening gradually towards clypeus, its margins obliterated below median ocellus. Foveolae square or slightly narrower apically; frontal margins rounded or angular. Vertex concave; margins rounded; median carina continuing on rugose occiput. Pronotum short, as long as or slightly longer than head; prozona much longer than metazona; lateral carinae distinct in anterior part of pronotum (Fig. 133). Mesosternal interspace wider than or as wide as mesosternal lobe.

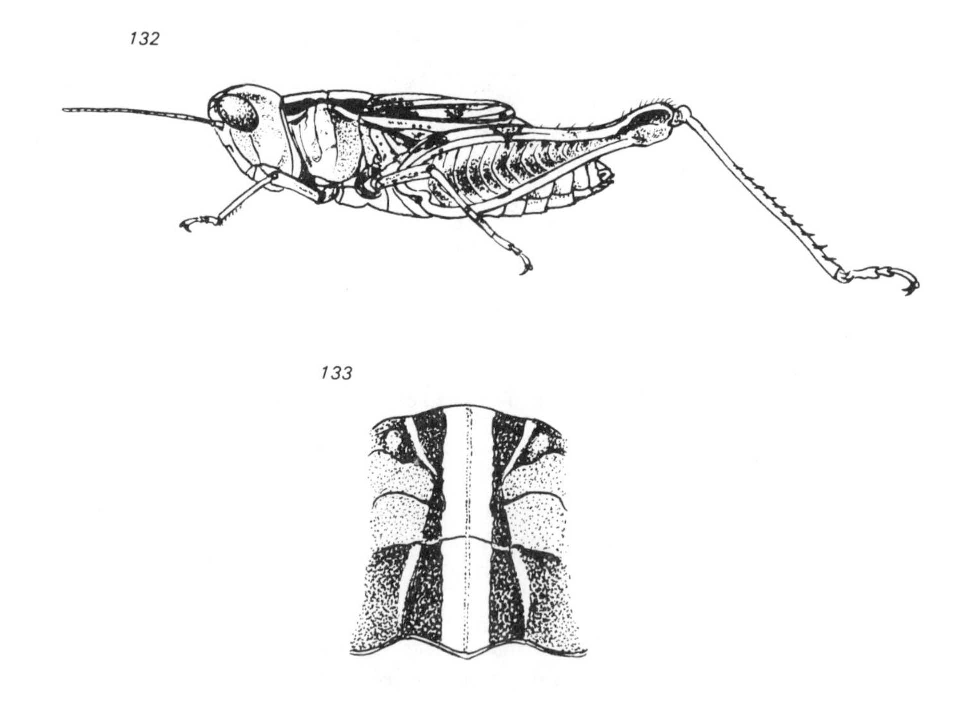

Figs. 132–133: *Notostaurus cephalotes* (Uvarov, 1923)
132. female; 133. pronotum

Tegmina abbreviate, in female covering 3–4 abdominal tergites (Fig. 132), in male extending almost to end of abdomen. Subcostal field much wider than median field, usually transparent, with regular, oblique veinlets ; median field reaching, or almost reaching, apical part of tegmina. Wings slightly shorter than tegmina. Hind femur thick, its length 2.2–2.5 times its maximum width ; dorsal carina at knee ends in a sharp spine.

Coloration : Brown or pale brown, with or without a pale line extending from vertex to apex of tegmina. Pronotum with or without X-shaped marking, usually brown on both sides of median carina (Fig. 133) and with a light marking on lateral lobes. Three dark blotches present on dorsal part of hind femur, the middle one usually triangular and extending to outer surface of femur ; the other blotches are either isolated or fused. Hind tibia greyish or greyish-violet.

Measurements (mm) : Body ♀ 16.0–18.0, ♂ 10.5–12.5 ; tegmina ♀ 5.5–7.5, ♂ 5.0–6.5.

Distribution : Endemic in Israel.

Israel : Upper Galilee (1), Golan Heights (18), Mount Hermon (19).

Genus DOCIOSTAURUS Fieber, 1853
Lotos, 3 : 118

Stauronotus Fischer L. H., 1853, *Orthoptera Europaea*, Leipzig, pp. 297, 351.
Dociostaurus —. Innes W., 1929, *Mém. Soc. R. ent. Égypte*, 3 (2) : 34.

Type Species : *Gryllus maroccanus* Thunberg, 1815.

Diagnosis : Small or medium sized, smooth. Head hypognathous or opisthognathous, often projecting above pronotum. Frontal ridge flat ; foveolae quadrangular, rectangular or slightly elongate (Figs. 136, 138) ; occiput without median carina. Antennae filiform, longer than or as long as head and pronotum together. Median carina of pronotum distinct ; lateral carinae in prozona obliterated or only slightly marked, converging towards transverse sulci, diverging in metazona. The light X-shaped marking on pronotum usually more pronounced on the posterior part (Figs. 135, 139).

Distribution : Europe, Canary Islands, Madeira, North Africa, Western Asia.

Four species in Israel.

Key to the Species of Dociostaurus in Israel

1. Wings short, not extending beyond hind knees (Fig. 134). White lines of X-shaped marking on metazona 2–3 times wider than those between transverse sulci (Fig. 135). Hind tibia red-orange. **D. hauensteini hauensteini** (Bolivar)

– Wings extending beyond hind knees (Fig. 137). White lines of X-shaped marking on prozona and metazona of nearly equal width. Hind tibia whitish-reddish or bluish 2

2. Size exceeding 20 mm. Prozona much shorter than metazona. Foveolae long and narrow. **D. maroccanus** (Thunberg)

- Size less than 20 mm. Prozona as long as or slightly longer than metazona. Foveolae square (Fig. 138). Hind tibia grey or bluish 3
3. Cerci of male curved inwards (Fig. 140). Anterior abdominal tergites of female dark brown. Hind tibia bluish. **D. curvicercus** Uvarov
- Cerci of male straight (Fig. 142). Anterior abdominal tergites of female light. Hind tibia grey. **D. genei** (Ocskay)

Dociostaurus hauensteini hauensteini (Bolivar, 1893)

Figs. 134–136; Plate III: 8

Type Locality: 'Palestine'.

Stauronotus hauensteini Bolivar I., 1892/93, *Rev. Biol. nord. Fr.*, 5: 481.
Stauronotus hauensteini —. Krauss H., 1909, in: Kneucker A., *Verh. naturw. Ver. Karlsruhe*, 21: 36.
Dociostaurus hauensteini —. Kirby W. F., 1910, *A synonymic catalogue of the Orthoptera*, Vol. III, Orthoptera Saltatoria, Part II, London, p. 158.
Dociostaurus crassiusculus Pantel. Uvarov B. P., 1921, *Bull. ent. Res.*, 11 (4): 401, 405.
Dociostaurus crassiusculus —. Buxton P. A. & B. P. Uvarov, 1923, *Bull. Soc. R. ent. Égypte*, p. 190.
Dociostaurus hauensteini —. Uvarov B. P., 1930, *Eos, Madr.*, 10: 359.
Dociostaurus hauensteini —. Bodenheimer F. S., 1935, *Animal Life in Palestine*, Jerusalem, pp. 89, 320, 323.
Dociostaurus hauensteini —. Bodenheimer F. S., 1935, *Arch. Naturgesch.*, 4 (2): 182.
Dociostaurus hauensteini —. Ramme W., 1951, *Mitt. zool. Mus. Berl.*, 27: 424.
Dociostaurus hauensteini —. Bei-Bienko G. Ya. & L. L. Mishchenko, 1951, *Locusts and Grasshoppers of the U. S. S. R. and Adjacent Countries*, Moscow, II: 442 [in Russian].

Short and robust. Head large, especially in female; slightly opisthognathous (Fig. 136). Frontal ridge of female flat and punctate, in male slightly concave, its margins diverging towards clypeus. Foveolae depressed, slightly elongate (Fig. 136). Vertex wide; horizontal diameter of eye equal to interocular space. Lateral carinae of pronotum distinct in anterior part of prozona and metazona, obliterated between the transverse sulci. Lines of X-shaped markings on metazona 2–3 times wider than those on posterior part of prozona (Fig. 135). Median carina of pronotum intersected only by third transverse sulcus. Prosternum protruding; mesosternal interspace 1.5–2 times as wide as long. Wings abbreviate, not reaching or just reaching hind knee. Length of hind femur 2.5–3 times its maximum width.
Coloration: Brown or pale brown with small brown spots on tegmina; lateral lobes of pronotum usually dark brown, the colour extending dorsally to border the X-shaped marking; ventral margins light. Two or three oblique brown bands present on outer surface of femur, often fused ventrally, usually continuing onto inner surface. Hind knees black; hind tibia orange-reddish.
Measurements (mm): Body ♀ 20.0–27.5, ♂ 17.0–18.5; tegmina ♀ 13.0–15.0, ♂ 8.0–9.0.

Distribution : Transcaucasus, Northern Iran, Turkey, Syria, Israel.
Israel : Mount Hermon (19), Jordan Valley (7), Judean Hills (11), Judean Desert (12), Southern Coastal Plain (9), Central Negev (17).
It inhabits areas with low sparse vegetation. Hoppers observed in March and April.

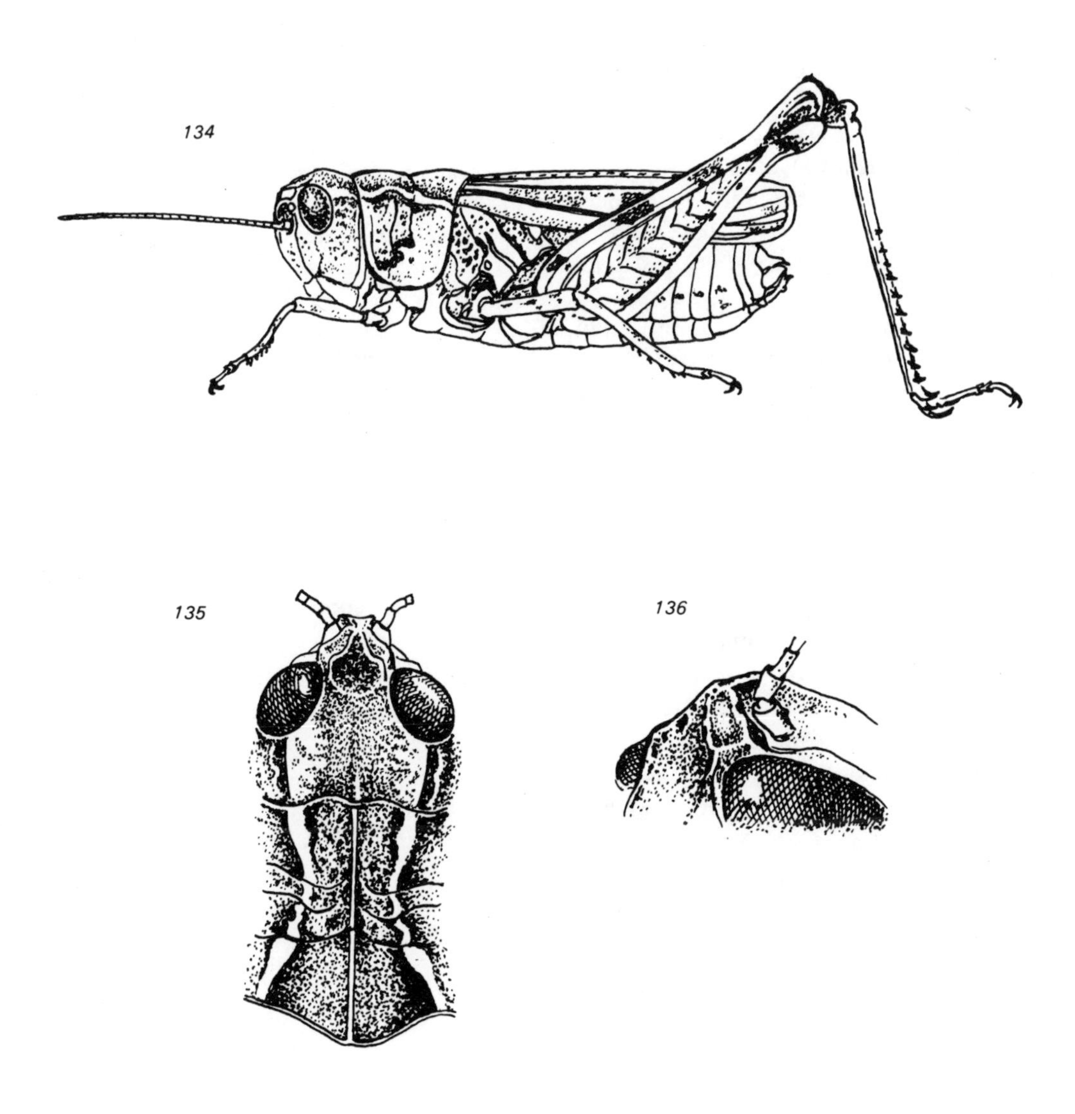

Figs. 134–136 : *Dociostaurus hauensteini hauensteini* (Bolivar, 1893)
134. female ; 135. head and pronotum ; 136. foveola

Dociostaurus maroccanus (Thunberg, 1815)

Type Locality : 'Barbary'. Type lost.

Gryllus maroccanus Thunberg C. P., 1815, *Mém. Acad. Sci. St.-Pétersb.*, 5 : 244.
Gryllus cruciatus Charpentier T. de, 1825, *Horae entomologicae*, Wratislaviae, p. 137.
Oedipoda cruciata —. Brullé A., 1832, in : Bory de St. V., *Expédition Scientifique de Morée*, 3 (1), Zool. Sect., 2 : 92.
Acridium cruciatum —. Brullé A. 1840, Orthoptera, in : Webb P. B. & S. Berthelot, *Histoire naturelle des Iles Canaries*, 2 (2) : 78.
Dociostaurus cruciatus —. Fieber F. X., 1853, *Lotos*, 3 : 118.
Stauronotus cruciatus —. Fischer L. H., 1853, *Orthoptera Europaea*, Leipzig, p. 352.
Epacromia oceanica Walker F., 1870, *Catalogue of the specimens of Dermaptera Saltatoria in the Collection of the British Museum*, London, Part IV, p. 779.
Stauronotus maroccanus —. Stål C., 1873, *Recensio Orthopterorum*, Stockholm, 1 : 111.
Stauronotus maroccanus —. Giglio-Tos E., 1893, *Boll. Musei Zool. Anat. comp. R. Univ. Torino*, 8 (164) : 5.
Dociostaurus maroccanus —. Werner F., 1908, *Zool. Jb.* (Syst.), 27 : 110.
Dociostaurus maroccanus —. Buxton P. A. & B. P. Uvarov, 1923, *Bull. Soc. R. ent. Égypte*, p. 186.
Dociostaurus maroccanus —. Bodenheimer F. S., 1935, *Animal Life in Palestine*, Jerusalem, pp. 28b, 152, 323.
Dociostaurus maroccanus —. Bodenheimer F. S., 1935, *Arch. Naturgesch.*, 4 (2) : 182.
Dociostaurus maroccanus —. Ramme W., 1951, *Mitt. zool. Mus. Berl.*, 27 : 424.

Medium sized, robust. Head short, slightly opisthognathous, not raised above pronotum. Frontal ridge flat, concave around median ocellus, constricted and densely punctate apically. Vertex concave, pentagonal. Foveolae contiguous in front or separated, 1.5–1.7 times longer than wide. Antennae as long as or shorter than head and pronotum together ; segments 6 to 8 square. Prozona of pronotum much shorter than metazona, slightly roof-shaped along median carina ; lateral carinae absent in prozona, very distinct in metazona. Margins of mesosternal interspace diverging. Cerci of male slightly curved, as long as subgenital plate. Wings extending markedly beyond hind knees. Median field of tegmina closed ; radius-sector with three branches at apex.

Coloration : Pale brown or ochraceous, with numerous brown blotches and spots on tegmina, which sometimes form transverse bands. Wings transparent. Inner surface of hind femur pale yellow ; three oblique dark bands on outer dorsal side. Hind tibia slightly reddish-orange.

Measurements (mm) : Body ♀ 28.0–38.0, ♂ 20.0–28.0 ; tegmina ♀ 25.0–36.0, ♂ 17.0–27.0.

Distribution : North Africa, Europe, Asia.

Israel : Upper Galilee (1), Northern and Central Coastal Plain (4, 8), Jordan Valley (7), Judean Hills (11) and Judean Desert (12), Dead Sea Area (13), Northern Negev (15).

The breeding grounds of *D. maroccanus* are in the Canary Islands and in Western

and Central Asia. This is the most important grasshopper pest in Southern Europe. It oviposits in loamy soil with stones and sparse vegetation. Such breeding grounds are also common in Jordan, Syria and Iraq. Utilization of this type of soil for agricultural purposes in the last century greatly reduced outbreaks of this species.
Only specimens of the solitary phase, on which this description is based, are found in our region. Specimens of ph. *gregaria* were also found in Israel in the past, differing from the solitary phase in the absence of oblique dark bands on the hind femur and a slightly higher projecting median carina of the pronotum.

Dociostaurus curvicercus Uvarov, 1942
Figs. 137–140

Type Locality : North Jaffa, Israel.

Dociostaurus genei (Ocskay). Buxton P. A. & B. P. Uvarov, 1923, *Bull. Soc. R. ent. Égypte*, p. 187.
Dociostaurus genei —. Bodenheimer F. S., 1935, *Animal Life in Palestine*, Jerusalem, pp. 89, 320, 323.
Dociostaurus genei —. Bodenheimer F. S., 1935, *Arch. Naturgesch.*, 4 (2) : 181.
Dociostaurus curvicercus Uvarov B. P., 1942, *Trans. Am. ent. Soc.*, 67 : 324 (revision).

Small, smooth. Head hypognathous or slightly opisthognathous, projecting above pronotum. Frontal ridge narrow, slightly depressed between bases of antennae, its margins gradually diverging, obliterated before clypeus. Vertex markedly concave, with sharp lateral and anterior margins, usually posteriorly open or with a rounded margin. Foveolae square or slightly elongate (Fig. 138). Antennae of female as long as those of male, longer than head and pronotum together ; segments 2 to 12 flattened and smooth ; distal segments cylindrical and densely punctate. Pronotum distinctly intersected by three transverse sulci ; prozona as long as or slightly shorter than metazona. Median carina raised, linear ; lateral carinae diverging posteriorly (Fig. 139). Mesosternal lobes twice as wide as long ; margins of mesosternal interspace gradually diverging. Cerci of male strongly curved inwards (Fig. 140). Wings developed, extending posteriorly beyond hind knees. Apex of tegmina rounded ; regular transverse veinlets in median field ; radius-sector with two or three branches. Hind femur slender. Hind tibia slightly shorter than femur ; ventral spur on inner line about twice as long as that on outer line.
Coloration : Brown or pale brown with a pattern of light and brown spots and blotches on body and tegmina. The X-shaped marking on pronotum usually very distinct, rarely obliterated on prozona. Inner surface of hind femur yellowish, with two oblique bands, also extending dorsally. Knee and base of tibia dark. Tibia and tarsus bluish or bluish-grey. First three abdominal terga usually dark brown.
Measurements (mm) : Body ♀ 17.5–21.5, ♂ 13.5–17.0 ; tegmina ♀ 18.0–21.0, ♂ 14.0–16.5 ; hind femur ♀ 14.0–15.0, ♂ 10.5–12.5.
Distribution : Eastern Mediterranean, Syria, Lebanon, Israel.

Israel: Golan Heights (18), Upper and Lower Galilee (1, 2), Jordan Valley (7), Coastal Plain (4, 8, 9), Judean Hills (11).

One of the most typical summer grasshoppers, found on uncultivated areas in the Coastal Plain. In areas where its distribution overlaps that of *D. genei* it apparently forms hybrids with intermediate characteristics. In the courting season, specimens of *D. curvicercus* aggregate in groups and emit strong chirping sounds.

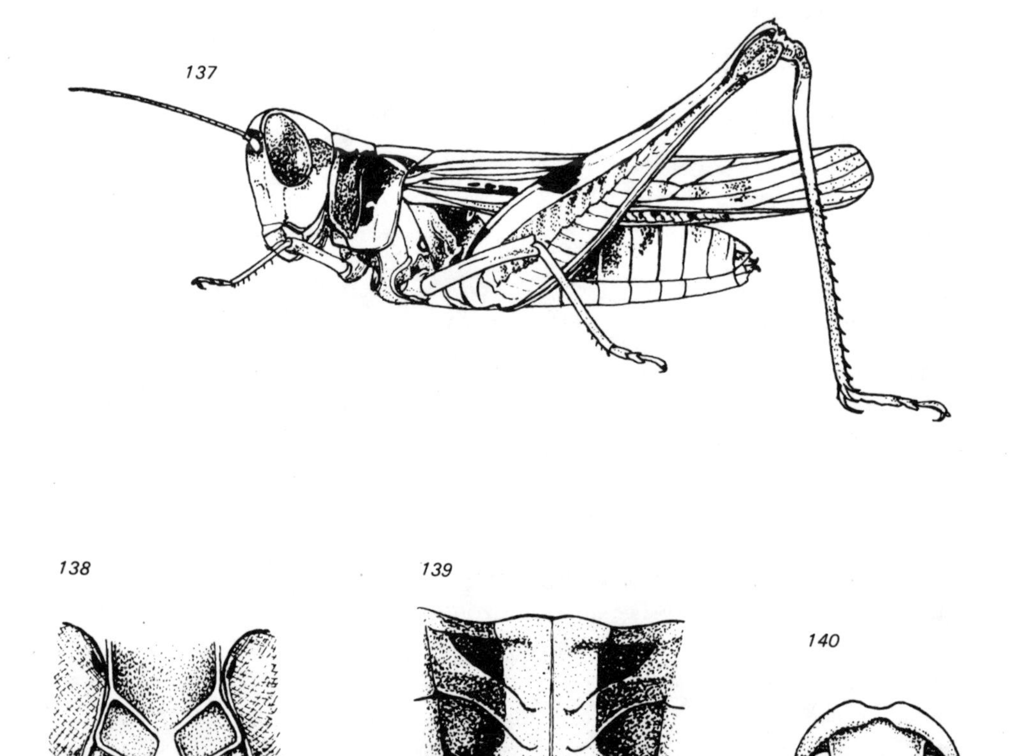

Figs. 137–140 : *Dociostaurus curvicercus* Uvarov, 1942
137. female ; 138. foveolae and frontal ridge ; 139. pronotum ; 140. cerci of male

Dociostaurus genei (Ocskay, 1833)
Figs. 141, 142

Type Locality : 'Italy'. Type lost.

Gryllus genei Ocskay F. L. B., 1833, *Nova Acta Acad. Caesar. Leop. Carol.*, 16 (2) : 961.
Stauronotus genei —. Fischer L. H., 1853, *Orthoptera Europaea*, Leipzig, p. 355.
Dociostaurus genei —. Fieber F. X., 1853, *Lotos*, 3 : 118.
Dociostaurus genei —. Buxton P. A. & B. P. Uvarov, 1923, *Bull. Soc. R. ent. Égypte*, p. 187.
Dociostaurus genei —. Innes W., 1929, *Mém. Soc. R. ent. Égypte*, 3 (2) : 34.
Dociostaurus genei —. Bodenheimer F. S., 1935, *Animal Life in Palestine*, Jerusalem, pp. 89, 320, 323.
Dociostaurus genei —. Bodenheimer F. S., 1935, *Arch. Naturgesch.*, 4 (2) : 181.
Dociostaurus genei —. Ramme W., 1951, *Mitt. zool. Mus. Berl.*, 27 : 424.

Small, smooth. Head hypognathous or opisthognathous. Frontal ridge densely punctate dorsally, obliterated below median carina. Vertex pentagonal, concave ; foveolae square, partly visible from above. Antennae longer than head and pronotum together. Hairs on pronotum sparse ; anterior part of prozona compressed ; median carina linear, raised, intersected by the posterior transverse sulcus. Cerci of male straight,

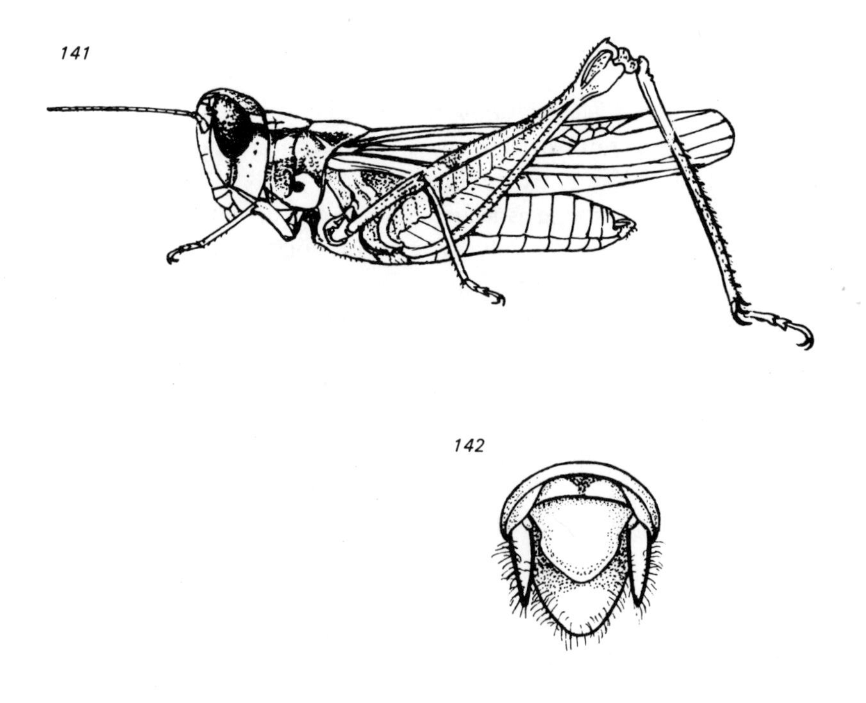

Figs. 141–142 : *Dociostaurus genei* (Ocskay, 1833)
141. male ; 142. cerci of male

slightly shorter than subgenital plate (Fig. 142). Wings extending to hind knees or slightly beyond them. Tegmina rounded apically; radius-sector with two or three branches.

Coloration: Generally pale brown, head and pronotum sometimes with white regions. The X-shaped marking on pronotum usually partly obliterated, especially in the southern populations. Numerous small brown blotches on tegmina. Two oblique brown bands present on inner surface of hind femur, only the apical one extending towards the ventral margin. Hind tibia grey.

Measurements (mm): Body ♀ 15.0–19.0, ♂ up to 13.0; tegmina ♀ 10.0–14.5, ♂ 7.5–12.0; hind femur ♀ 9.0–9.5, ♂ 11.5–12.5.

Distribution: North Africa, Europe, Asia Minor, Israel.

Israel & Sinai: Golan Heights (18), Jordan Valley (7), Judean Hills (11), Coastal Plain (8, 9), Northern and Central Negev (15, 17), Northern Sinai (20).

Genus CHORTHIPPUS Fieber, 1852

In: Kelch A., *Grundlage zur Kenntnis der Orthopteren Oberschlesiens, etc.*, Ratibor, Bögner, pp. 1, 4

Stenobothrus Jacobson G. G. & V. L. Bianchi, 1902, *Orthoptera and Odonata of the Russian Empire*, pp. 165, 177, 219 [in Russian].

Type Species: *Acridium albomarginatum* De Geer, 1773.

Diagnosis: Small or medium sized, compressed laterally. Head slightly opisthognathous (Fig. 143). Foveolae narrow, elongate, only partly visible from above (Fig. 146). Antennae filiform, as long as head and pronotum together, or longer in males. Pronotum intersected distinctly only by the posterior transverse sulcus (Fig. 144). Lateral carinae of pronotum very distinct, nearly parallel or converging anteriorly, diverging posteriorly (Fig. 147). Metasternal lobes contiguous or separated. Tegmina and wings fully developed or residual. Precostal field of tegmina widened at the base; costal and subcostal veins straight. Inner side of hind femur with or without a longitudinal black line near the base.

Distribution: More than 80 species throughout Europe, North Africa, Asia and North America.

Four species in our region, more to be expected.

Key to the Species of Chorthippus in Israel

1. Tegmina vestigial, situated laterally, extending to posterior margin of fourth abdominal segment (Fig. 143). **C. dirshi** Fishelson
– Tegmina fully developed, usually extending beyond hind knees 2

2. Pronotum distinctly intersected by one transverse sulcus beyond its middle ; lateral carinae in prozona almost parallel (Figs. 150, 151) 3
- Pronotum intersected by two transverse sulci, the posterior sulcus situated in middle of pronotum or before it. Lateral carinae of prozona distinctly converging towards median carina (Fig. 147). **C. peneri** Fishelson
3. Anterior part of lateral lobes of pronotum with numerous carinulae (especially in males) (Fig. 150). Length of foveolae 2.2–2.5 times their width. Mesosternal interspace of males long and narrow. Median field of tegmina closed ; intercalary vein present. **C. loratus** Fischer-Waldheim
- Lateral lobes of pronotum punctate or slightly rugose (Fig. 151). Length of foveolae 3.5–4.0 times their width. Mesosternal interspace nearly square. Median field of tegmina usually open ; intercalary vein absent. **C. dorsatus palaestinus** Uvarov

Chorthippus dirshi Fishelson, 1969

Figs. 143–145

Type Locality : Mount Hermon (Zoological Museum, Tel Aviv University).

Chorthippus dirshi Fishelson L., 1969, *Israel J. Ent.*, 4 : 235–237.

Small. Head large, opisthognathous. Fastigium slightly oblique ; median carina absent or present. Foveolae straight, moderately concave, twice as long as wide, margins rounded. Eyes as long as or slightly longer than subocular groove. Metazona of pronotum shorter than prozona ; median carina slightly raised ; lateral carinae distinct, converging anteriorly towards first transverse sulcus (Fig. 144).
Tegmina vestigial, extending to posterior margin of second or fourth abdominal terga, their length twice their maximum width (Fig. 145). Wings shorter than tegmina, membranous. Hind femur thick ; in male almost three times longer than wide. Hind tibia with 13–15 spines on the outer line.
Coloration : Brown, with a pattern of blackish and whitish lines and spots on head, pronotum and exposed areas of abdominal terga. Dark basal line present on inner side of hind femur and black spots usually on outer side. Hind tibia grey.
Measurements (mm) : Body ♀ 16.0–16.5, ♂ 12.5–13.5 ; vestigial tegmina ♀ 3.0–3.5, ♂ 4.5–5.0.
Distribution : Mount Hermon (19).
Found at an altitude of 1,200 m in areas with dense grass and bush-like vegetation.

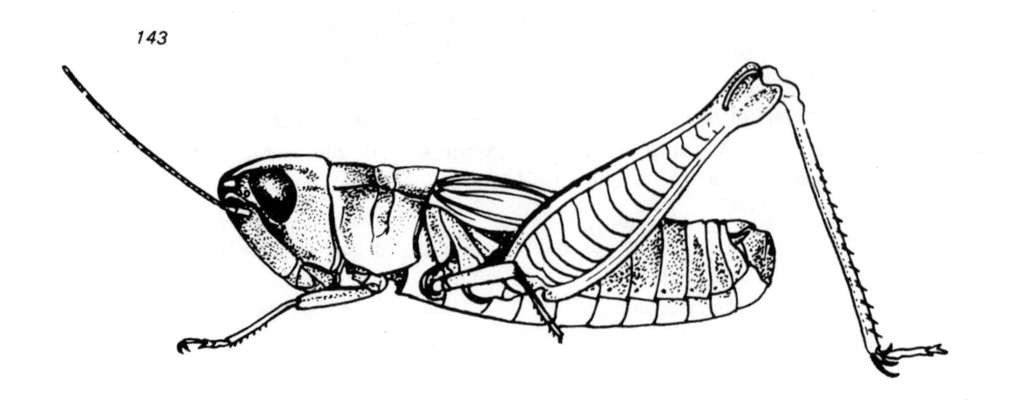

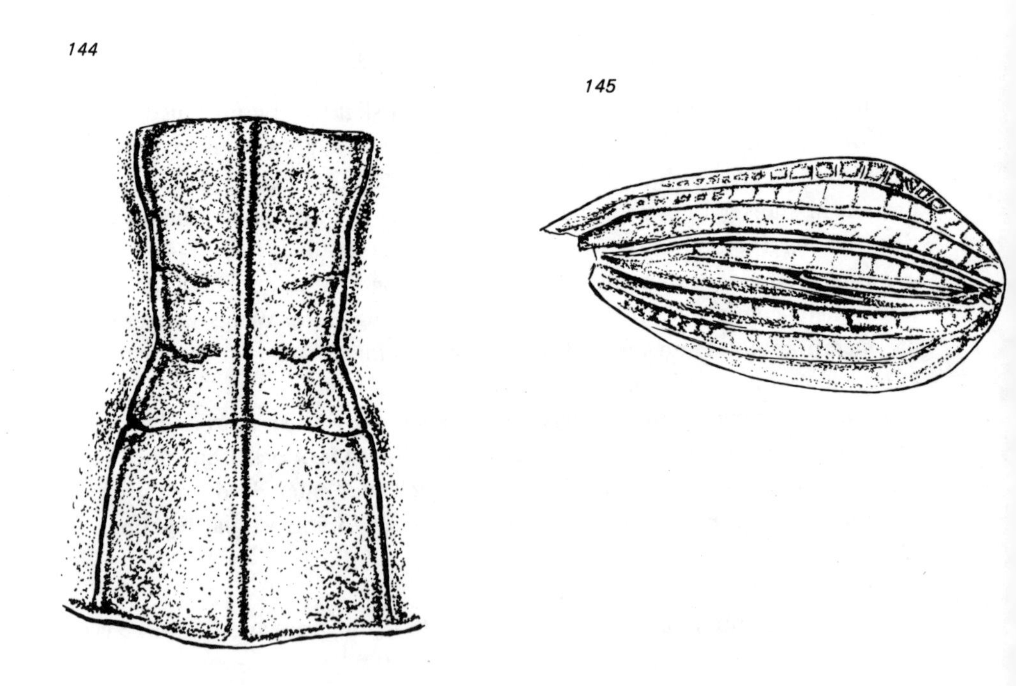

Figs. 143–145 : *Chorthippus dirshi* Fishelson, 1969
143. male ; 144. pronotum, male ; 145. tegmen, female

Chorthippus peneri Fishelson, 1969

Figs. 146–149

Type Locality: Mount Hermon (1,800 m altitude) (Zoological Museum, Tel Aviv University).

Chorthippus peneri Fishelson L., 1969, *Israel J. Ent.*, 4 : 237–242.

Small, almost smooth, hairs sparse. Head opisthognathous. Frontal ridge concave from median field downwards, widening towards clypeus. Vertex horizontal ; fastigium pentagonal, angular apically. Foveolae concave, slightly curved, at least twice as long as wide (Fig. 146). Posterior transverse sulcus situated before or in middle of pronotum (Fig. 147). Lateral carinae of prozona converging towards the median carina before the second transverse sulcus, distinctly diverging posteriorly, ending close to posterior margin of metazona. Mesosternal interspace wide and short, its margins parallel or slightly diverging posteriorly. Metasternal interspace square (Fig. 148) or twice as long as wide in male.

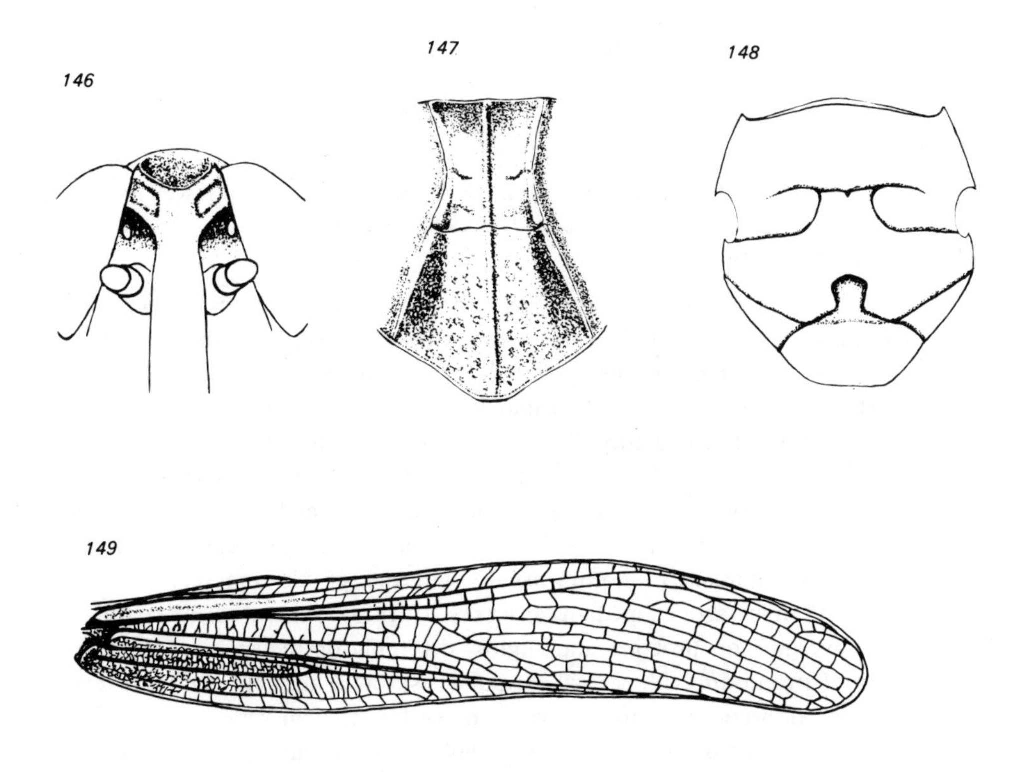

Figs. 146–149 : *Chorthippus peneri* Fishelson, 1969
146. foveolae and frontal ridge ; 147. pronotum ; 148. sternum ; 149. tegmen

Tegmina extending beyond hind knees, slightly curved, with rounded apex; median field, as wide as narrowest part of costal field (Fig. 149). Abdominal terga of female with numerous lateral sulci. Hind femur slender, its length four times its maximum width. Hind tibia with 12 or 13 spines on the inner line. Arolium more than half as long as claws.

Coloration: Brown, with numerous small black spots; face sometimes dark and dorsal part of head and pronotum pale. Lateral carinae of pronotum usually whitish, bordered by dark brown. Posterior margin of tegmina sometimes reddish-brown. Wings smoky in distal field. Abdominal terga dorsally and apically yellowish-red or reddish, the first three or four segments blackish laterally. Black basal line present on inner side of hind femur. Hind tibia and tarsus reddish or reddish-grey.

Measurements (mm): Body ♀ 20.0–23.5, ♂ 17.0–19.5; tegmina ♀ 18.0–19.5, ♂ 15.0–16.5.

Distribution: Northern Israel, Golan Heights.

Israel: Mount Hermon (19), Judean Hills (11), Upper Galilee (1; Mount Meron).

Chorthippus loratus Fischer-Waldheim, 1846
Fig. 150

Type Locality: 'Russia'.

Chorthippus lorata Fischer de Waldheim G., 1846, *Orthoptera Imperii Russici*, Moscow, p. 307.
Oedipoda lorata —. Jacobson G. G. & V, L. Bianchi, 1902, *Orthoptera and Odonata of the Russian Empire*, p. 234 [in Russian].

Small, smooth. Head prominently opisthognathous; in male slightly swollen and longer than pronotum. Frontal ridge flat between bases of antennae, then concave, margins diverging towards clypeus and slightly converging near median ocellus. Fastigium short, not wider than interocular space. Foveolae 2.2–2.5 times longer than wide; anterior margin slightly blurred. Eyes large, in female twice as long as subocular groove, more than twice as long in male. Occiput usually with a median carina. Antennae of male as long as those of female, shorter than head and pronotum together; almost all segments square and flattened in female; distal segment of male slightly narrow or rounded. Pronotum usually with one visible transverse sulcus; prozona longer than metazona; median carina slightly raised; lateral carinae parallel or slightly converging on prozona and slightly diverging on metazona (Fig. 150). Lateral lobes of pronotum with several carinulae, especially in male. Sternum very hairy; mesosternal interspace usually narrow, much longer than wide.

Tegmina extending to or beyond hind knee, usually with a spurious intercalary vein; median field closed; radius-sector with a single branch. Hind femur slender, its length 3.5–4.5 times its maximum width. Hind tibia with 13 spines on both inner and outer lines; outer spurs of almost equal length. Arolium two-thirds to three-quarters the length of claws.

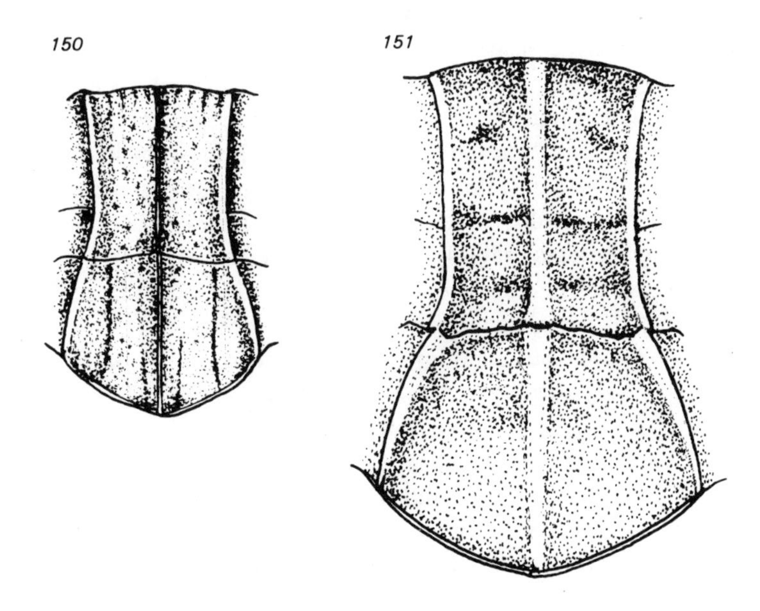

Figs. 150–151 : Pronotum
150. *Chorthippus loratus* Fischer-Waldheim, 1846, male;
151. *Chorthippus dorsatus palaestinus* Uvarov, 1933, female

Coloration : Greyish-brown to brown ; dorsal parts of head, pronotum and tegmina pale. Lateral carinae of pronotum white with brown borders. Tegmina of uniform colour laterally or with white and brown lines at the anterior margin. Anterior abdominal terga blackish. Basal line on inner side of hind femur distinct and long. Hind knees black.
Measurements (mm) : Body ♀ 19.0 ; tegmina ♀ 16.0.
Distribution : Europe, Caucasus, Persia, Israel.
Israel : Golan Heights (18), Upper Galilee (1 ; Mount Meron).

Chorthippus dorsatus palaestinus Uvarov, 1933
Fig. 151

Type Locality : Caesarea, Israel.

Acrydium albomarginatum De Geer C., 1773, *Mémoires pour servir à l'histoire des insectes*, 3 : 480.
Chorthippus albomarginatus —. Buxton P. A. & B. P. Uvarov, 1923, *Bull. Soc. R. ent. Égypte*, p. 186.
Chorthippus palaestinus Uvarov B. P., 1933, *Ann. Mag. nat. Hist.*, Ser. 10, 11 : 671, fig. 5.
Chorthippus dorsatus palaestinus —. Bodenheimer F. S., 1935, *Arch. Naturgesch.*, 4 (2) : 180.
Chorthippus dorsatus palaestinus —. Bodenheimer F. S., 1935, *Animal Life in Palestine*, Jerusalem, pp. 28b, 39, 320, 323.

Head opisthognathous, in male much shorter than pronotum. Frontal ridge sparsely punctate, flat apically, concave above median ocellus; margins slightly raised, almost parallel, diverging before clypeus. Fastigium longer than interocular space; foveolae narrow, 3.6–4 times longer than wide; margins sometimes obliterated. Antennae of male much longer than head and pronotum together; segments, from the 7th onwards, narrow in both sexes. Pronotum intersected by the posterior transverse sulcus, situated beyond its middle; traces of first sulcus visible on the roof-shaped prozona (Fig. 151). Lateral carinae slightly curved inwards in prozona, diverging in metazona; posterior part of lateral lobe granulate. Hairs on sternum sparse; mesosternal interspace almost square, its margins slightly diverging.
Tegmina extending beyond hind knees; median field usually open; intercalary vein either absent or only traces present. Hind tibia usually with 13 spines on inner and outer lines.
Coloration: Usually brown or pale brown; face and lateral lobe of pronotum paler or all dorsal parts of pronotum and wings pale, straw-yellow. Anterior abdominal terga pale or with dispersed brown pigmentation. Apical parts of wing greyish, especially in male. Dorsal black line on inner side of femur short, sometimes obliterated. Hind tibia light brown.
Measurements (mm): Body ♀ 20.0–24.5, ♂ 15.0–16.5; tegmina ♀ 16.5–19.5, ♂ 12.5–14.0.
Distribution, Northern Israel: Upper and Lower Galilee (1, 2), Central Coastal Plain (8), Judean Hills (11).

Genus OCHRILIDIA Stål, 1873

Recensio Orthopterorum, Stockholm, 1: 92, 104

Platypterna (nom preoc.) Fieber F. X., 1853, *Lotos*, 3: 98.

Type Species: *Ochrilidia tryxalicera* Stål, 1873.
Diagnosis: Medium sized or small, smooth, body cylindrical. Head short, opisthognathous. Frontal ridge with distinct marginal carinae. Vertex triangular, usually with a wide base and a median carina present on its dorsal surface. Foveolae narrow and long, usually curved, not visible from above (Fig. 155). Interocular space shorter than the horizontal diameter of eye. Antennae more or less ensiform, in females almost filiform, but basal segments slightly wider than apical ones (Fig. 160). Three longitudinal carinae present on pronotum, the median carina only intersected by the posterior transverse sulcus (Figs. 152, 158). Mesosternal interspace narrow or lobes contiguous.
Tegmina extending beyond hind knees, rounded apically; intercalary vein usually present in median field. Apical part of inner surface of hind femur with or without a black blotch. Arolium large, usually two-thirds the length of the claws.
Coloration: Usually mimetic of dry grass.

Ochrilidia

Distribution : Arid areas of Southern Europe, North Africa and Asia.
About 30 species in this genus. According to the revision of Jago (1977), four species in our region.

Key to the Species of Ochrilidia in Israel and Sinai

1. Pronotum intersected by posterior transverse sulcus beyond its middle, so that the prozona is much longer than the metazona (Fig. 158). Lateral carinae of pronotum straight or slightly diverging in metazona 2
– Pronotum intersected by posterior transverse sulcus in its middle or slightly anterior to it ; prozona as long as or slightly shorter than metazona. Lateral carinae of pronotum gradually diverging towards posterior margin (Fig. 152). Hind tibia violet. **O. geniculata** (Bolivar)
2. Hind femur with a black subapical spot on the inner surface. Foveolae partly obliterate, their length 1.5–2.0 times their maximum width (Fig. 153). **O. tibialis** (Fieber)
– Hind femur without subapical black spot. Foveolae distinct, their length 3.0–4.5 times their maximum width 3
3. Pronotum with straight, lateral carinae which gradually diverge towards the posterior margin (Fig. 154). Hind tibia blue-grey to brown. **O. gracilis** (Krauss)
– Lateral carinae of pronotum converging in anterior part of prozona, parallel in middle, between transverse sulci, and very slightly diverging in metazona (Fig. 158). Occiput with numerous rugulae, especially in male. Hind tibia pale grey. **O. persica** (Salfi)

Ochrilidia geniculata (Bolivar, 1913).

Fig. 152 ; Plate IV : 2

Type Locality : 'Algeria' (Madrid Museum).

Platypterna geniculata Bolivar I., 1913, *Novit. zool.*, 30 : 608.
Platypterna pruinosa agedabiae Salfi M., 1928, *Boll. Soc. Nat. Napoli*, 39 : 244.
Platypterna ladakiae Salfi M., 1931, *Eos, Madr.*, 7 (3) : 300, 301.
Platypterna ladakiae —. Bodenheimer F. S., 1935, *Animal Life in Palestine*, Jerusalem, p. 320.
Platypterna ladakiae —. Bodenheimer F. S., 1935, *Arch. Naturgesch.*, 4 (2) : 181.
Ochrilidia geniculata —. Jago N. D., 1977, *Acrida*, 6 : 177.

Large. Head large, slightly swollen dorsally. Frontal ridge with almost parallel margins, usually slightly wider between the base of antennae. Vertex with raised margins anteriorly, swollen in the middle, with a median carina. Foveolae distinctly curved, punctate with rounded or obliterated margins anteriorly. Antennae distinctly ensiform. Lateral carinae of pronotum almost parallel, slightly diverging in metazona ; third transverse sulcus situated beyond middle of pronotum (Fig. 152). Sternum sparsely pitted anteriorly. Mesosternal interspace narrow, its length three times its maximum width.

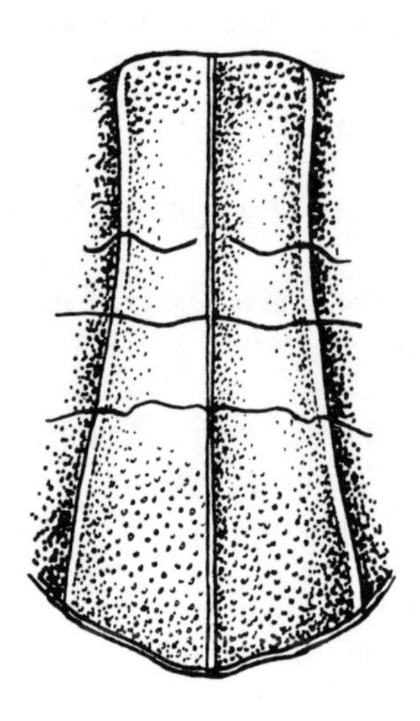

Fig. 152.: *Ochrilidia geniculata* (Bolivar, 1913), pronotum

Tegmina long, extending for one-third of their length beyond hind knees. Male cerci and subgenital plate equal in length, both sharp apically.
Coloration: Straw-yellow, greyish or greenish-grey, sometimes with a pattern of pale bands on head and body. Inner surface of hind femur with a black blotch apically. Hind tibia violet.
Measurements (mm): Body ♀ 29.0–35.0, ♂ 18.0–23.5; tegmina ♀ 28.0–31.0, ♂16.0–19.5.
Distribution: North Africa, Syria, Israel, Iran.
Israel: Central and Southern Coastal Plain (8, 9), Central Negev (17), Dead Sea Area (13).
The most common species along the Coastal Plain; found on stems of *Ammophila arenaria, Artemisia monosperma* and *Arundo donax.* Hoppers observed from August to January; oviposition in October.

Ochrilidia tibialis (Fieber, 1853)
Fig. 153

Type Locality : Crete. Type lost.

Platypterna tibialis Fieber F. X., 1853, *Lotos*, 3 : 98.
Ochrilidia tryxalicera Stål. Giglio-Tos E., 1893, *Boll. Musei Zool. Anat. comp. R. Univ. Torino*, 8 (164) : 5.
Platypterna filicornis judaica Salfi A., 1931, *Eos, Madr.*, 7 (3) : 279.
Platypterna filicornis judaica —. Bodenheimer F. S., 1935, *Arch. Naturgesch.*, 4 (2) : 181.
Ochrilidia tibialis —. Jago N. D., 1977, *Acrida*, 6 : 171.

Size variable. Head acute ; face sparsely punctate. Frontal ridge slightly concave, usually wider between the antennae than near the median ocellus, its margins diverging towards clypeus. Vertex triangular, shorter than the interocular space, its anterior carinae only slightly raised. Foveolae, if present, have a length not exceeding 1.5–2.0 times their maximum width, always indistinct anteriorly, distinct posteriorly. Antennae slightly ensiform, the widest segments 2.0–2.5 times wider than long. Lateral carinae of pronotum almost parallel along their entire length, or diverging in metazona. Sternum sparsely punctate ; mesosternal interspace twice as long as wide. Tegmina extending beyond hind knees for one-quarter of their length ; median field closed, usually with one spurious vein. Male cercus as long as subgenital plate.
Coloration : Pale or pale brown, usually with a brown band on the lateral lobes of the pronotum and the median part of tegmina. Prominent black spot present on inner apical part of hind femur. Hind tibia pale bluish.
Measurements (mm) : Body ♀ 27.0–31.2, ♂ 18.0–20.5 ; tegmina ♀ 23.5–29.5, ♂ 14.0–16.0.
Distribution, Israel & Sinai : Dead Sea Area (13), Coastal Plain (4, 8, 9), Judean Hills (11), Central Negev (17), ‘Arava Valley (14), Northern Sinai (20).

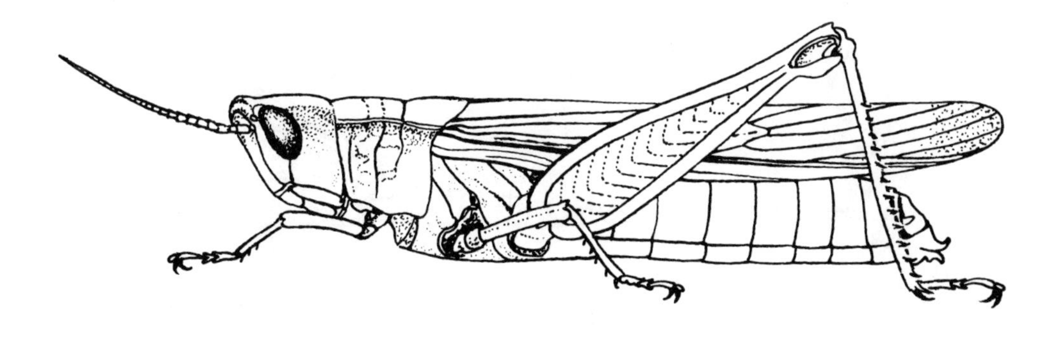

Fig. 153 : *Ochrilidia tibialis* (Fieber, 1853), female

Ochrilidia gracilis (Krauss, 1902)
Figs, 154, 155

Type Locality : Sahara (Algeria). Type lost.

Platypterna gracilis Krauss H. A., 1902, *Verh. zool.-bot. Ges. Wien*, 52 : 236, fig. 2.
Platypterna acuta Bolivar I., 1908, *Bull. Soc. ent. Fr.*, 1908 : 244.
Platypterna tibialis —. Innes W., 1929, *Mém. Soc. R. ent. Égypte*, 3 (2) : 20.
Platypterna obtusa Salfi M., 1931, *Eos, Madr.*, 7 (3) : 324–326.
Platypterna acuta —. Bodenheimer F. S., 1935, *Arch. Naturgesch.*, 4 (2) : 180.
Ochrilidia gracilis —. Jago N. D., 1977, *Acrida*, 6 : 186.

Large. Head distinctly pointed. Frontal ridge concave, its margins gradually diverging towards clypeus. Vertex longer than the interocular space, swollen ; its median carina raised (Fig. 155). Length of foveolae 3.5–5 times their maximum width. Diameter of eye of female 1.5 times as long as subocular groove, twice as long in male. Antennae wide at the base, widest segment 3.5–5 times wider than long. Lateral carinae of pronotum slightly diverging ; first transverse sulcus dividing prozona into two equal parts ; third sulcus situated beyond middle of pronotum ; metazona between lateral carinae almost as wide as long. Sternum long and narrow ; mesosternal

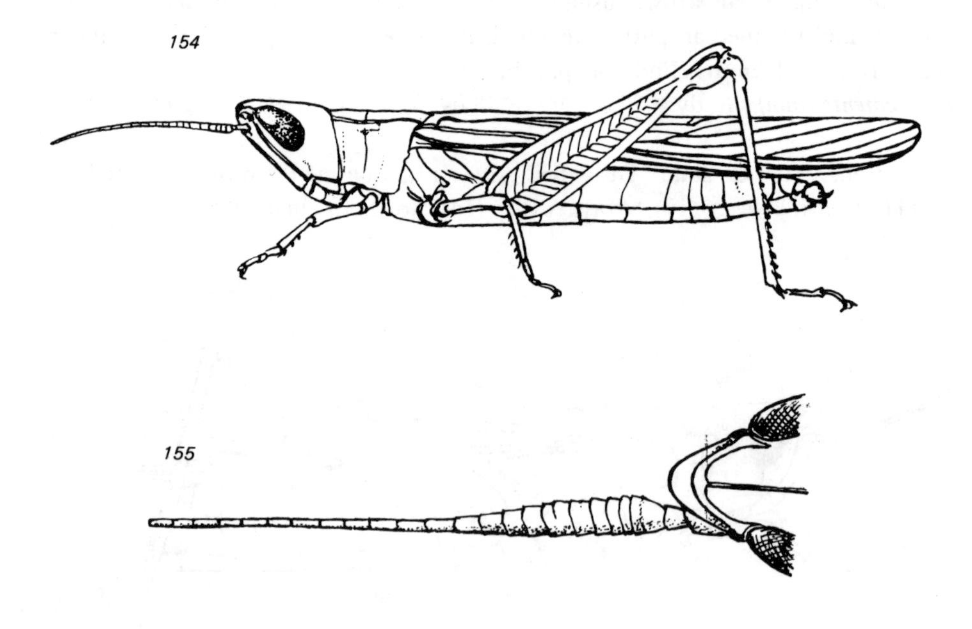

Figs. 154–155 : *Ochrilidia gracilis* (Krauss, 1902)
154. female ; 155. head, from above

lobes contiguous or slightly separated anteriorly. Cerci of male much shorter than subgenital plate. Tegmina membranous, extending far beyond hind femur; median field with long intercalary vein.
Coloration: Usually straw-yellow, sometimes light greenish with brown or white lines on head and pronotum. Hind femur without apical black blotch. Hind tibia pale.
Measurements (mm): Body ♀ 33.5–39.0, ♂ 24.0–26.0; tegmina ♀ 26.0–29.5, ♂ 19.0–20.0.
Distribution: Iraq, Syria, Israel.
Israel: Jordan Valley (7), Judean Hills (11), Central Negev (17), 'Arava Valley (14), Central Sinai Foothills (21).
Usually resting on stems of *Juncus* and *Ammophila.*

Ochrilidia persica (Salfi, 1931)
Figs. 156–161

Type Locality: Northern Iran (Leningrad Museum).

Platypterna persica Salfi M., 1931, *Eos, Madr.*, 7 (3): 228, 230.
Ochrilidia marmorata Uvarov P. B., 1952 (Holotype), *J. Linn. Soc.* (Zool), 42 (284): 178.

Robust. Head short, with numerous transverse carinulae, especially in male, slightly swollen towards occiput (Fig. 156). Frontal ridge concave; margins above median ocellus parallel or slightly converging below ocellus, diverging towards clypeus. Foveolae curved, their length 3.5–4.5 times their maximum width (Fig. 157). Antennae only slightly ensiform. Lateral carinae of pronotum nearly parallel, slightly converging anteriorly, diverging posteriorly (Fig. 158); prozona longer than metazona (Fig. 159) or in some isolated populations shorter (Fig. 161). Mesosternal interspace twice as long as wide; metasternal lobes contiguous. Tegmina narrow, rounded apically. Cerci of male shorter than subgenital plate.
Coloration: Greyish-brown; face and lateral parts of body usually whitish. Hind femur without black blotch. Hind tibia greyish.
Measurements (mm): Body ♀ 23.0–24.5, ♂ 15.0–21.0; tegmina ♀ 19.0–22.5, ♂ 14.0–17.5.
Distribution: East Africa, Israel, Northern Iran.
Israel: Central Negev (17).
Found on stems of *Ammophila arenaria* and *Juncus maritimus.*

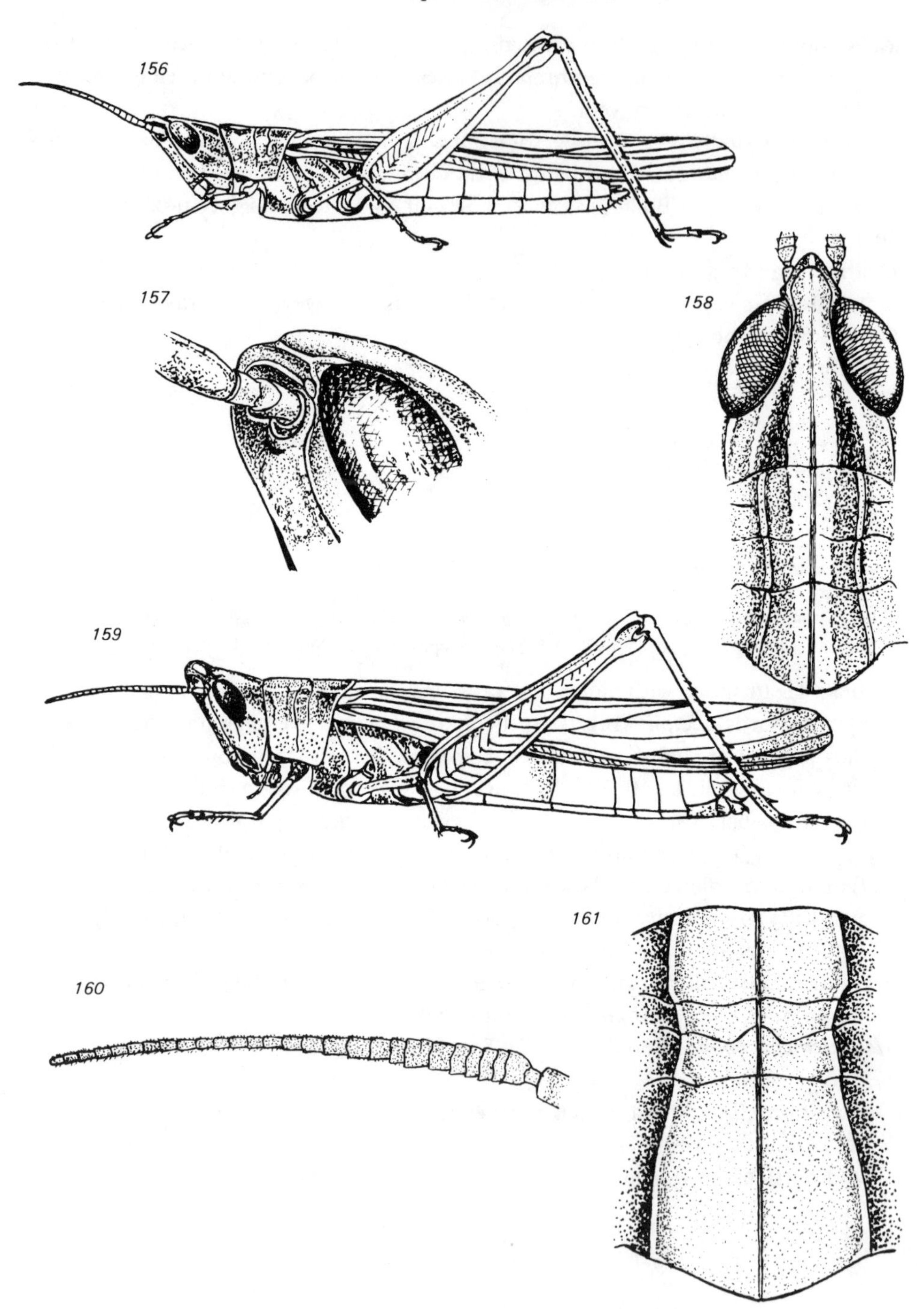

Figs. 156–161 : *Ochrilidia persica* (Salfi, 1931)
156. male ; 157. head and foveolae ; 158. head and pronotum ;
159. female ; 160. antenna ; 161. pronotum

Genus XEROHIPPUS Uvarov, 1942

Trans. Am. ent. Soc., 67 : 325

Type Species : *Stauroderus savigny* Buxton & Uvarov, 1923.
Diagnosis : Small. Head hypognathous, projecting above pronotum. Frontal ridge punctate, concave ; margins usually raised, diverging gradually towards clypeus. Foveolae elongate, partly visible or not visible from above. Eyes of female as long as or shorter than subocular groove. Antennae ensiform. Carinae of pronotum distinct ; lateral carinae slightly undulate, converging in middle of prozona, diverging in posterior part and in metazona (Fig. 163).
Tegmina delicate, slender, usually transparent, either not reaching hind knees or extending to or slightly beyond them. Arolium as long as claws.
Distribution : Near East and Central Asia to Mongolia.
Two species in our region.

Key to the Species of Xerohippus in Israel

1. Small ; length of female 16.0–18.5 mm. Upper valves of ovipositor longer than their maximum width. Tegmina of female 12.0–13.0 mm long. **X. palaestinae** Uvarov
– Large ; length of female 23.0–24.5 mm. Upper valves of ovipositor not longer than their maximum width. Tegmina of female 18.0–19.5 mm long. **X. savignyi** (Krauss)

Xerohippus palaestinae Uvarov, 1942

Type Locality : Nazareth.

Xerohippus palaestinae Uvarov B. P., 1942, *Trans. Am. ent. Soc.*, 57 : 328.

Small, compressed laterally. Head opisthognathous, especially in male. Frontal ridge protruding, concave, its margins parallel to the median ocellus, then gradually diverging, usually slightly blurred above clypeus. Vertex pentagonal, slightly depressed, with signs of a median carina. Foveolae oblong, wider posteriorly than anteriorly, in males not visible from above. Eyes 1.2–1.4 times longer than subocular groove. Antennae slightly ensiform, in female 1.2 times as long as head and pronotum together, in male 1.6 times as long. Pronotum slightly roof-shaped, usually intersected by the first and third transverse sulci ; metazona shorter than prozona ; lateral carinae parallel to anterior margin of prozona, gradually converging towards the first sulcus and then diverging, ending at posterior margin of metazona. Mesosternal interspace of female as wide as mesosternal lobe, that of male narrower. Metasternal interspace present, in female square with rounded angles.
Tegmina not reaching or extending to hind knees, membranous ; median field open or closed, narrower than the subcostal field in its widest part ; spurious intercalary

vein present or absent. Wings narrow, triangular, slightly shorter than tegmina. Hind femur with compressed and raised dorsal carina. Ventro-lateral lobe of hind knees slightly pointed. Hind tibia with 11 or 12 spines on each line. Arolium two-thirds the length of claws.

Coloration : Pale brown, usually darker laterally ; dorsum sometimes very pale ; head and pronotum dark brown. Inner side of hind femur with a definite, dark basal line.

Measurements (mm) : Body ♀ 16.0–19.5, ♂ 12.5–13.5 ; tegmina ♀ 12.0–13.5, ♂ 9.0–9.5.

Distribution, Israel : Upper Galilee (1), Judean Hills (11).

Apparently endemic in our region, developed from *X. savignyi* described below. One of the most common phytophilous species in late summer, found among dense grassy vegetation, hidden by its cryptic coloration.

Xerohippus savignyi (Krauss, 1890)

Figs. 162, 163

Type Locality : 'Cyprus'.

Duronia savigny Krauss H. A., 1890, *Verh. zool.-bot. Ges. Wien*, 40 : 227–272.
Duronia savigny —. Giglio-Tos E., 1894, *Boll. Musei Zool. Anat. comp. R. Univ. Torino*, 8 (164) : 2.
Stauroderus savigny —. Buxton P. A. & B. P. Uvarov, 1923, *Bull. Soc. R. ent. Égypte*, p. 185.
Eremippus savignyi—. Bodenheimer F. S., 1935, *Animal Life in Palestine*, Jerusalem, pp. 86, 320.
Eremippus savignyi —. Bodenheimer F. S., 1935, *Arch. Naturgesch.*, 4 (2) : 184.
Xerohippus savignyi —. Uvarov B. P., 1942, *Trans. Am. ent. Soc.*, 67 : 328.

Larger than *X. palaestinae*, compressed laterally. Head distinctly opisthognathous (Fig. 162). Frontal ridge concave, flat and punctate apically, its margins gradually diverging. Vertex flat or slightly concave. Foveolae either not visible or partly visible from above, usually slightly curved, their margins blurred anteriorly, their length twice their maximum width. Eyes of female shorter than the subocular groove, of male as long or slightly longer. Median carina on occiput almost always present. Antennae of female shorter, in male 1.5 times longer than head and pronotum together. Prozona of pronotum longer than metazona ; usually one transverse sulcus on dorsum ; lateral carinae parallel on anterior part of prozona, converging towards the first transverse sulcus and then diverging towards posterior margin of metazona (Fig. 163). Mesosternal interspace narrower than the lobes ; metasternal lobes of male contiguous.

Tegmina reaching hind knees or extending slightly beyond them, membranous ; subcostal field of male much wider than median field. Upper valves of ovipositor wider than long.

Coloration : Usually straw-yellow or pale brown, especially the tegmina ; sometimes greyish-brown with a pale area extending from head and pronotum to dorsal part of wings. Basal dark line on inner side of hind femur usually wide and extending to middle of femur.

Measurements (mm) : Body ♀ 23.0–24.5, ♂ 15.0–16.5 ; tegmina ♀ 18.0–19.5, ♂ 11.0–11.5.

Distribution : Lebanon, Syria, Israel.

Israel : Upper and Lower Galilee (1, 2), Central and Southern Coastal Plain (8, 9), Judean Hills (11).

Common from May to July, inhabiting areas of dry, grassy vegetation. Usually occurring together with *Notostaurus anatolicus* and *Pyrgomorphella granosa*. Hoppers found from March to May.

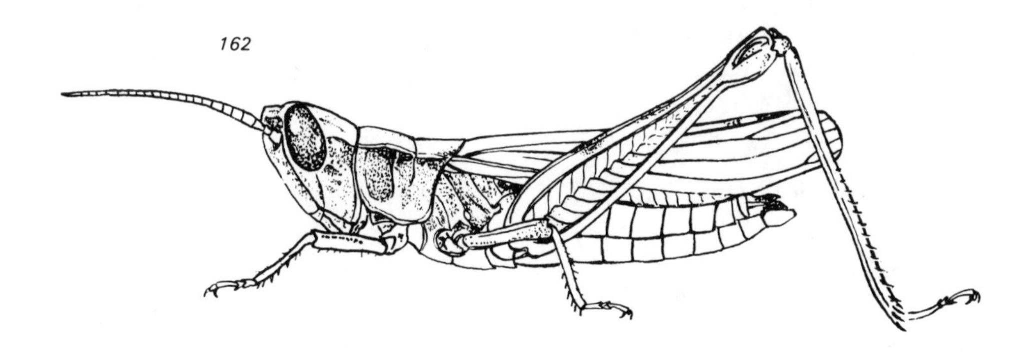

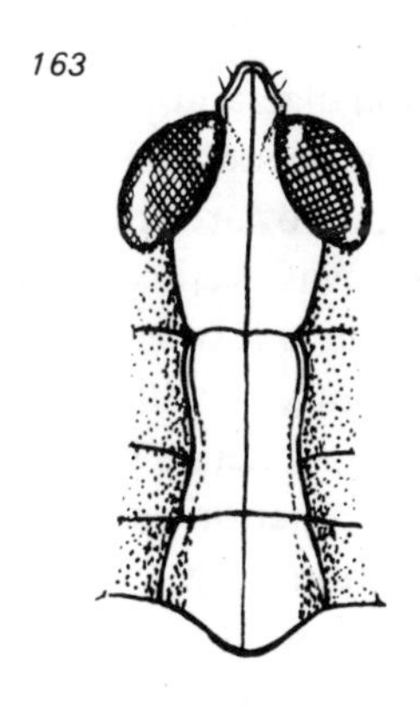

Figs. 162–163 : *Xerohippus savignyi* (Krauss, 1890)
162. male ; 163. head and pronotum

Genus NOTOPLEURA Krauss, 1902

Verh. zool.-bot. Ges. Wien, 52 (4) : 240

Notopleura —. Innes W., 1929, *Mém. Soc. R. ent. Égypte*, 3 (2) : 35.
Notopleura —. Dirsh V. M., 1961, *Bull. Br. Mus. nat. Hist. (Ent.)*, 10 (9) : 413.

Type Species : *Notopleura saharica* Krauss, 1902.
Diagnosis : Small, robust. Head hypognathous or slightly opisthognathous, rugose (Fig. 164). Vertex short, triangular, concave, with sharp margins ; foveolae elongate, rectangular, partly visible from above. Pronotum short, intersected before its middle by one transverse sulcus. Prozona shorter than metazona. Median carina slightly raised ; lateral carinae parallel in prozona, strongly diverging in metazona. Mesosternal interspace much wider than mesosternal lobe.
Tegmina fully developed, extending only slightly beyond hind knees. Hind femur wide and hairy (Fig. 165). Arolium very small.
Distribution : North Africa — Morocco to Egypt.
Five species known ; one in our region.

Notopleura saharica Krauss, 1902

Figs, 164, 165

Type Locality : 'Algeria' (Berlin Museum).

Notopleura saharica Krauss H. A., 1902, *Verh. zool.-bot. Ges. Wien*, 52 (4) : 231, 241.
Notopleura saharica —. Uvarov B. P., 1924, *Tech. Bull. Minist. Agric. Egypt*, 41 : 20.
Notopleura saharica —. Dirsh V. M., 1961, *Bull. Br. Mus. nat. Hist. (Ent.)*, 10 (9) : 413.
Notopleura saharica —. Pener M. P., 1966, *Israel J. Ent.*, 1 : 191, 192.

Small, robust, with numerous rugulae and tubercles on head and pronotum. Head almost hypognathous or slightly opisthognathous in male. Frontal ridge raised ; margins diverging, then converging towards median ocellus and again diverging and becoming obliterated towards clypeus. Vertex concave ; anterior margins sharp (Fig. 164). Foveolae trapezoidal, their length twice their maximum width. Eyes globular, in male their vertical diameter longer than subocular groove, in female shorter. Pronotum short, rugose anteriorly, punctate posteriorly. Prozona shorter than metazona ; lateral carinae parallel in anterior part of prozona (Fig. 164), diverging in metazona. Lateral lobes of pronotum slightly raised, their ventro-posterior margin rounded. Prosternum swollen between forelegs ; mesosternal interspace 2.0–2.5 times wider than mesosternal lobe. Wings extending beyond hind knees or shorter.
Tegmina wide ; median field closed ; intercalary vein absent ; radial vein projecting. Hind femur hairy, its length 2.5–2.8 times its maximum width (Fig. 165). Hind tibia with eight or nine spines on outer line and 11 or 12 on inner line. Inner spurs of hind tibia about half the length of first tarsal segment. Arolium minute.

Coloration : Sandy-brown ; face and dorsal part of pronotum pale ; tegmina with sparse brown blotches. Inner surface of hind femur yellowish, with or without a brown band ; upper part with a dark band, sometimes triangular in form. Hind tibia blackish on inner surface close to the knee, the rest pale bluish.
Measurements (mm) : Body ♀ 15.0, ♂ 10.5 ; tegmina ♀ 11.5, ♂ 8.5.
Distribution : Tunisia, Algeria, Egypt, Israel.
Israel : Central Negev (17 ; Hamachtesh Hagadol, 1962, three specimens collected).

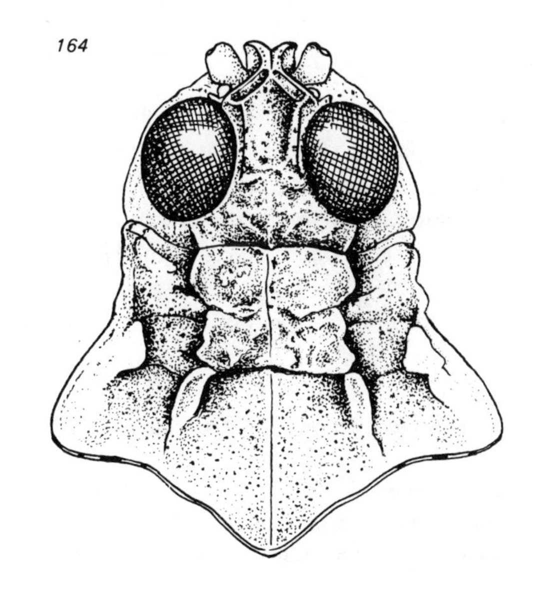

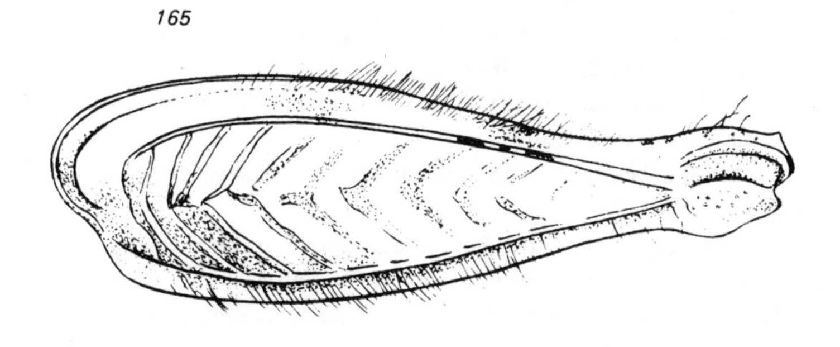

Figs. 164–165 : *Notopleura saharica* Krauss, 1902, female
164. head and pronotum ; 165. hind femur

Subfamily OEDIPODINAE

Diagnosis: Body strong, mostly granulate and hairy or nearly smooth. Head hypognathous, rarely moderately opisthognathous *(Aiolopus)*. Antennae always filiform. Foveolae usually absent; if present, they are lateral and never contiguous in front, triangular or trapezoidal in form, never square (Figs. 215, 218). Lateral carinae of pronotum absent or indistinct, forming shoulders only in metazona. Median carina always present, often raised and roof-shaped. Wings fully developed, extending beyond hind knees in most species, usually bright red, blue or yellow, and frequently with a black or brown band. Intercalary vein of tegmina always present, often serrate. Hind legs usually short and strong. Arolium small, half the length of the claws or shorter.

Coloration: Generally cryptic, varying in different habitats, but rarely with green spots. Mainly geophilous species, typical of arid zones, where they constitute the largest population of Acridoidea, particularly the genera *Crinita* Dirsh *Sphingonotus* Fieber, *Leptopternis* Saussure and *Scintharista* Saussure.

Many species are found in specific habitats, e.g., sand *(Hyalorrhipis)* or rocky areas *(Crinita)*.

Sixteen genera in our region, most of them of East African origin.

Key to the Genera of Oedipodinae in Israel

1. Median carina of pronotum intersected by two or three transverse sulci. Carina often obliterated in the middle, so that the pronotum is saddle-shaped (Fig. 186) — 2
- Median carina intersected by one transverse sulcus. Carina usually complete, raised above pronotum, often forming a roof-like structure (Fig. 224) — 7
2. Median carina intersected by two transverse sulci — 3
- Median carina intersected by three transverse sulci — 4
3. Posterior margin of pronotum rounded (Fig. 167); posterior transverse sulcus situated in centre of pronotum. Hind femur with two black triangular spots dorsally. Basal parts of wings red, with a black band. Body and legs hairy. — **Acrotylus** Fieber
- Pronotum with acute posterior margin (Fig. 170); posterior transverse sulcus situated beyond the middle. Wings transparent. — **Crinita** Dirsh
4. Spurs of hind tibia very long; inner pair longer than half the length of first tarsal segment (Fig. 173) — 5
- Spurs not very long; shorter than half the length of first tarsal segment — 6
5. Face rectangular, widening ventrally (Fig. 172). Spurs of hind leg longer than first tarsal segment (Fig. 173). Coloration of body sandy, with small brown spots. — **Hyalorrhipis** Saussure
- Head narrow in front, markedly higher than wide, not widening ventrally (Fig. 175). Spurs of hind leg not longer than first tarsal segment. Light and dark lines present on body. — **Leptopternis** Saussure

6. Tympanal lobe small, covering less than a third of tympanal opening. Abdominal terga with dense vertical carinulae (Fig. 178). **Pseudoceles** Bolivar
– Tympanal lobe distinct, covering at least a third of tympanal opening, sharply isolated. Abdominal terga without vertical carinulae 7
7. Wings bluish, without dark band ; main veins thickened ; veins of second lobe of wing distinctly converging ; third lobe with an accessory vein (Fig. 180). **Helioscirtus** Saussure
– Wings coloured or colourless ; if coloured a black band is always present, sometimes covering a large part of the wing (Fig. 191) ; venation regular, without thickened veins and accessory vein. **Sphingonotus** Fieber
8. Pronotum with numerous longitudinal sulci on both sides of the median carina (Fig. 208). Wings yellowish with a dark area. **Morphacris** Walker
– Dorsum of pronotum without longitudinal sulci 9
9. Median carina of pronotum distinctly and deeply intersected by posterior transverse sulcus (Fig. 210) 10
– Median carina complete, sometimes intersected slightly and indistinctly by a transverse sulcus 12
10. Dorsal carina of hind femur indented in its posterior third, forming a ledge (Fig. 212). **Oedipoda** Latreille
– Hind femur not indented, gradually narrowing towards apex 11
11. Foveolae distinct, rectangular, trapezoidal, nearly contiguous in front (Fig. 218). Head opisthognathous. Wings transparent, without black band. **Aiolopus** Fieber
– Foveolae reduced, very small. Base of wings coloured, with black band (Figs. 219, 220). **Mioscirtus** Saussure
12. Abdominal segments 1–5, black (Fig. 221). Hind tibia black, much shorter than femur. Three distinct black spots on tegmina. Eyes round. **Hilethera** Uvarov
– All abdominal segments of the same colour. Hind tibia not shorter than femur 13
13. Median carina of pronotum strongly raised, helmet-like, projecting anteriorly above head (Fig. 224). Wings with a black band branching radially towards base. **Pyrgodera** Fischer-Waldheim
– Median carina lower, not forming a helmet-like structure. Black band on wing, if present, without radial branch towards base 14
14. Wings transparent. Median carina of pronotum distinct, intersected slightly by transverse sulcus (Fig. 226). Sternum densely covered with short hairs. Large, mainly brown with green parts. **Locusta** L.
– Wings coloured ; a black band present 15
15. An X-shaped marking present on pronotum, sometimes extending in two pale lines on head to base of eyes (Fig. 230). Inner side of hind femur light or with two or three black bands. **Oedaleus** Fieber
– Pronotum without X-shaped marking (Fig. 232). Inner side of hind femur black, with a white band before knee. **Scintharista** Saussure

Genus ACROTYLUS Fieber, 1853

Lotos, 3 : 125

Acrotylus —. Innes W., 1929, *Mém. Soc. R. ent. Égypte*, 3 (2) : 45.

Type Species : *Gryllus insubricus* Scopoli, 1786.

Diagnosis : Small to medium sized, hairy. Head hypognathous (Fig. 166). Eyes round. Frontal ridge with raised margins, gradually extending ventrally. Foveolae triangular, granulate. Pronotum short, saddle-shaped, intersected by two transverse sulci (Fig. 167). Metazona larger than prozona ; median carina distinct along its entire length.

Tegmina narrow and long, transparent, with a few dark spots on the apical part. Basal part of wing red, with a dark semi-circular band in the middle (Fig. 168).

Distribution : Eastern and Northern Africa, Southern and Eastern Europe (U. S. S. R. and Central Asia to Kashmir).

Two species in Israel.

Key to the Species of Acrotylus in Israel

1. Body delicate, smooth. Antennae 1.5 times as long as head and pronotum together ; basal and middle segments flat, elongate, 2–3 times longer than wide. Dark band on wing nearly reaching inner margin. Posterior margin of pronotum rounded (Fig. 167). . **A. patruelis** (Herrich-Schäffer)

– Body granulate. Antennae shorter ; in female shorter than head and pronotum together, in male slightly longer ; basal and middle segments square or only slightly longer than wide. Posterior margin of pronotum obtuse-angled. **A. insubricus** (Scopoli)

Acrotylus patruelis (Herrich-Schäffer, 1838)

Figs. 166–168 ; Plate IV : 3 ; colour plate : 6

Type Locality : Unknown.

Gryllus patruelis Herrich-Schäffer G. A. W., 1838, *Fauna insectorum Germanica initia oder Deutschlands-Insecten, etc.*, Heft 157, pl. 18.

Oedipoda (Acrotylus) patruelis —. Fieber F. X., 1853, *Lotos*, 3 : 126.

Acrotylus patruelis —. Walker F., 1871, *Catalogue of the specimens of Dermaptera Saltatoria in the Collection of the British Museum*, London, Suppl., Part V, p. 74.

Acrotylus patruelis —. Buxton P. A. & B. P. Uvarov, 1923, *Bull. Soc. R. ent. Égypte*, p. 194.

Acrotylus patruelis —. Innes W., 1929, *Mém. Soc. R. ent. Égypte*, 3 (2) : 50.

Acrotylus patruelis —. Bodenheimer F. S., 1935, *Arch. Naturgesch.*, 4 (2) : 192.

Acrotylus patruelis —. Bei-Bienko G. Ya. & L. L. Mishchenko, 1951, *Locusts and Grasshoppers of the U. S. S. R. and Adjacent Countries*, Moscow, II : 595 [in Russian].

Acrotylus patruelis —. Ramme W., 1951, *Mitt. zool. Mus. Berl.*, 27 : 426.

Slender, hairy, especially ventrally and on hind legs. Head hypognathous ; foveolae triangular, visible from above. Eyes round, as long as subocular groove. Antennae longer than head and pronotum together ; most segments 2–3 times longer than wide. Pronotum saddle-shaped ; metazona 1.5 times as long as prozona, with distinct median carina, raised shoulders and rounded posterior margin (Fig. 167). Lateral lobes rounded ventrally. Mesosternal and metasternal interspaces twice as wide as long.

Tegmina slightly curved, usually extending beyond middle of hind tibia ; intercalary vein in middle of median field straight. Arolium relatively large, half the length of the claws.

Coloration : Dark brown or partly pale, with a black and white blotch on lateral lobe of pronotum. Three dark bands on upper and inner side of femur ; knees dark. Base of wings reddish, with a dark band which usually reaches the inner margin ; two or three dark specks at the apex (Fig. 168).

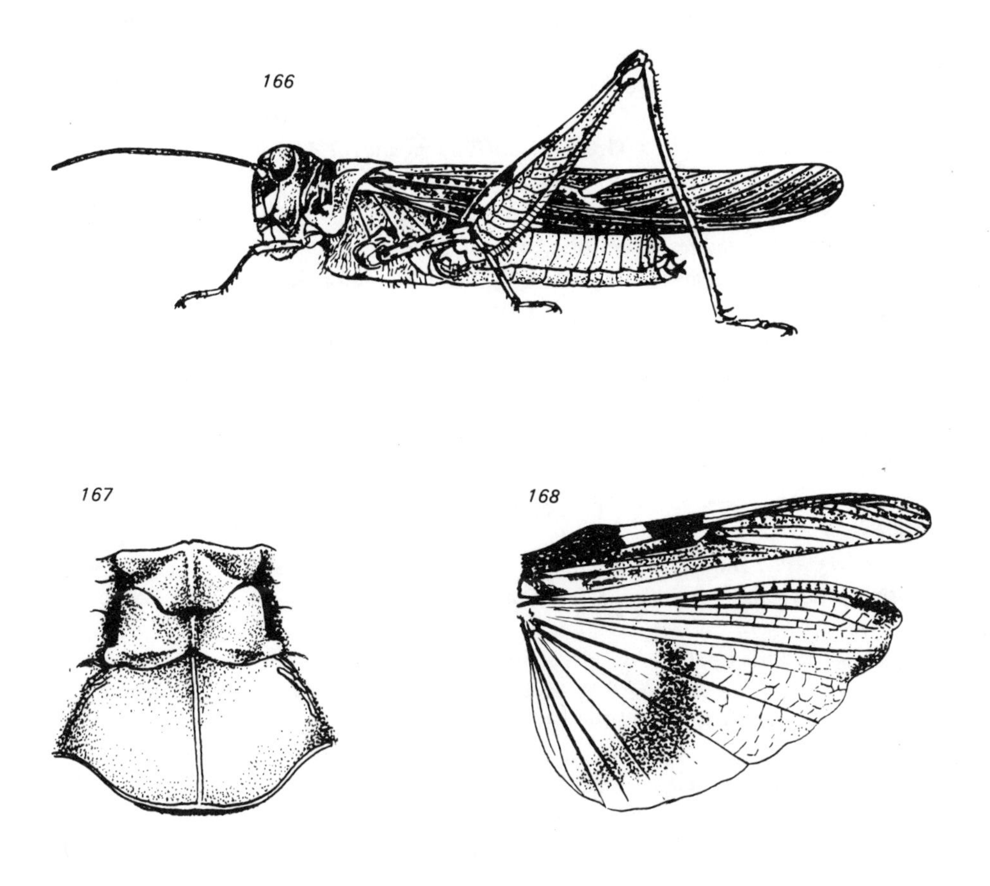

Figs. 166–168 : *Acrotylus patruelis* (Herrich-Schäffer, 1838)
166. female ; 167. pronotum ; 168. wings

Measurements (mm): Body ♀ 20.0–25.0, ♂ 16.5–17.5; tegmina ♀ 23.0–25.0, ♂ 19.0–21.0.
Distribution: North Africa, South Europe, Asia Minor, Caucasus.
Israel: Jordan Valley (7), Coastal Plain (4, 8, 9).
It inhabits areas with low, sparse, grassy vegetation. A typical geophile, resting on the soil, digging with hind legs. During courting season, it emits sounds while on ground. Copulation and oviposition from June to August.

Acrotylus insubricus (Scopoli, 1786)
Fig. 169

Type Locality: 'Northern Italy'.

Gryllus insubricus Scopoli J. A., 1786, *Deliciae faunae et florae insubricae*, fasc. 1, p. 64, pl. 24, fig. D.
Acrydium maculatum Olivier G. A., 1791, *Encyclopédie Méthodique, Histoire naturelle, Dictionnaire des Insectes,* Paris, 6: 224.
Gryllus fasciatus Fabricius J. C., 1793, *Entomologia Systematica*, 2: 58.
Gryllus maculatus —. Thunberg C. P., 1815, *Mém. Acad. Sci. St.-Pétersb.*, 5: 231.
Oedipoda maderae (?) Serville J. G. A., 1838, *Histoire naturelle des Insectes*, in: Roret, *Collection des Suites à Buffon. Orthoptères*, Paris, p. 730.
Acridium insubricum —. Brullé A., 1840, Orthoptera, in: Webb P. B. & S. Berthelot, *Histoire naturelle des Iles Canaries*, 2 (2): 78, pl. 5, figs. 11, 11a.
Oedipoda (Acrotylus) insubrica —. Fieber F. X., 1853, *Lotos*, 3: 125.
Oedipoda variegata Walker F., 1870, *Zoologist*, (2) 5: 2301.
Oedipoda maderae —. Walker F., 1870, *Catalogue of the specimens of Dermaptera Saltatoria in the Collection of the British Museum*, London, Part IV, p. 736.
Acrotylus insubricus —. Walker F., 1871, *ibid.*, suppl., Part V, p. 74.
Oedipoda insubrica —. Hart H. C., 1891, *Fauna and Flora of Sinai, Petra and Wadi Arabah*, London, p. 183.
Acrotylus insubricus —. Giglio-Tos E., 1893, *Boll. Musei Zool. Anat. comp. R. Univ. Torino*, 8 (164): 6.
Acrotylus versicolor Burr M., 1898, *Trans. R. ent. Soc. Lond.*, 1898: 50.
Acrotylus insubricus —. Krauss H. A.,1909, in: Kneucker A., *Verh. naturw. Ver. Karlsruhe*, 21: 105.
Thalpomena maderae —. Burr M., 1912, *Entomologist's Rec. J. Var.*, 24: 31.
Acrotylus insubricus —. Buxton P. A. & B. P. Uvarov, 1923, *Bull. Soc. R. ent. Égypte*, p. 194.
Acrotylus insubricus —. Innes W., 1929, *Mém. Soc. R. ent. Égypte*, 3 (2): 46, 47.
Sphingonotus variegatus —. Innes W., 1929, *ibid.*, 3 (2): 56, 70.
Acrotylus insubricus —. Bodenheimer F. S., 1935, *Animal Life in Palestine*, Jerusalem, pp. 86, 88, 89, 311, 320, 322, 323.
Acrotylus insubricus —. Bodenheimer F. S., 1935, *Arch. Naturgesch.*, 4 (2): 191.
Acrotylus insubricus —. Ramme W., 1951, *Mitt. zool. Mus. Berl.*, 27: 426.

Medium sized, slightly rugose and hairy. Head hypognathous or slightly opisthognathous. Antennae flat, shorter or slightly longer than head and pronotum to-

together ; basal segments square or slightly elongate. Pronotum rugose, raised along the median carina in prozona, roof-shaped, with low shoulders in metazona, obtuse-angled posteriorly.

Tegmina narrow ; intercalary vein slightly obliterated. Arolium minute, less than half the length of the claws.

Coloration : Cryptic. Wings with a dark band, usually not reaching the inner margin ; dark blotches absent or rarely present on apex. On lateral lobe of pronotum a black shiny blotch, with a white spot inside, extending anteriorly over the gular region of the head. Hind femur with two or more triangular dark blotches on upper margin, and traces of dark bands on inner margin.

Measurements (mm) : Body ♀ 19.5–22.5, ♂ 15.0–16.0 ; tegmina ♀ 21.0–24.0, ♂ 16.0–17.5.

Distribution : North Africa, South Europe to Asia Minor, Caucasus, South Central Asia.

Israel : Upper Galilee (1), Jordan Valley (7), Coastal Plain (4, 8, 9), Northern Sinai (20), Judean Hills (11), Northern Negev (15).

A very euryoecous geophilous species, occurring in bare habitats with very sparse low vegetation and in cultivated fields. Apparently passes the winter as adults ; first hoppers found from April to May. During the summer, it is the most common species in cultivated habitats.

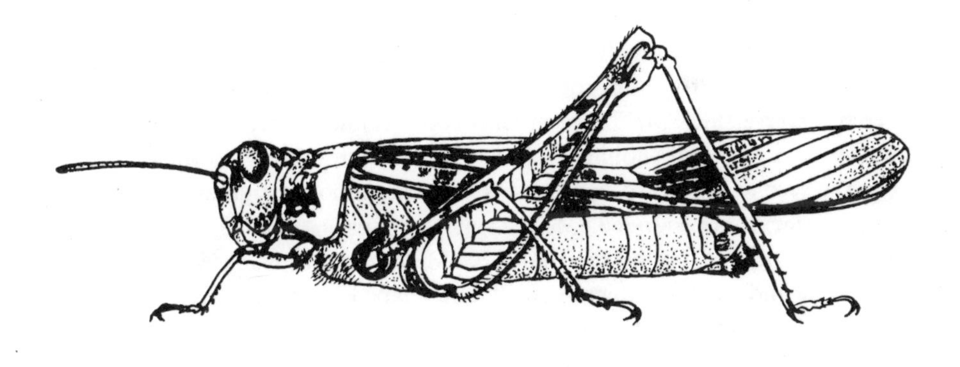

Fig. 169 : *Acrotylus insubricus* (Scopoli, 1786), female

Genus CRINITA Dirsh, 1949

Trans. R. ent. Soc. Lond., 100 (13) : 390, 391

Thalpomena —. Uvarov B. P., 1923, *Entomoligst's mon. Mag.*, 59 : 84.

Type Species : *Thalpomena hirtipes* Uvarov, 1923.
Diagnosis : Short, hairy. Head large, slightly opisthognathous, projecting above pronotum. Antennae filiform. Eyes round ; interocular distance shorter than diameter of eye. Pronotum with a small prozona compressed laterally in the anterior part ; metazona more than twice as long as prozona and with a right-angled posterior margin.
Distribution : Africa and Asia.
One species in our region.

Crinita hirtipes (Uvarov, 1923)
Fig. 170 ; Plate IV : 4

Type Locality : Jericho.

Thalpomena hirtipes Uvarov B. P., 1923, *Entomologist's mon. Mag.*, 59 : 84.
Thalpomena hirtipes —. Buxton P. A. & B. P. Uvarov, 1923, *Bull. Soc. R. ent. Égypte*, p. 193.
Thalpomena hirtipes —. Bodenheimer F. S., 1935, *Arch. Naturgesch.*, 4 (2) : 191.
Crinita hirtipes —. Dirsh V. M., 1949, *Trans. R. ent. Soc. Lond.*, 100 (13) : 391.

Medium sized, short ; very hairy, especially thorax and hind femur. Head slightly opisthognathous. Eyes oval, in female as long as subocular groove, in male 1.5 times longer. Frontal ridge with a distinct carina depressed from median ocellus downwards. Fastigium of vertex with a sharp carina, at least twice as long as the minimum interocular distance. Pronotum roof-shaped, angular posteriorly, intersected by two or three transverse sulci. Metazona at least twice as long as prozona, with a distinct median carina and raised shoulders. A deep depression present on the median line, between the second and third transverse sulci. Lateral lobes high and narrow, their ventro-posterior angle rounded. Mesosternal interspace twice as wide as long, its width equal to that of the mesosternal lobe. Metasternal interspace square.
Tegmina wide and rounded apically, with an S-shaped intercalary vein and a radius-sector with two branches. Hind tibia with short spurs ; inner spur as long as or slightly longer than the external one. Arolium one-half or one-third the length of the claws.
Coloration : Generally pale brown, with brown blotches on tegmina. Face usually whitish. Inner side of hind femur with two black bands ; knee and adjacent part of tibia black with two additional black rings on the tibia.
Measurements (mm) : Body ♀ 19.5–22.5, ♂ 15.0–16.5 ; tegmina ♀ 20.5–21.5, ♂ 15.5–16.5.
Distribution : Endemic in Israel.

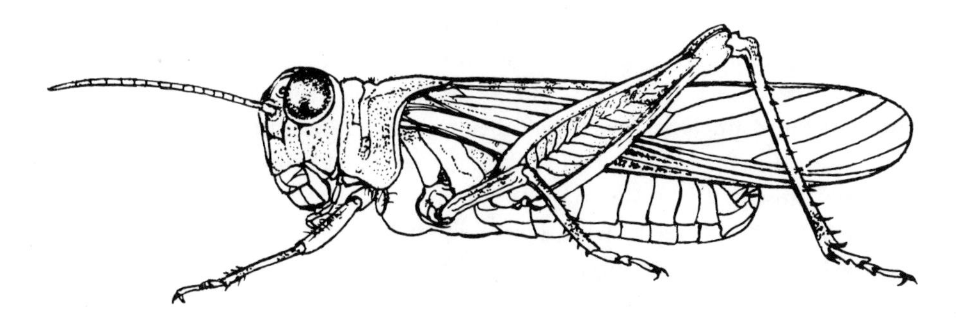

Fig. 170 : *Crinita hirtipes* (Uvarov, 1923), male

Israel : Jordan Valley (7), Dead Sea Area (13), Judean Hills (11), Central Negev (17). A typical cryptic-coloured petrobiont, found usually from May to October in bare, stony areas. It emits loud, short chirps during flight.

Genus HYALORRHIPIS Saussure, 1884

Mém. Soc. Phys. Hist. nat. Genève, 28 (9) : 198, 210

Hyalorhipis —. Kirby W. F., 1910, *A synonymic catalogue of the Orthoptera*, Vol. III, Orthoptera Saltatoria, Part II, London, p. 279.
Hyalorrhipis —. Innes W., 1929, *Mém. Soc. R. ent. Égypte*, 3 (2) : 41, 83.

Type Species : *Oedipoda clausii* Kittary, 1849.
Diagnosis : Slender, smooth. Head widening slightly towards the mouth (Fig. 172). Eyes round ; interocular distance less than their diameter.
Tegmina narrow ; wings delicate, transparent. Spurs of hind tibia long, as long as or longer than first tarsal segment (Fig. 173).
Distribution : North Africa, South Europe, Central Asia, Iran. Inhabits sandy deserts.
One species in our region.

Hyalorrhipis calcarata (Vosseler, 1902)

Figs. 171–173 ; colour plate : 8

Type Locality : Bou Saada, Algeria (Stuttgart Museum).

Leptopternis calcarata Vosseler J., 1902, *Zool. Jb.* (Syst.), 16 : 382, pl. 18, figs. 9a, 9b, 10.
Hyalorhipis calcarata —. Kirby W. F., 1910, *A synonymic catalogue of the Orthoptera*, Vol. III, Orthoptera Saltatoria, Part II, London, p. 280.
Hyalorrhipis calcarata —. Salfi M., 1929, *Mém. Soc. ent. ital.*, 6 (1927) : 161, fig. 1.

Medium sized, smooth, slightly flattened. Head of female hypognathous, that of male slightly opisthognathous. Ventral part of face wider than dorsal part (Fig. 172). Antennae filiform, long; length of segments at least twice their width. Eyes round; interocular distance much narrower than their diameter. Pronotum narrow in prozona, saddle-shaped in metazona; median carina intersected by three transverse sulci; lateral carinae absent. Lateral lobe of pronotum with projecting angles (Fig. 171). Mesosternal interspace twice as wide as mesosternal lobes.

Tegmina narrow, rounded apically, with a strong intercalary vein in the median field and two branches on the radius-sector. Wings transparent. Spurs of hind tibia very long, as long as or longer than first tarsal segment (Fig. 173). Arolium absent.

Coloration: Sandy with numerous brown spots, rarely with brown bands on the pronotum.

Measurements (mm): Body ♀ 22.0–22.5, ♂ 14.5–15.5; tegmina ♀ 21.5–22.0, ♂ 15.0–16.0.

Distribution: North Africa, Israel.

Israel & Sinai: Southern Coastal Plain (9), 'Arava Valley (14), Northern Sinai (20).

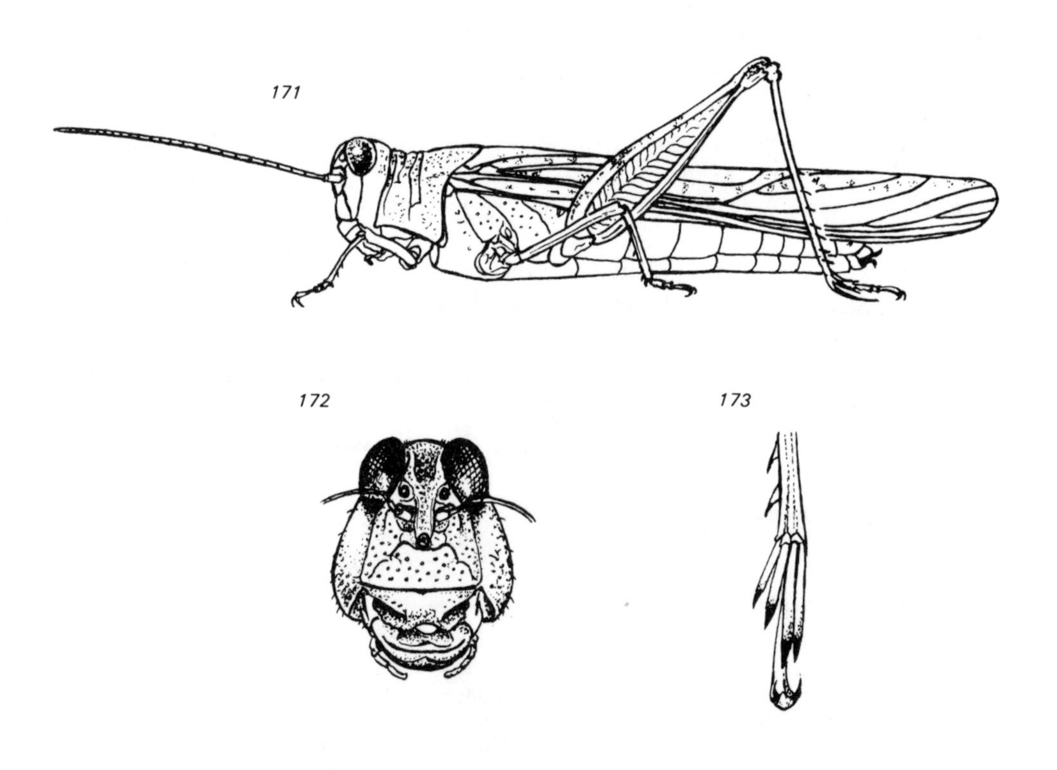

Figs. 171–173: *Hyalorrhipis calcarata* (Vosseler, 1902)
171. female; 172. head, frontal view; 173. posterior spurs and tarsus

It occurs only on sand dunes, blending well with its habitat. On sensing danger, this grasshopper immediately flies away, but usually does not cover a distance of more than one or two metres. When alighting on sand, it digs itself a groove in which to hide by throwing sand backwards and sideways with its hind legs and by executing wriggling body movements. When digging *H. calcarata* positions itself parallel to the sun's rays, thus minimizing the length of its shadow and preventing overheating. It advances on the sand by short jumps of 10–15 cm according to the following procedure : digging in slightly with the hind tarsi, jumping, digging in again, etc., thus leaving tracks like those of a miniature hopping bird or a miniature jerboa. Found during the hot summer months, from May to September ; hoppers during April and May.

Genus LEPTOPTERNIS Saussure, 1884

Mém. Soc. Phys. Hist. nat. Genève, 28 (9) : 193, 196, 198, 209

Sphingonotus (Leptopternis) Saussure H. de, 1884, *Mém. Soc. Phys. Hist. nat. Genève*, 28 (9) : 193, 196, 198, 209.
Leptopternis —. Saussure H. de, 1888, *ibid.*, 30 (1) : 24, 88.

Type Species : *Oedipoda gracilis* Eversmann, 1848.
Diagnosis : Smooth, slender, laterally compressed. Head hypognathous or slightly opisthognathous, narrow and high, dorsal part not wider than the ventral, projecting above pronotum (Fig. 174). Antennae filiform, longer than head and pronotum together. Pronotum compressed in middle ; prozona shorter than metazona. Median carina obliterated in the middle, distinct in metazona. Posterior margin of pronotum obtuse (Fig. 176). Mid femur 1.5 times as long as front femur. Spurs of hind tibia not longer than first tarsal segment.
Coloration : Sandy-grey with brown spots ; often with a white marking on head and wings.
Distribution : Sandy deserts throughout East Asia and North Africa.
Two species in our region.

Key to the Species of Leptopternis in Israel and Sinai

1. Wings with a black blotch or band which sometimes extends to the anterior margin (Fig. 176). **L. maculata** Vosseler
– Wings transparent, without markings (Fig. 177). **L. gracilis** (Eversmann)

Leptopternis maculata Vosseler, 1902
Figs. 174–176

Type Locality : Bou Saada, Algeria (Stuttgart Museum).

Leptopternis maculata Vosseler J., 1902, *Zool. Jb.* (Syst.), 16 : 380, pl. 17, figs. 14a, 14b, 15.
Sphingonotus acrotyloides Werner F., 1908, *Zool. Anz.*, 32 : 715.
Hyalorrhipis maculata —. Salfi M., 1929, *Mém. Soc. ent. ital.*, 6 (1927) : 151.
Hyalorrhipis maculata —. Bodenheimer F. S., 1935, *Animal Life in Palestine*, Jerusalem, p. 320.
Hyalorrhipis maculata —. Bodenheimer F. S., 1935, *Arch. Naturgesch.*, 4 (2) : 196.

Smooth, small. Head slightly opisthognathous, raised above pronotum. Frontal ridge concave, with a raised carinula (Fig. 175). Eyes round, as long as subocular groove. Antennae filiform, longer than head and pronotum together. Pronotum with short prozona, slightly compressed anteriorly along the median carina. Between the second and third transverse grooves there are two small tubercles. Posterior margin of pronotum obtuse-angled ; posterior angle of lateral lobe obtuse or with a short process.

Tegmina rounded at apex, with a more or less straight intercalary vein and a radius-sector with one branch. Wings transparent, with a large black blotch or a band extending to the anterior margin (Fig. 176). Hind tibia with strong spines ; spurs as long as first tarsal segment.

Coloration : Sandy or pale brown, matching colour of environment.

Measurements (mm) : Body ♀ 16.5–17.5, ♂ 12.0–12.5 ; tegmina ♀ 16.5–17.0, ♂ 11.5–12.0.

Distribution : North Africa, Israel.

Israel : Northern and Central Coastal Plain (4, 8).

Like *Hyalorrhipis*, this species also digs well, using its hind legs and spending the hot hours of the day almost completely buried in the sand. A typical summer species ; first hoppers found during June–July. Oviposition in July and August, usually near stems or bushes at a depth of 3–5 cm. Eight to 15 eggs in egg-pod. In captivity a female deposits 4–6 egg-pods.

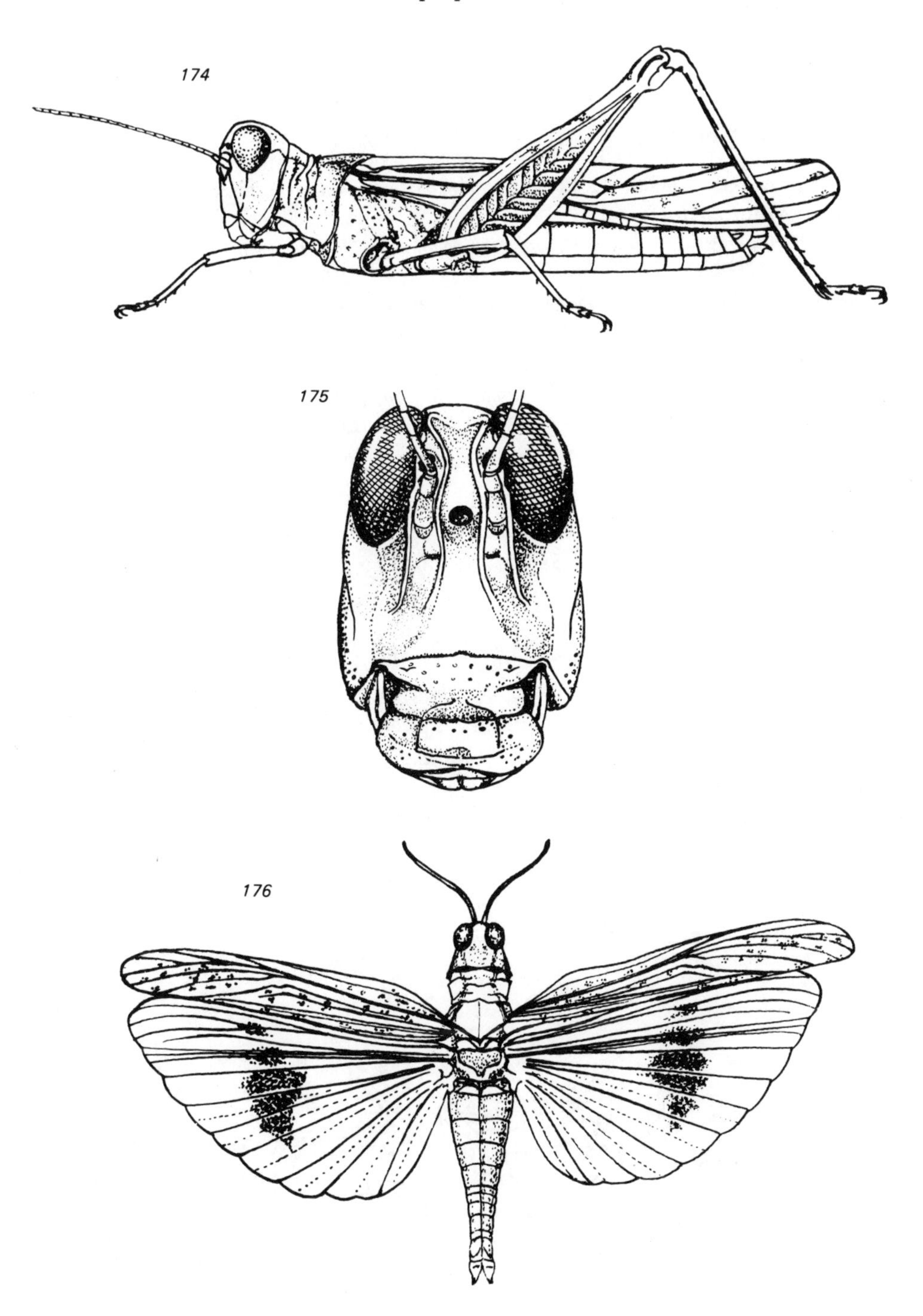

Figs. 174–176 : *Leptopternis maculata* Vosseler, 1902
174. female ; 175. head, frontal view ; 176. female with open wings

Leptopternis gracilis (Eversmann, 1848)
Fig. 177 ; Plate IV : 5

Type Locality : 'Songoria'.

Oedipoda gracilis Eversmann E., 1848, *Additamenta quaedam levia ad Fischeri de Waldheim Orthoptera rossica*, Moscow, p. 10.
Sphingonotus gracilis —. Saussure H. de, 1884, *Mém. Soc. Phys. Hist. nat. Genève*, 28 (9) : 198, 210.
Leptopternis gracilis —. Saussure H. de, 1888, *ibid.*, 30 (1) : 88.
Sphingonotus grobbeni Werner F., 1905, *Sber. Akad. Wiss. Wien*, 114 (1) : 361, 418.
Leptopternis gracilis —. Innes W., 1929, *Mém. Soc. R. ent. Égypte*, 3 (2) : 87.
Sphingonotus grobbeni —. Innes W., 1929, *ibid.*, 3 (2) : 55, 59.
Leptopternis gracilis —. Bodenheimer F. S., 1935, *Animal Life in Palestine*, Jerusalem, p. 320.
Leptopternis gracilis —. Bodenheimer F. S., 1935, *Arch. Naturgesch.*, 4 (2) : 196.

Smooth, small to medium sized. Resembling *L. maculata*, but without dark coloration on the wings. With pale brown lines on head, pronotum and tegmina.
Measurements (mm) : Body ♀ 16.5–17.5, ♂ 12.0–12.5 ; tegmina ♀ 16.5–17.0, ♂ 11.5–12.0.
Distribution : North Africa, Asia.
Israel : Negev (15, 16, 17), Coastal Plain (8, 9), Northern and Central Sinai (20, 21). Found on loose sand together with *Hyalorrhipis* in the Southern Coastal Plain and with *L. maculata* in the Northern Coastal Plain.
The three species together with *Tenuitarsus angustus* constitute a typical group of psammobiotic forms. *L. gracilis* is also a typical summer species ; the first hoppers are found in March and April.

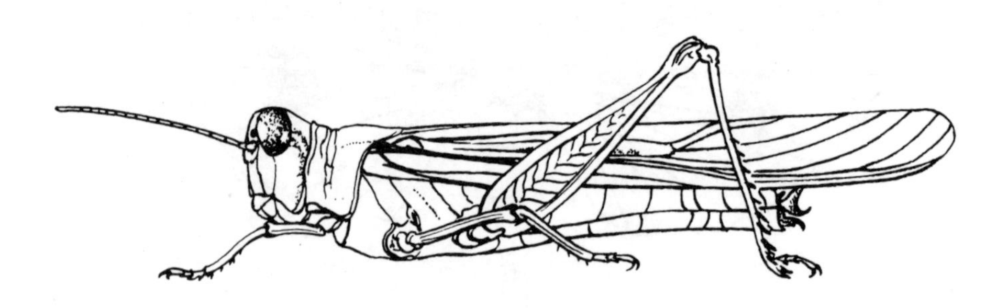

Fig. 177 : *Leptopternis gracilis* (Eversmann, 1848), female

Genus PSEUDOCELES Bolivar, 1899

Annls Soc. r. ent. Belg., 43 : 593

Thalpomena Saussure H. de, 1884, *Mém. Soc. Phys. Hist. nat. Genève*, 28 (9) : 184.

Type Species : *Pseudoceles oedipodioides* Bolivar, 1899.
Diagnosis : Medium sized. Head hypognathous or slightly opisthognathous, raised above pronotum (Fig. 178). Frontal ridge flat, punctate on upper part. Vertex depressed, with distinct lateral carinae ; foveolae triangular, flat. Eyes oval, their vertical diameter as long as subocular groove. Pronotum with a low median carina usually intersected by three transverse sulci.
Tegmina wide, usually monochromatic ; venation at apex sparse. Dark band rarely present on wings. Tympanal opening large, with a weakly separated tympanal membrane which covers less than a third of the opening. Last abdominal sternite of female dark bluish.
The genus includes about 10 species distributed in Iran, Central Asia and in the Caucasus. Only one species in our region.

Pseudoceles ebneri Dirsh, 1949
Fig. 178.

Type Locality : 'Syria, Lebanon'.

Pseudoceles ebneri Dirsh V. M., 1949, *Trans. R. ent. Soc. Lond.*, 100 (13) : 377.
Pseudoceles palaestinus Dirsh, V. M., 1949, *ibid.*, 100 (13) : 380.

Medium sized, robust, granulate. Head of female hypognathous, of male slightly opisthognathous ; frontal ridge flat or slightly sunken around the median ocellus, densely punctate dorsally. Fastigium pentagonal, elongate, concave with raised margins, its median carinulae not continuous on the vertex. Foveolae irregularly triangu-

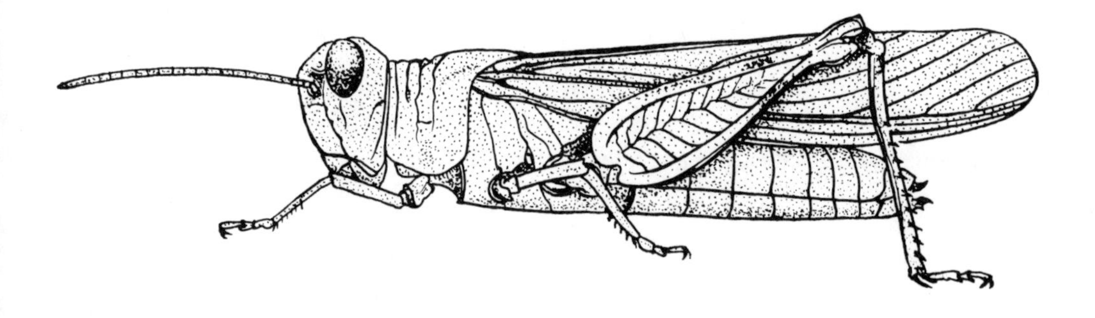

Fig. 178 : *Pseudoceles ebneri* Dirsh, 1949, female

lar. Antennae of female longer than head and pronotum together. Anterior part of pronotum laterally compressed, with a slightly raised median carina, especially in the prozona. Median carina intersected by three transverse sulci, all situated anterior to middle of pronotum. Metazona densely granulate, its posterior margin angular. Sternum hairy; mesosternal interspace 1.5 times longer than wide; metasternal interspace usually square.

Tegmina of uniform colour; intercalary vein straight; radius-sector with two or three branches. Hind femur thick, its length 3.0–3.5 times its maximum width; inner surface black with a single light band before the apex.

Coloration: Usually uniformly brown, sometimes blackish. Wings bluish, with or without traces of a dark marginal area. Inner side of hind tibia dark bluish, with a light ring before the knee, or bluish-grey in males; the upper side rose or light bluish.

Measurements (mm): Body ♀ 21.5–25.0, 15.0–17.5; tegmina ♀ 20.0–21.5, ♂ 14.5–16.0.

Distribution: Syria, Lebanon, Israel.

Israel: Upper Galilee (1), Golan Heights (18), Mount Hermon (19).

Genus HELIOSCIRTUS Saussure, 1884

Mém. Soc. Phys. Hist. nat. Genève, 28 (9): 59, 193, 194

Vosseleria Uvarov B. P., 1923, *Konowia*, 2: 30.

Helioscirtus —. Innes W., 1929, *Mém. Soc. R. ent. Égypte*, 3 (2): 41, 81.

Type Species: *Helioscirtus moseri* Saussure, 1884.

Diagnosis: Medium sized, strong. Head hypognathous, projecting above pronotum. Eyes elongate. Antennae filiform; segments cylindrical, at least twice as long as wide. Metazona of pronotum at least twice as long as prozona, intersected by three transverse sulci; median carina distinct.

Tegmina long and wide, extending beyond hind knees. Wings transparent, with two anal veins distinctly converging (Fig. 180).

Distribution: Somalia, North Africa, Near East.

One species in our region.

Helioscirtus moseri tichomirovi Shchelkanovtsev, 1909

Figs. 179, 180; Plate V: 2

Type Locality: 'Persia'.

Helioscirtus moseri Saussure. Giglio-Tos E., 1893, *Boll. Musei Zool. Anat. comp. R. comp. R. Univ. Torino*, 8 (164): 6.

Helioscirtus moseri tichomirovi Shchelkanovtsev J. P., 1909, *Proc. Warsaw Univ.*, p. 40 (separate publication) [in Russian].

Helioscirtus tichomirovi ebneri Bodenheimer F. S., 1935, *Arch. Naturgesch.*, 4 (2) : 187.
Helioscirtus moseri —. Bodenheimer F. S., 1935, *ibid.*, 4 (2) : 187.
Helioscirtus moseri —. Ramme W., 1951, *Mitt. zool. Mus. Berl.*, 27 : 426.

Medium sized, strong. Head hypognathous ; frontal ridge distinct only on its upper part ; lower part of face flat. Antennae filiform, as long as or shorter than head and pronotum together ; segments cylindrical, their length twice or three times their width. Eyes round ; shorter than subocular groove ; carinulae present on fastigium of vertex. Pronotum flat, its lateral lobes higher than long, their ventro-posterior margin rounded (Fig. 179). Disc of pronotum flat, intersected by three transverse sulci, the posterior one situated before middle of pronotum, so that the metazona is three times as long as the prozona.
Tegmina wide, usually extending for a third of their length beyond the hind knees ; intercalary vein S-shaped and granulate ; radius-sector with three branches (Fig. 180). Wings transparent, with typical convergence of the two anal veins and an additional vein in the third field which does not reach the base of the wing. Mesosternal and metasternal interspaces wider than the bordering lobes. Arolium very short.

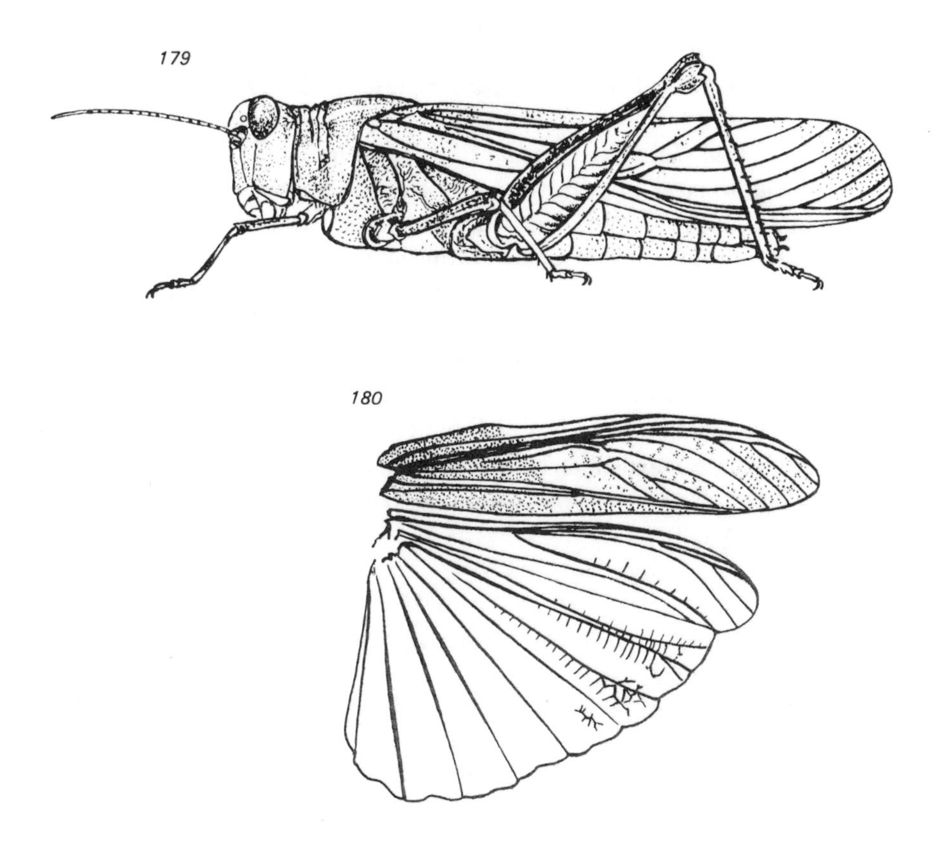

Figs. 179–180 : *Helioscirtus moseri tichomirovi* Shchelkanovtsev, 1909
179. female ; 180. wings

Coloration: Reddish-brown or dark brown, resembling the rocky environment which the species inhabits. Inner median field of hind femur brown; hind tibia yellowish.
Measurements (mm): Body ♀ 27.0–29.0, ♂ 27.5–28.0; tegmina ♀ 28.5, ♂ 27.0–28.5.
Distribution, Israel: Northern and Central Negev (15, 17), Dead Sea Area (13).
A typical geophilous species inhabiting the rocky deserts in the Dead Sea Area, Central Negev and continuing to the Sinai mountains. Active in the hot summer months. In flight it emits a loud sound resembling that of species of *Sphingonotus*.

Genus SPHINGONOTUS Fieber, 1852

In: Kelch A., *Grundlage zur Kenntnis der Orthopteren Oberschlesiens, etc.*, Ratibor, Bögner, p. 2.

Oedipoda (Sphingonotus) Fieber F. X., 1852, in: Kelch A., *Grundlage zur Kenntnis der Orthopteren Oberschlesiens, etc.*, Ratibor, Bögner, p. 2.
Sphingonotus —. Fieber F. X., 1853, *Lotos*, 3: 124.
Sphingonotus —. Fischer L. H., 1853, *Orthoptera Europaea*, Leipzig, pp. 52, 297, 401.
Sphingonotus —. Innes W., 1929, *Mém. Soc. R. ent. Égypte*, 3 (2): 41, 54–57.
Sphingonotus —. Mishchenko L. L., 1936, *Eos, Madr.*, 12 (1–2): 65–192.

Type Species: *Gryllus (Locusta) caerulans* Linnaeus, 1767.
Diagnosis: Small to large, robust, strong. Head hypognathous or slightly opisthognathous, raised above pronotum. Eyes slightly elliptical, the interocular space as long as their diameter. Pronotum with median carina intersected by three transverse sulci; metazona wider than prozona, its posterior margin obtuse or rounded. Wings colourless or coloured and with a black band. Intercalary vein distinct, straight or S-shaped. Subgenital plate of male conical, sharp. Valves of ovipositor widened at base, usually with a tooth-like process on the ventral valves.
Distribution: The genus, which includes about 115 taxa, is distributed throughout the world, including isolated islands like Galapagos, Cuba, Jamaica, usually in arid and semi-arid habitats with low and sparse vegetation.
There are 16 species in Israel.

Key to the Species of Sphingonotus in Israel and Sinai

1. Tegmina, between radius and media, with dense projecting cross-veins (Fig. 182). Spurious intercalary vein smooth. — **S. pictus onerosus** Mishchenko
– Dense projecting cross-veins between radius and media absent. Intercalary vein usually granulate — 2
2. Wings with distinct dark band (Fig. 191) — 3
– Wings without dark band — 9

3. Base of wings bluish or transparent 4
– Base of wings pinkish, yellow, green or red 7
4. Dark band on wings narrow (3–4 mm wide), covering less than one-third of wing. Base of wings colourless. Hind tibia yellow 5
– Band wide, especially in centre, covering more than one-third wing. Base of wings bluish 6
5. Large species (25.0–35.0 mm), with distinct black spot on apical part of wing (Fig. 183). Anterior part of prozona smooth ; inner part of hind femur entirely yellow or with a diffuse dark blotch. **S. savignyi** Saussure
– Smaller (15.0–20.0 mm), without black spot on apical part of wing. Prozona raised anteriorly along median carina (Fig. 186). Inner side of hind femur usually dark or with black band. lobe of hind knee usually black. **S. eurasius** Mishchenko
6. Wings with large apical black area (Fig. 187). Inner side of hind femur black, except the knee. Hind tibia dark blue. **S. obscuratus** (Walker)
– Wing without apical black area (Figs. 189, 191). Inner side of hind femur and hind tibia yellowish-grey. **S. angulatus** Uvarov
7. Base of wings yellow or greenish-yellow. Pronotum with strongly raised median carina on prozona (Fig. 192). **S. satrapes** Saussure
– Base of wings violet or red. Median carina not raised on anterior part of pronotum 8
8. Base of wings red, with a large dark spot on the apical part, visible through the transparent part of the tegmina (Fig. 193). Two distinct black transverse bands on tegmina. **S. octofasciatus** (Serville)
– Base of wings violet ; no dark region on the apical part. Tegmina without distinct bands, only darker in the basal part (Fig. 194). **S. balteatus latifasciatus** (Walker)
9. Inner side of hind femur mainly black with one or two light bands 10
– Inner side of hind femur light yellow with one or two dark bands or spots 11
10. Foveolae triangular or obliterated. Radius-sector of tegmina with three branches ; intercalary vein smooth (Fig. 195). Two light bands present on inner side of hind femur. **S. rubescens** (Walker)
– Foveolae trapezoidal. Radius-sector with two branches ; intercalary vein granulate. Inner side of hind femur black with one light band. **S. coerulans** (Linnaeus)
11. Mesosternal and metasternal interspaces 2.5–3 times wider than long (Fig. 197). **S. carinatus** Saussure
– Mesosternal and metasternal interspaces much narrower 12
12. Radius-sector of tegmina with three branches (Figs. 198, 200). Hind tibia sulphurous-yellow with a shiny round black spot near knee. **S. theodori** Uvarov
– Radius-sector with two branches. Hind tibia different 13
13. Hind femur thick, its length approximately three times its maximum width at the base. Cerci of male swollen at the base and curved (Fig. 204). **S. femoralis** Uvarov
– Hind femur slender, its length at least four times its maximum width. Cerci of male elongate, sharp, not curved 14
14. Tegmina with one or two distinct transverse dark bands, the anterior band oblique and with distinct boundaries ; intercalary vein in the middle of the median field, more or less straight 15
– Tegmina without transverse bands, ochraceous only at base. Intercalary vein near apical part of radius S-shaped, and below it a row of large irregular cells. **S. vosseleri** Krauss

15. Prozona of pronotum divided by first transverse sulcus into a wider anterior part and a narrower posterior part (Fig. 205). Regular square cells below apical part of intercalary veins. **S. hierichonicus** Uvarov
– Prozona divided by first transverse sulcus into two equal parts (Fig. 206). Irregular cells below the apical part of the intercalary vein. **S. laxus** Uvarov

Sphingonotus pictus onerosus Mishchenko, 1936
Figs. 181, 182

Type Locality : Mokattam Hills, Egypt.

Sphingonotus niloticus var. *picta* Werner F., 1905, *Sber. Akad. Wiss. Wien*, 114 (1) : 418 (partim).
Sphingonotus pictus onerosus Mishchenko L. L., 1936, *Eos, Madr.*, 12 (1–2) : 108.
Vosseleriana picta onerosa —. Bei-Bienko G. Ya. & L. L. Mishchenko, 1951, *Locusts and Grasshoppers of the U. S. S. R. and Adjacent Countries*, Moscow, II : 634 [in Russian].

Small, smooth ; hairs sparse. Head hypognathous, raised above pronotum. Eyes round, protruding laterally. Frontal ridge obliterated below median ocellus, especially in male. Antennae delicate ; segments 2.5–3 times longer than wide. Pronotum saddle-shaped, forming a small crest on the median carina, which is obtuse posteriorly and angular anteriorly. Transverse sulci distinct, the first situated in middle of

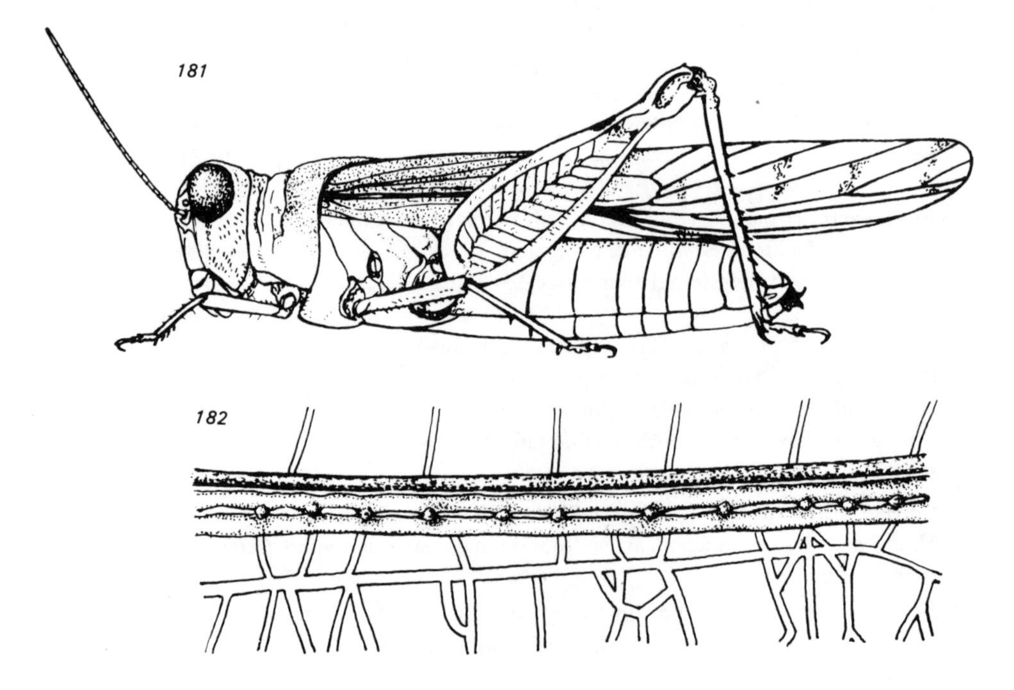

Figs. 181–182 : *Sphingonotus pictus onerosus* Mishchenko, 1936
181. female ; 182. field of cross-veinlets on tegmen

prozona, curved posteriorly. Transverse depression present between second and third sulci, often with raised margins. Postero-ventral angle of lateral lobes with short spiny processes. Sternum sparsely punctate. Mesosternal and metasternal interspaces more than twice as wide as long.

Tegmina narrow, their length more than five times their width; intercalary vein S-shaped. Protruding transverse veinlets between radius and media form the typical chirping organ (Fig. 182). Hind femur short and thick.

Coloration : Generally pale brown. Antennae with light rings ; tegmina ochraceous at the base, with or without transverse bands. Hind femur yellow on inner side and with a dark band on the posterior third. Wings transparent, sometimes bluish at the base with traces of a very diffuse dark band.

Measurements (mm) : Body ♀ 25.5–28.0, ♂ 15.5–16.5 ; tegmina ♀ 24.5–29.5, ♂ 18.0–18.5.

Distribution : From Pakistan, Persia, to Sinai.

Israel : Southern and Central Negev (16, 17), Dead Sea Area (13).

Sphingonotus savignyi Saussure, 1884

Figs. 183, 184

Type Locality : 'Egypt'.

Sphingonotus savignyi Saussure H. de, 1884, *Mém. Soc. Phys. Hist. nat. Genève*, 28 (9) : 198, 208.
Sphingonotus savignyi stirps *apicalis* Saussure H. de, 1884, *ibid.*, 28 (9) : 208.
Sphingonotus savignyi var. *major* Saussure H. de, 1888, *ibid.*, 30 (1) : 84.
Sphingonotus savignyi —. Krauss H. A., 1909, *Verh. naturw. Ver. Karlsruhe*, 21 : 36.
Sphingonotus savignyi —. Innes W., 1929, *Mém. Soc. R. ent. Égypte*, 3 (2) : 56, 69.
Sphingonotus savignyi —. Bodenheimer F. S., 1935, *Arch. Naturgesch.*, 4 (2) : 195.
Sphingonotus savignyi —. Mishchenko L. L., 1936, *Eos, Madr.*, 12 (1–2) : 95.
Sphingonotus savignyi —. Bei-Bienko G. Ya. & L. L. Mishchenko, 1951, *Locusts and Grasshoppers of the U. S. S. R. and Adjacent Countries*, Moscow, II : 626 [in Russian].
Sphingonotus savignyi —. Ramme W., 1951, *Mitt. zool. Mus. Berl.*, 27 : 425.

Large, slender ; hairs sparse. Head hypognathous, laterally compressed, projecting above pronotum (Fig. 183). Eyes small, round, their diameter as wide as the interocular distance. Foveolae minute, flat, pentagonal. Segments of antennae long and slender, punctate. Pronotum with distinct transverse sulci, constricted in prozona ; metazona densely punctate, more than twice as long as prozona, its posterior angle obtuse and shoulders rounded. Lateral lobes slightly attenuate at their ventro-posterior angle (Fig. 184). Sternum square ; mesosternal and metasternal interspaces twice as long as wide.

Tegmina with sparse venation ; radius-sector with three branches ; intercalary vein S-shaped. Hind femur thick, its length 3.5–4.0 times its maximum width. Hind tibia shorter than femur, with eight or nine spines on the outer side and 11 spines on the inner side.

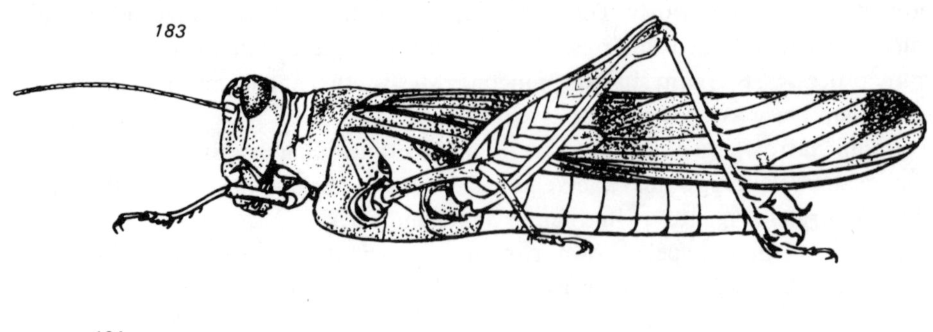

Figs. 183–184 : *Sphingonotus savignyi* Saussure, 1884
183. female ; 184. ventral margin of pronotum

Coloration : Pale to dark brown, sometimes reddish with white spots. Basal third of tegmina brown ; apical third transparent, showing the dark apical spot on the wings. Wings colourless, transparent, with a narrow dark band which reaches the anterior margin. Inner side of hind femur yellowish with or without a diffuse dark band near the apex. Hind tibia sulphurous-yellow ; inner base black.
Measurements (mm) : Body ♀ 29.0–38.0, ♂ 22.0–32.0 ; tegmina ♀ 29.0–37.0, ♂ 23.0–31.0.
Distribution : Canary Islands, North and East Africa, Arabia to West India.
Israel & Sinai : Along the shores of the Sinai Peninsula (16, 22, 23).

Sphingonotus eurasius Mishchenko, 1936
Figs. 185, 186

Type Locality : Tedjen, Turkmenistan (Academy, Leningrad).

Sphingonotus callosus (Fieber). Brunner von Wattenwyl C., 1882, *Prodromus der europäischen Orthopteren*, Leipzig, p. 154.
Sphingonotus scabriusculus Finot A., 1895, *Annls Soc. ent. Fr.*, 64 : 468, 474.
Sphingonotus callosus —. Bodenheimer F. S., 1935, *Animal Life in Palestine*, Jerusalem, p. 322.
Sphingonotus callosus —. Bodenheimer F. S., 1935, *Arch. Naturgesch.*, 4 (2) : 193.
Sphingonotus eurasius eurasius Mishchenko L. L., 1936, *Eos, Madr.*, 12 (1–2) : 86.
Sphingonotus eurasius eurasius —. Bei-Bienko G. Ya. & L. L. Mishchenko, 1951, *Locusts and Grasshoppers of the U. S. S. R. and Adjacent Countries*, Moscow, II : 625 [in Russian].
Sphingonotus eurasius —. Ramme W., 1951, *Mitt. zool. Mus. Berl.*, 27 : 426.

Small, smooth, hairy. Head hypognathous, raised above pronotum, densely punctate. Facial carinae raised, almost parallel, nearly reaching the clypeus ventrally. Foveolae visible from above ; margins irregular. Horizontal diameter of eye as long as interocular distance. Antennae filiform, 1.5 times as long as head and pronotum together. Pronotum saddle-shaped ; prozona constricted anteriorly and raised along the median carina (Fig. 186) ; metazona obtuse-angled, 1.5 times as long as prozona. Processes on posterior angle of lateral lobe of variable size. Sternum square, sparsely punctate. Width of mesosternal interspace 1.5 times its length ; metasternal interspace twice as wide as long.

Tegmina not reaching apex of hind tibia, rounded apically with regular venation. Hind femur slender, its length 3.5–4 times its maximum width. Hind tibia with eight spines on the outer side and 10 on the inner ; inner spurs 1.5 times as long as outer spurs.

Coloration : Pale brown, sandy or slightly reddish with a pattern of brown blotches on tegmina. Inner side of hind femur yellowish with a single dark band near the apex. Wings transparent, colourless, with a dark band extending to the anterior margin, sometimes with a dark reticulate apex.

Measurements (mm) : Body ♀ 19.0–25.0, ♂ 13.0–17.0 ; tegmina ♀ 17.0–25.0, ♂ 16.0–17.5.

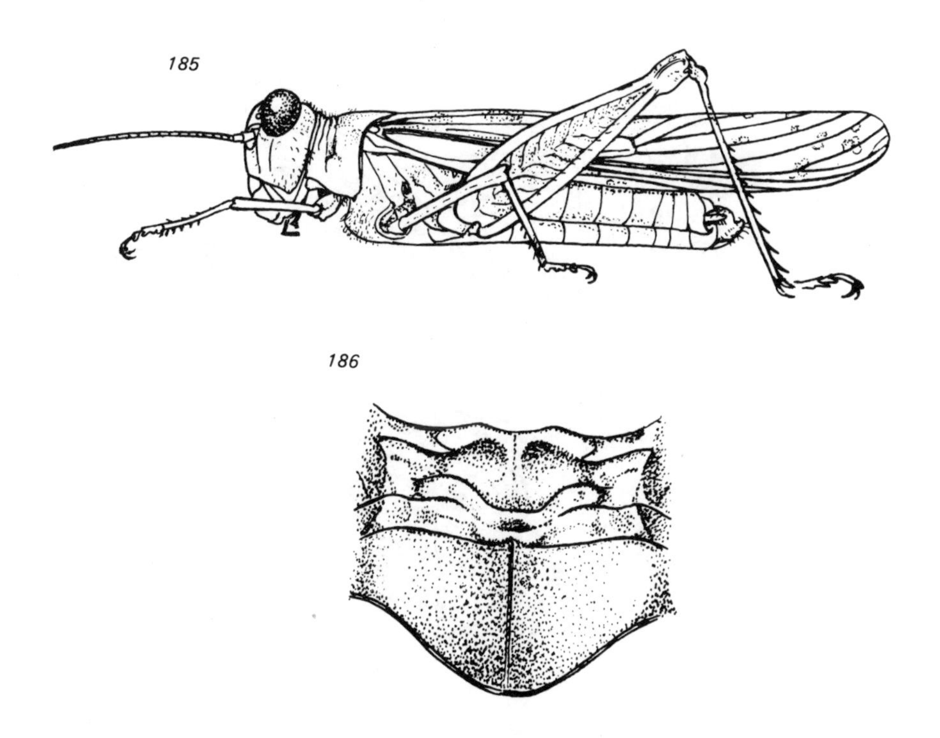

Figs. 185–186 : *Sphingonotus eurasius* Mishchenko, 1936
185. male ; 186. pronotum

Distribution: Central Asia, Asia Minor, North Africa.
Israel & Sinai: Coastal Plain (4, 8, 9), Northern Negev (15), Northern Sinai (20). It occurs mainly on sandy soils.

Sphingonotus obscuratus (Walker, 1870)
Figs. 187, 188; Plate IV: 6

Type Locality: Wadi Genneh, Sinai. Type lost.

Oedipoda obscurata Walker F., 1870, *Zoologist*, (2) 5: 2300.
Sphingonotus brunneri Saussure H. de, 1884, *Mém. Soc. Phys. Hist. nat. Genève*, 28 (9): 206.
Sphingonotus obscuratus —. Kirby W. F., 1910, *A synonymic catalogue of the Orthoptera*, Vol. III, Orthoptera Saltatoria, Part II, London, p. 272.
Sphingonotus quadrifasciatus Innes W., 1919, *Bull. Soc. R. ent. Égypte*, 5 (3) (1918): 37, 38.
Sphingonotus obscuratus —. Bodenheimer F. S., 1935, *Arch. Naturgesch.*, 4 (2): 195.
Sphingonotus obscuratus —. Mishchenko L. L., 1937, *Eos, Madr.*, 12 (3–4): 266.

Large, strong. Head slightly opisthognathous. Eyes oval, their diameter shorter than subocular groove, their horizontal diameter as long as interocular space. Fastigium

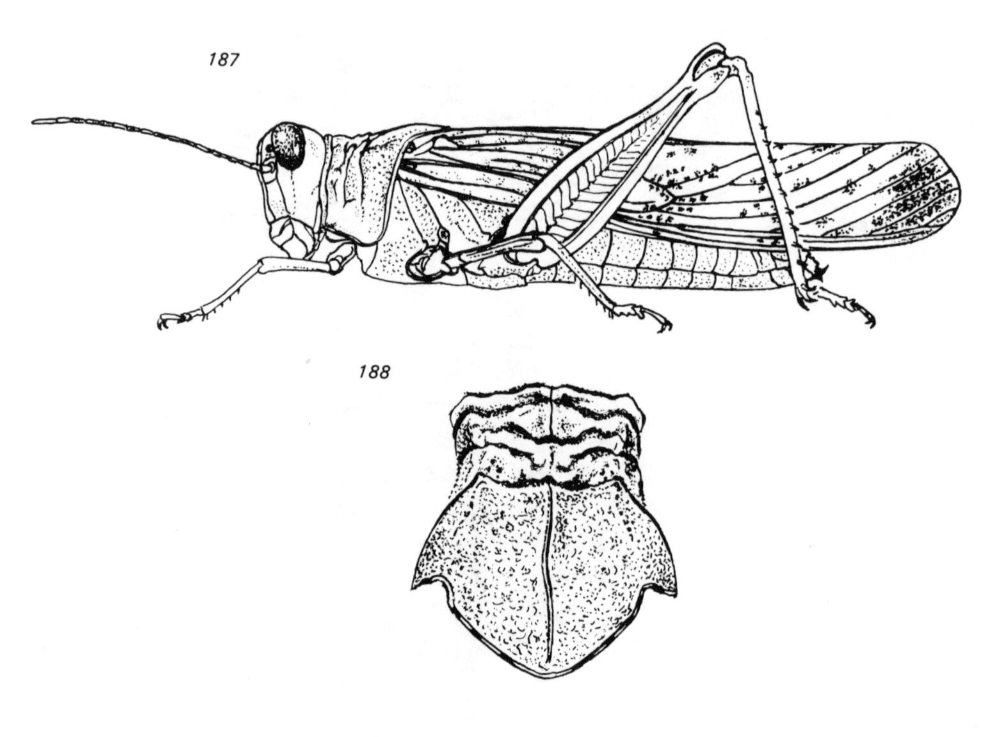

Figs. 187–188: *Sphingonotus obscuratus* (Walker, 1870)
187. female; 188. pronotum

wide with raised margins and fastigial carinulae ; foveolae punctate. Antennae slightly longer than head and pronotum together. Pronotum rounded ; prozona divided in the middle by the first sulcus, swollen before the third one (Fig. 188) ; metazona 1.2 times as long as prozona, densely punctate, its posterior angle obtuse. Ventral margin of lateral lobe undulate, with rounded posterior angle. Sternum longer than wide ; mesosternal interspace 1.5 times wider than long ; metasternal interspace twice as wide as long.

Tegmina rounded apically, their length five times their maximum width ; venation dense and regular. Radius-sector with three branches ; intercalary vein granulate and S-shaped. Two-thirds of wings black, with bluish base and wide black apical area. Inner side of hind femur mostly black with one or two light areas at the apex. Hind tibia dark violet ; tarsus pale or reddish.

Coloration : Pale brown with dark punctations and blotches. In most specimens the apical dark area of the wing is visible through the transparent tegmen.

Measurements (mm) : Body ♀ 36.0, ♂ 31.0 ; tegmina ♀ 39.0, ♂ 39.0.

Distribution, Israel & Sinai : Central and Southern Negev (15, 17), Sinai (22) along the shoreline.

Inhabits rocky sites with very sparse vegetation.

Sphingonotus angulatus Uvarov, 1922

Figs. 189–191 ; Plate V : 4

Type Locality : Haifa, 'Akko, Israel.

Sphingonotus angulatus Uvarov B. P., 1922, *Entomologist's mon. Mag.*, 8 : 84.
Sphingonotus tricinctus angulatus —. Bodenheimer F. S., 1935, *Arch. Naturgesch.*, 4 (2) : 195.
Sphingonotus tricinctus tricinctus Walker. Mishchenko L. L., 1936, *Eos. Madr.*, 12 (1–2) : 86.

Medium sized to small. Head hypognathous, projecting above pronotum. Frontal ridge concave, with raised parallel margins which diverge ventrally. Foveolae triangular, deep ; interorbital space less wide than horizontal diameter of eyes. Antennae 1.5 times as long as head and pronotum together. Pronotum punctate ; hairs sparse with numerous carinulae, especially in the female. Median carina slightly raised in anterior part of prozona, obliterated between the transverse sulci, distinct in metazona. Posterior margin of pronotum obtuse-angled. Lateral lobes acute anteriorly and with a sharp process on the postero-ventral margin (Fig. 190). Sternum square, punctate anteriorly ; hairs sparse. Metasternal interspace less than twice as wide as long.

Tegmina narrow, their length 5.5–6.0 times their maximum width, extending to apex of hind tibia ; apex rounded, transparent. Intercalary vein usually straight (Fig. 191), sometimes slightly S-shaped. Radius-sector with two branches. Wings bluish at the base, with a dark band covering one-third of the wing membrane (Fig. 191).

Coloration : Sandy-brown with a pattern of white and brown dots. Inner side of hind femur yellowish-grey, usually with a dark band extending onto the outer side. Hind tibia yellowish-bluish.

Measurements (mm) : Body ♀ 22.5–25.0, ♂ 18.0–20.5 ; tegmina ♀ 24.0–28.0, ♂ 19.0–22.0.

Distribution, Israel : Coastal Plain (4, 8, 9), Northern Negev (20).

Occurs in the Coastal Plain from Sinai north to Caesarea. During the summer, especially in July–August, this is the most typical species on loose sand. Oviposition takes place in these months and the first hoppers are observed in March.

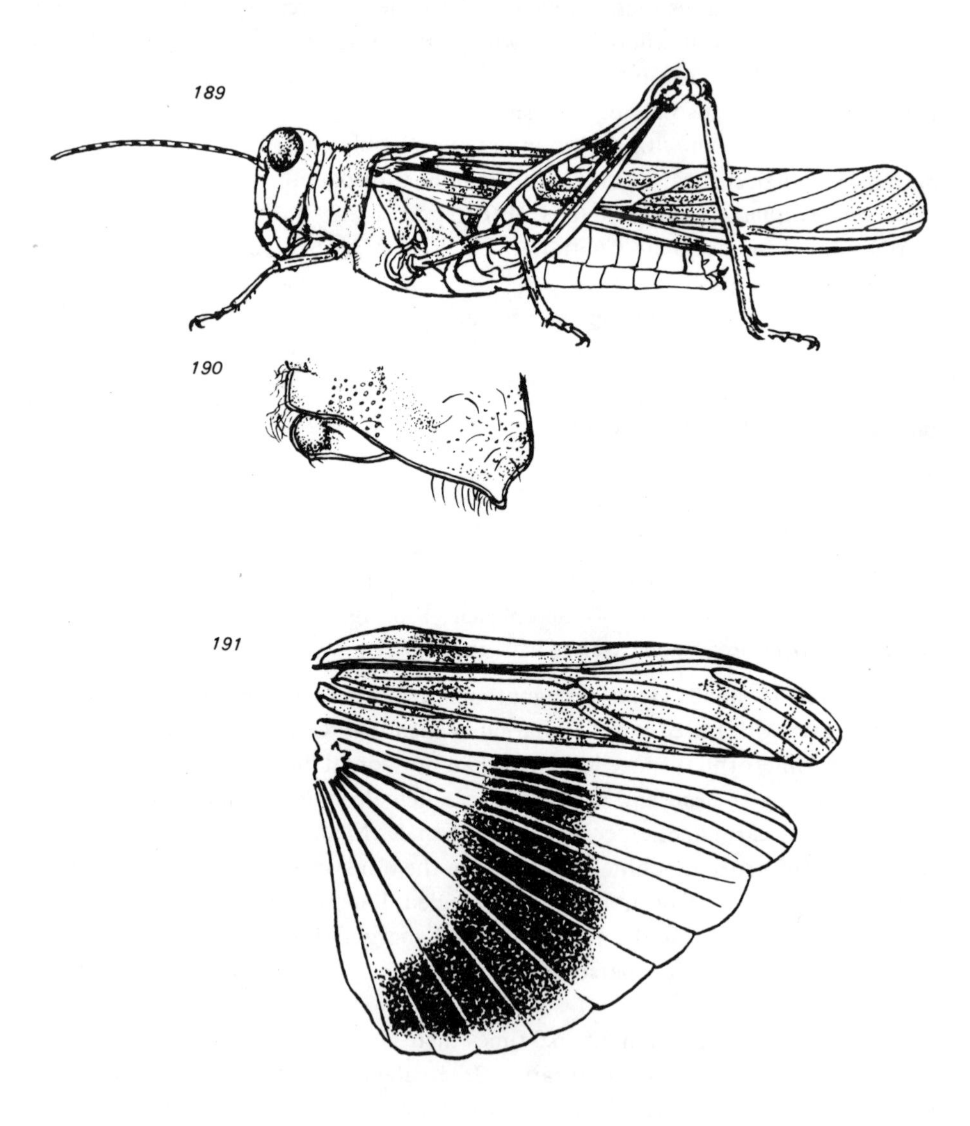

Figs. 189–191 : *Sphingonotus angulatus* Uvarov, 1922
189. female ; 190. ventral margin of pronotum ; 191. wings

Sphingonotus satrapes Saussure, 1884
Fig. 192

Type Locality : 'Turkestan'.

Sphingonotus satrapes Saussure H. de, 1884, *Mém. Soc. Phys. Hist. nat. Genève*, 28 (9) : 199.
Sphingonotus satrapes —. Buxton P. A. & B. P. Uvarov, 1923, *Bull. Soc. R. ent. Égypte*, p. 195.
Sphingonotus satrapes —. Bodenheimer F. S., 1935, *Animal Life in Palestine*, Jerusalem, pp. 322, 323.
Sphingonotus satrapes —. Bodenheimer F. S., 1935, *Arch. Naturgesch.*, 4 (2) : 193.
Sphingonotus satrapes satrapes —. Mishchenko L. L., 1936, *Eos, Madr.*, 12 (1–2) : 92, 276.
Sphingonotus satrapes —. Bei-Bienko G. Ya. & L. L. Mishchenko, 1951, *Locusts and Grasshoppers of the U. S. S. R. and Adjacent Countries*, Moscow, II : 632 [in Russian].
Sphingonotus satrapes —. Ramme W., 1951, *Mitt. zool. Mus. Berl.*, 27 : 426.

Large. Head slightly opisthognathous, raised above pronotum. Margins of frontal ridge converging towards median ocellus, diverging ventrally. Fastigium of vertex short, flat, wider than horizontal diameter of eye, with a longitudinal carina in its middle. Foveolae triangular, flat, visible from above. Antennae pale yellow, as long as head and pronotum together. Pronotum granulate, with a distinct crest in prozona (Fig. 192), sharp shoulders in metazona and a rounded posterior margin. Space between second and third sulci swollen, with a transverse depression. Ventral angles of lateral lobes rounded or slightly pointed. Sternum longer than wide. Mesosternal interspace 1.3 times wider than long ; margins parallel. Large tooth-like process present on the ovipositor, near the apex of the dorsal valve.
Tegmina transparent at apex, not reaching the end of hind tibia. Intercalary vein S-shaped ; radius-sector with three or four branches.
Coloration : Generally pale brown, sometimes whitish on head and thorax, with brown blotches and traces of two brown diffuse bands on the tegmina. Wings milky-yellowish, with a wide dark band which narrows in the posterior half and reaches the

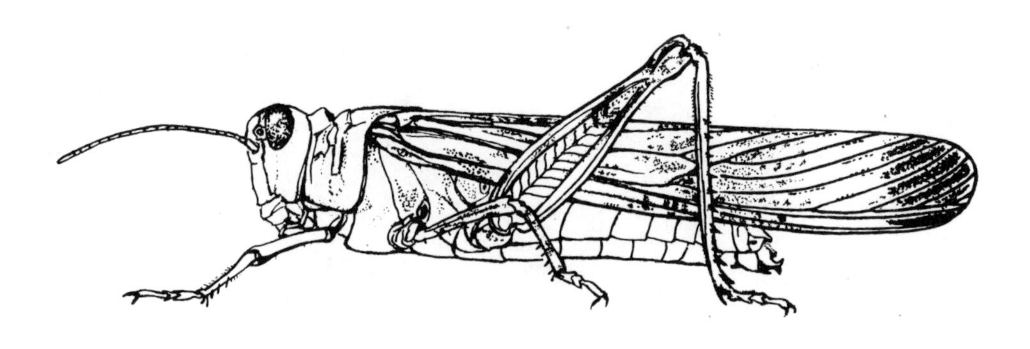

Fig. 192 : *Sphingonotus satrapes* Saussure, 1884, female

posterior margin ; large black spot at the apex visible through the transparent tegmina. Inner side of hind femur dark, with two light bands. Hind tibia pale yellowish, with short dense hairs distally.

Measurements (mm) : Body ♀ 37.0–45.0, ♂ 27.0–35.0 ; tegmina ♀ 33.0–42.0, ♂ 31.0–37.5.

Distribution : Central Asia, Caucasus, Persia, Iraq, Israel.

Israel : Central and Southern Coastal Plain (8, 9), Northern Negev (20), Jordan Valley (7).

Sphingonotus octofasciatus (Serville, 1838)
Fig. 193

Type Locality : 'Egypt' (Paris Museum).

Oedipoda octofasciata Serville J. G. A., 1838, *Histoire naturelle des Insectes*, in : Roret, *Collection des Suites à Buffon. Orthoptères*, Paris, p. 728.

Sphingonotus kittaryi Saussure H. de, 1884, *Mém. Soc. Phys. Hist. nat. Genève*, 28 (9) : 207.

Acrotylus octofasciatus —. Bonnet E. & A. Finot, 1884, *Revue Sci. nat. Montpellier* (3), 4 : 215.

Sphingonotus kittaryi —. Giglio-Tos E., 1893, *Boll. Musei Zool. Anat. comp. R. Univ. Torino*, 8 (164) : 6.

Sphingonotus octofasciatus —. Buxton P. A. & B. P. Uvarov, 1923, *Bull. Soc. R. ent. Égypte*, p. 201.

Sphingonotus octofasciatus —. Innes W., 1929, *Mém. Soc. R. ent. Égypte*, 3 (2) : 57, 87.

Sphingonotus octofasciatus —. Bodenheimer F. S., 1935, *Animal Life in Palestine*, Jerusalem, pp. 320, 323.

Sphingonotus octofasciatus —. Bodenheimer F. S., 1935, *Arch. Naturgesch.*, 4 (2) : 194.

Sphingonotus octofasciatus —. Mishchenko L. L., 1936, *Eos, Madr.*, 12 (1–2) : 91, 264.

Sphingonotus octofasciatus —. Ramme W., 1951, *Mitt. zool. Mus. Berl.*, 27 : 426.

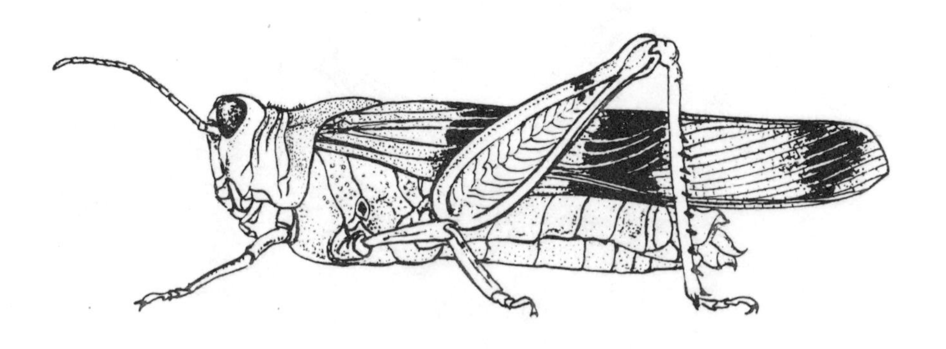

Fig. 193 : *Sphingonotus octofasciatus* (Serville, 1838), female

Medium sized, robust. Head opisthognathous. Margins of frontal ridge diverging from median ocellus downwards. Fastigium rhomboidal, with a raised median carina and sharp, raised anterior margins. Foveolae triangular, with distinct margins, not visible from above. Antennae with alternating light and dark rings, in females as long as head and pronotum together, in males 1.5 times as long. Pronotum anteriorly narrower than head; first transverse sulcus situated beyond middle of prozona. Metazona more than twice as long as prozona; posterior margin rounded.
Tegmina wide, crossed by two distinct dark brown transverse bands. Intercalary vein straight, raised; radius-sector with three branches. Base of wing red, with a dark band reaching the posterior margin and a large dark apical area. Inner side of hind femur greyish-yellow with a dark band extending onto outer side.
Measurements (mm): Body ♀ 25.0–34.5, ♂ 16.0–24.5; tegmina ♀ 25.0–37.0, ♂ 19.0–27.5.
Distribution: Central to Western Asia, North Africa.
Israel: Jordan Valley (7), Dead Sea Area (13), Judean Hills (11), Northern Negev (20).
Inhabits stony slopes in arid areas, near water, e.g., in the Dead Sea Area. In summer forms the most common population of *Sphingonotus* in these areas.

Sphingonotus balteatus latifasciatus (Walker, 1870)
Fig. 194

Type Locality: Rafah, Sinai. Type lost.

Oedipoda latifasciata Walker F., 1870, *Zoologist*, (2) 5 : 2299.
Oedipoda terminalis Walker F., 1870, *ibid.*, (2) 5 : 2300.
Sphingonotus balteatus (Serville). Saussure H. de, 1884, *Mém. Soc. Phys. nat. Genève*, 28 (9) : 197, 202.
Sphingonotus latifasciatus —. Kirby W. F., 1910, *A synonymic catalogue of the Orthoptera*, Vol. III, Orthoptera Saltatoria, Part II, London, p. 278.
Sphingonotus bifasciatus Innes W., 1919, *Bull. Soc. R. ent. Égypte*, 5 (3) (1918) : 44, 48.
Sphingonotus balteatus latifasciatus —. Uvarov B. P., 1923, *J. Bombay nat. Hist. Soc.*, 29 : 646.
Sphingonotus balteatus latifasciatus —. Bodenheimer F. S., 1935, *Arch. Naturgesch.*, 4 (2) : 195.

Large, robust. Head hypognathous, raised above pronotum. Foveolae obliterated; frontal ridge flat. Fastigium wider than horizontal diameter of eye, concave with a carinula extending over the vertex. Prozona of pronotum divided in the middle by the first transverse sulcus; metazona of pronotum markedly granulate; shoulders of pronotum distinct. Ventral angles of lateral lobes rounded. Tegmina wide. Intercalary vein dentate, S-shaped (Fig. 194); radius-sector with three branches.
Coloration: Pale brown with wide brown bands on tegmina. Base of wings pinkish-purple, the colour extending along the posterior margin; dark band very wide, narrowing slightly towards the posterior and anterior margins (Fig. 194). Inner side

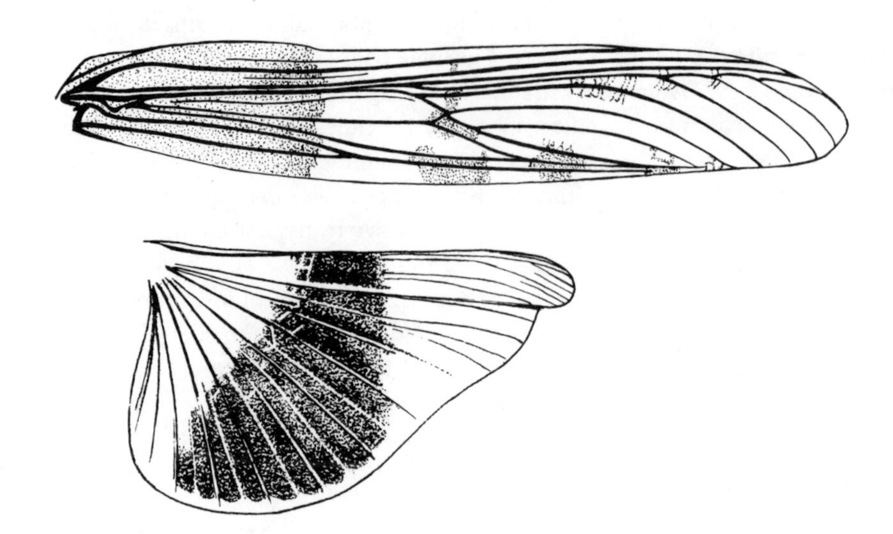

Fig. 194 : *Sphingonotus balteatus latifasciatus* (Walker, 1870), wings of female

of hind femur dark, with a wide apical band ; inner side of hind tibia purple-pink. Parts of body are often pinkish, like the wings.
Measurements (mm) : Body ♀ 36.0–40.0, ♂ 21.0–22.0 ; tegmina ♀ 36.0–38.0, ♂ 30.0–33.0.
Distribution : North and East Africa, Israel and Sinai.
Israel & Sinai : Southern Negev (16), Northern Sinai (20).
Rare ; usually found in deltas of dry wadis.

Sphingonotus rubescens (Walker, 1870)
Fig. 195

Type Locality : Sinai. Type lost.

Oedipoda rubescens Walker F., 1870, *Zoologist*, (2) 5 : 2301.
Sphingonotus coerulans var. *aegyptiaca* Saussure H. de, 1884, *Mém. Soc. Phys. Hist. nat. Genève*, 28 (9) : 200.
Sphingonotus rubescens —. Kirby W. F., 1910, *A synonymic catalogue of the Orthoptera*, Vol. III, Orthoptera Saltatoria, Part II, London, p. 274 (partim).
Sphingonotus rubescens —. Uvarov B. P., 1922, *Entomologist's mon. Mag.*, 8 : 84.
Sphingonotus rubescens —. Buxton P. A. & B. P. Uvarov, 1923, *Bull. Soc. R. ent. Égypte*, p. 201.
Sphingonotus rubescens —. Bodenheimer F. S., 1935, *Animal Life in Palestine*, Jerusalem, p. 322.

Sphingonotus rubescens —. Bodenheimer F. S., 1935, *Arch. Naturgesch.*, 4 (2) : 194.
Sphingonotus rubescens rubescens —. Mishchenko L. L., 1936, *Eos, Madr.*, 12 : 84, 168.
Sphingonotus rubescens —. Ramme W., 1951, *Mitt. zool. Mus. Berl.*, 27 : 425.

Very variable in size ; body slender, compressed laterally, hairy. Head opisthognathous, especially in male, slightly raised above pronotum. Fastigium narrow, concave with margins converging between eyes. Interocular distance small, less than half the eye diameter. Foveolae obliterated. Antennae banded, 1.2–1.4 times as long as head and pronotum together. Pronotum constricted anteriorly ; first transverse sulcus situated anteriorly to its middle. Metazona punctate, more than twice as long as prozona ; carinulae sometimes present ; posterior margin obtuse. Ventro-posterior line of lateral lobe with a sharp process. Sternum hairy ; mesosternal and metasternal interspaces twice as wide as long.

Tegmina long, extending posteriorly to end of hind tibia. Intercalary vein S-shaped, extending along the whole median field ; radius-sector with three branches.

Coloration : Generally brownish with groups of spots and blotches, usually without dark cross-bands on tegmina. Wings transparent, colourless. Inner side of hind femur dark with a complete light band and often also with an incomplete one. Hind tibia bluish.

Measurements (mm) : Body ♀ 20.0–33.0, ♂ 15.0–23.0 ; tegmina ♀ 24.0–35.0, ♂ 17.5–28.0.

Distribution : North Africa, Asia.

Israel & Sinai : Lower Galilee (2), Coastal Plain (4, 8, 9), Judean Hills (11), Negev (15, 16, 17), Dead Sea Area (13), ‘Arava Valley (14), Sinai (21, 22, 23).

The most common and euryoecous species, inhabiting stony hammadas and mountains. It is an excellent flier, and individuals can also be found north of their range of distribution, in the Coastal Plain and Judean Hills. They emit chirping sounds during flight. Strongly attracted by light during the night.

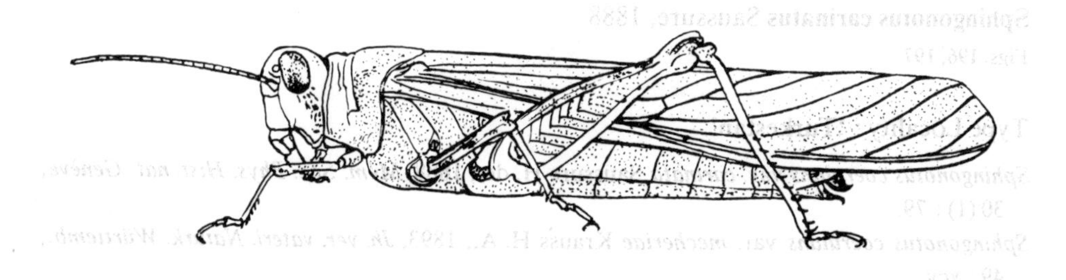

Fig. 195 : *Sphingonotus rubescens* (Walker, 1870), female

Sphingonotus coerulans (Linnaeus, 1767)

Type Locality : 'Sweden'.

Gryllus (Locusta) caerulans Linnaeus C., 1767, *Systema Naturae*, 12th ed., p. 701.
Acridium coerulans —. Brullé A., 1840, Orthoptera, in : Webb P. B. & S. Berthelot, *Histoire naturelle des Iles Canaries*, 2 (2) : 78.
Oedipoda coerulans —. Lucas H., 1849, Insectes, in : *Exploration scientifique de l'Algérie pendant les années 1846–1849*, 3 (3) : 35.
Oedipoda (Sphingonotus) coerulans —. Fieber F. X., 1852, in : Kelch A., *Grundlage zur Kenntnis der Orthopteren Oberschlesiens, etc.*, Ratibor, Bögner, p. 2.
Sphingonotus coerulans —. Walker F., 1871, *Catalogue of the specimens of Dermaptera Saltatoria in the Collection of the British Museum*, London, Suppl., Part V, p. 74.
Gryllus coerulans —. Uvarov B. P., 1948, *Eos, Madr.*, 24 : 384.

Medium sized, smooth. Head hypognathous. Frontal ridge slightly concave ; margins parallel. Foveolae rectangular-trapezoidal, situated laterally. Fastigium of vertex concave, with raised margins ; interocular distance shorter than half the diameter of the eye ; vertical diameter of eye shorter than or nearly as long as subocular groove. Base of antennae flattened ; segments 7–10 square. Pronotum slightly raised anteriorly in the median line, angular posteriorly. Metazona twice as long as prozona ; shoulders sharp, granulate. Lateral lobes higher than wide, with a small process situated anterior to their posterior angle.
Tegmina membranous, wide, reaching posteriorly to middle of hind tibia. Radius-sector with two branches ; intercalary vein straight proximally, markedly S-shaped apically.
Coloration : Wings colourless, wide. Inner surface of hind femur dark bluish, with a light apical area and dark knee. Hind tibia bluish ; spines strong, short.
Measurements (mm) : Body ♀ 22.5 ; tegmina ♀ 25.0.
Distribution : European U.S.S.R., Sweden, Italy, North-Eastern Mediterranean.
Israel : Dead Sea Area (13 ; 'En Gedi).

Sphingonotus carinatus Saussure, 1888
Figs. 196, 197

Type Locality : 'Turkestan'.

Sphingonotus coerulans var. *carinata* Saussure H. de., 1888, *Mém. Soc. Phys. Hist. nat. Genève*, 30 (1) : 79.
Sphingonotus coerulans var. *mecheriae* Krauss H. A., 1893, *Jh. ver. vaterl. Naturk. Württemb.*, 49 : xcv.
Sphingonotus mecheriae —. Buxton P. A. & B. P. Uvarov, 1923, *Bull. Soc. R. ent. Égypte*, p. 195.
Sphingonotus mecheriae —. Bodenheimer F. S., 1935, *Animal Life in Palestine*, Jerusalem, p. 322.
Sphingonotus mecheriae —. Bodenheimer F. S., *Arch. Naturgesch.*, 4 (2) : 193.

Sphingonotus carinatus —. Mishchenko L. L., 1936, *Eos, Madr.*, 12 (1–2) : 85, 186.
Sphingonotus carinatus —. Ramme W., 1951, *Mitt. zool. Mus. Berl.*, 27 : 426.

Medium sized or small. Head opisthognathous (especially in male), slightly raised above pronotum. Frontal ridge concave ; margins elevated, reaching clypeus. Eyes oval, their vertical diameter in male longer than subocular groove, as long in female. Interocular distance in male shorter than horizontal diameter of eye, in female equal in length to it. Vertex concave ; margins sharp ; median carinulae present, sharp in male. Foveolae triangular, distinct. A crest formed by the raised median carina on anterior part of pronotum ; first transverse sulcus situated beyond middle of prozona ; all three sulci deep. Metazona rugose, granulate, 1.5 times as long as prozona ; posterior angle obtuse. Ventral margin of lateral lobes distinctly projecting at their anterior and posterior angles. Sternum wider than long. Mesosternal and metasternal interspaces 2.5–3.0 times wider than long.
Tegmina extend over middle of hind tibia ; venation sparse ; radius-sector with two or three branches ; intercalary vein slightly curved apically. Hind femur slender, its length almost four times its maximum width.

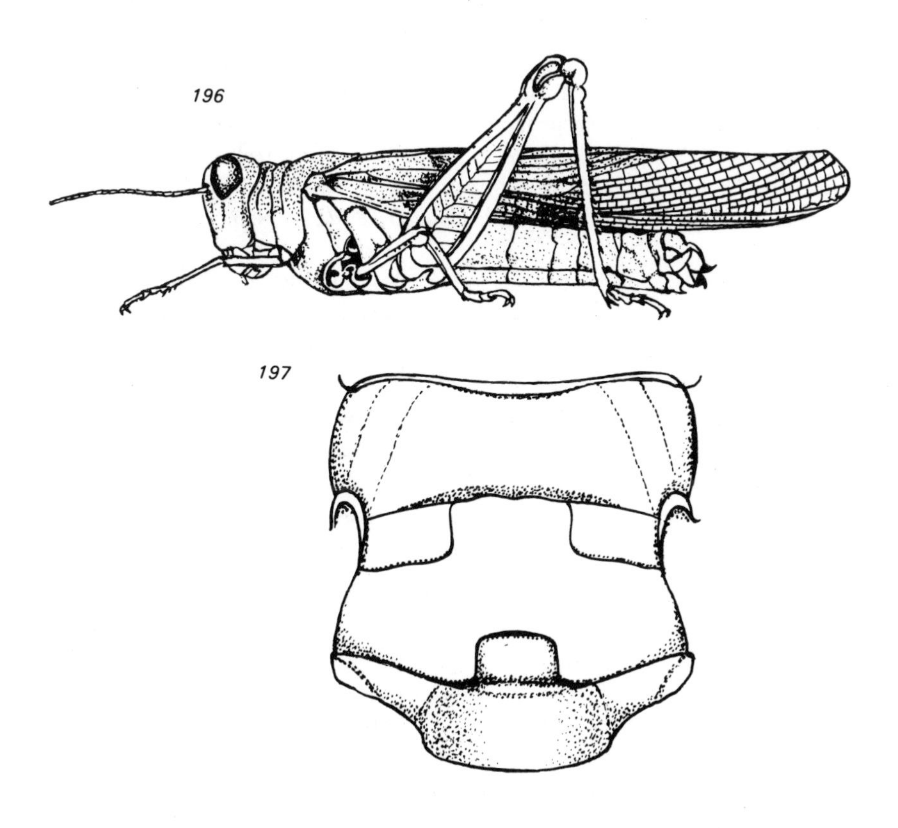

Figs. 196–197 : *Sphingonotus carinatus* Saussure, 1888
196. female ; 197. sternum

Coloration: Brown or pale brown, sometimes with whitish parts. Tegmina with or without indistinct brown bands, usually only with brown blotches. Wings bluish at the base; veins black. Inner surface of hind femur bluish-grey with two light bands intersected by a dark band. Hind tibia yellowish.
Measurements (mm): Body ♀ 21.0–32.0, ♂ 14.0–21.0; tegmina ♀ 20.0–30.0, ♂ 14.0–20.5.
Distribution: North Africa to the Near East, Iraq, Central Asia to Western Mongolia.
Israel & Sinai: Jordan Valley (7), Dead Sea Area (13), 'Arava Valley (14), Sinai Mountains (22).

Sphingonotus theodori Uvarov, 1923

Figs. 198–200

Type Locality: Sinai.

Sphingonotus coerulans theodori Uvarov B. P., 1923, in: Buxton P. A. & B. P. Uvarov, *Bull. Soc. R. ent. Égypte*, p. 195, figs. 2, 3.
Sphingonotus theodori —. Uvarov B. P., 1929, in: *Ergebnisse Sinai-Expedition*, Leipzig, p. 95.
Sphingonotus theodori —. Bodenheimer F. S., 1935, *Arch. Naturgesch.*, 4 (2): 193.
Sphingonotus theodori theodori —. Mishchenko L. L., 1936, *Eos, Madr.*, 12 (1–2): 83, 159.
Sphingonotus theodori theodori —. Bei-Bienko G. Ya. & L. L. Mishchenko, 1951, *Locusts and Grasshoppers of the U. S. S. R. and Adjacent Countries*, Moscow, II: 616 [in Russian].
Sphingonotus theodori —. Ramme W., 1951, *Mitt. zool. Mus. Berl.*, 27: 425.

Medium sized, smooth; hairs sparse. Head hypognathous, projecting above pronotum. Face punctate; frontl ridge flat, markedly concave around median ocellus. Foveolae triangular; margins indistinct. Eyes irregularly oval; horizontal diameter as long as interocular distance; vertical diameter as long as subocular groove. Antennae 1.2 times as long as head and pronotum together; segments rounded. Pronotum compressed anteriorly, slightly raised; first transverse sulcus situated beyond middle of prozona (Fig. 199); metazona 2.5 times as long as prozona, punctate, with numerous short carinulae; posterior margin angular. Lateral lobes with straight ventral margins, ventro-posterior angle rounded or slightly pointed. Sternum square, punctate; hair sparse. Mesosternal and metasternal interspaces 1.5 times wider than long.
Tegmina straight, with rounded apex, extending almost to end of hind tibia; irregular venation at base; radius-sector with two branches; intercalary vein almost straight, situated in middle of median field (Fig. 200). Wings triangular, colourless. Hind femur widened basally, its length 3.2–3.5 times its maximum width.
Coloration: Generally pale brown or sandy, face sometimes white; pale brown or brown bands present on tegmina. Inner surface of posterior femur pale yellow with a small indistinct dark spot. Posterior tibia sulphurous-yellow with a round dark spot on the knee.

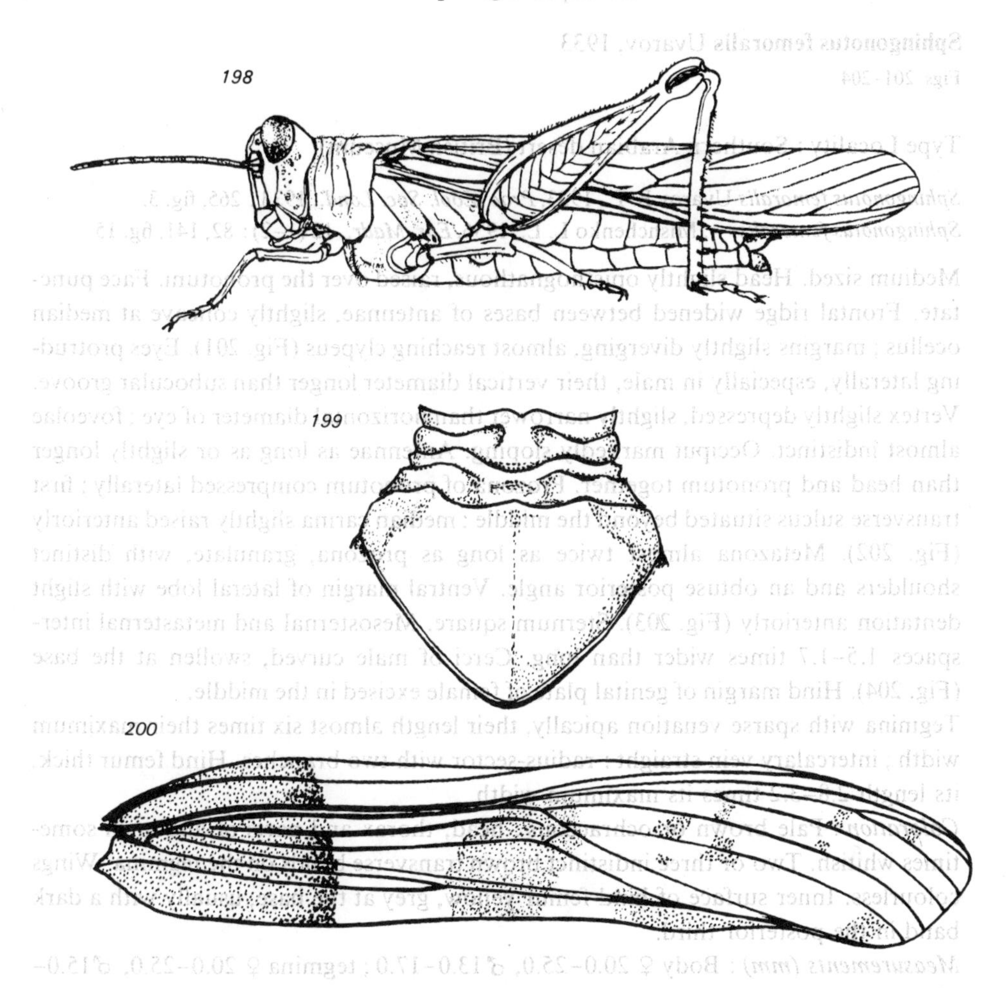

Figs. 198–200 : *Sphingonotus theodori* Uvarov, 1923
198. female ; 199. pronotum ; 200. tegmen

Measurements (mm) : Body ♀ 22.0–27.0, ♂ 14.5–19.0 ; tegmina ♀ 24.0–26.0, ♂ 17.0–22.0.

Distribution : Asia Minor, Syria, Israel, North-West Persia.

Israel : Judean Hills (11), Jordan Valley (7), Dead Sea Area (13), Southern and Central Negev (16, 17), ‘Arava Valley (13).

Inhabits lower parts of rocky mountain slopes and dry wadis ; typical on loose soil.

Sphingonotus femoralis Uvarov, 1933
Figs. 201–204

Type Locality : Southern Arabian desert (British Museum).

Sphingonotus femoralis Uvarov B. P., 1933, *Proc. Zool. Soc. Lond.*, 1933 : 265, fig. 3.
Sphingonotus femoralis —. Mishchenko L. L., 1936, *Eos, Madr.*, 12 (1–2) : 82, 141, fig. 15.

Medium sized. Head slightly opisthognathous, raised over the pronotum. Face punctate. Frontal ridge widened between bases of antennae, slightly concave at median ocellus ; margins slightly diverging, almost reaching clypeus (Fig. 201). Eyes protruding laterally, especially in male, their vertical diameter longer than subocular groove. Vertex slightly depressed, slightly narrower than horizontal diameter of eye ; foveolae almost indistinct. Occiput markedly sloping. Antennae as long as or slightly longer than head and pronotum together. Prozona of pronotum compressed laterally ; first transverse sulcus situated beyond the middle ; median carina slightly raised anteriorly (Fig. 202). Metazona almost twice as long as prozona, granulate, with distinct shoulders and an obtuse posterior angle. Ventral margin of lateral lobe with slight dentation anteriorly (Fig. 203). Sternum square. Mesosternal and metasternal interspaces 1.5–1.7 times wider than long. Cerci of male curved, swollen at the base (Fig. 204). Hind margin of genital plate of female excised in the middle.
Tegmina with sparse venation apically, their length almost six times their maximum width ; intercalary vein straight : radius-sector with two branches. Hind femur thick, its length 2.8–3.2 times its maximum width.
Coloration : Pale brown or ochraceous ; head, thorax and parts of abdomen sometimes whitish. Two or three indistinct brown transverse bands on the tegmina. Wings colourless. Inner surface of hind femur yellow, grey at the base, usually with a dark band in the posterior third.
Measurements (mm) : Body ♀ 20.0–25.0, ♂ 13.0–17.0 ; tegmina ♀ 20.0–25.0, ♂ 15.0–18.0.
Distribution : North Africa (Sudan and Egypt), Israel, Arabia, Iran, Baluchistan. Israel : Dead Sea Area (13).
It forms very large populations in May–June in the southern part of the Dead Sea Area (‘En Gedi).

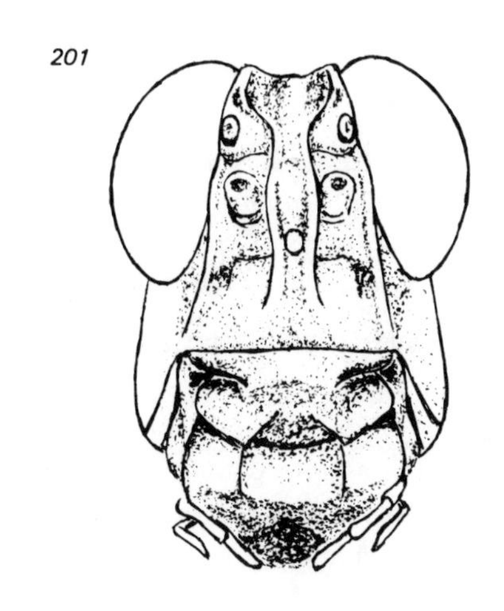

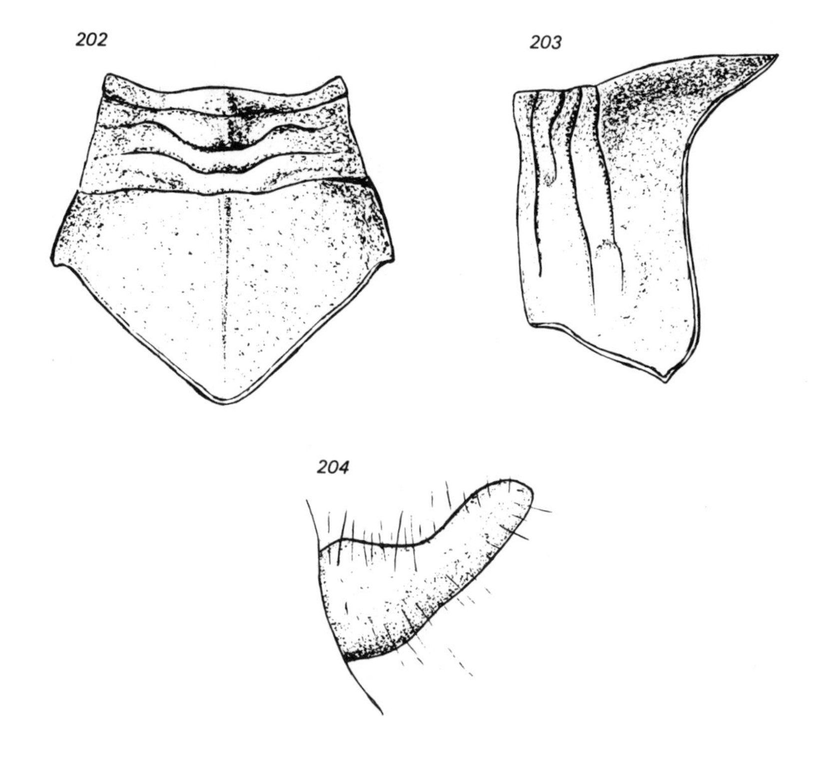

Figs. 201–204 : *Sphingonotus femoralis* Uvarov, 1933
201. head, frontal view ; 202. pronotum, dorsal ; 203. same, lateral ; 204. cercus of male

Sphingonotus vosseleri Krauss, 1902

Type Locality : Biskra, Algeria (Berlin Museum).

Sphingonotus vosseleri Krauss H. A., 1902, *Verh. zool.-bot. Ges. Wien*, 52 (4) : 231, 242, fig. 10.
Sphingonotus desertorum Vosseler J., 1902, *Zool. Jb.* (Syst.), 16 : 372, pl. 17, figs. 12a, b, 13.
Sphingonotus vosseleri —. Innes W., 1929, *Mém. Soc. R. ent. Égypte*, 3 (2) : 55, 57.
Sphingonotus vosseleri —. Mishchenko L. L., 1936, *Eos, Madr.*, 12 (1–2) : 178.

Small, smooth ; hairs sparse. Head slightly opisthognathous, sparsely punctate, in female slightly projecting above pronotum, in male markedly projecting. Eyes oval, large, projecting laterally, their vertical diameter longer than subocular groove, their horizontal diameter twice as long as interocular space. Frontal ridge slightly concave ; margins thick, diverging towards clypeus. Vertex depressed ; margins distinctly raised. Foveolae indistinct. Antennae longer than head and pronotum together. Pronotum saddle-shaped, constricted, slightly raised in anterior part of prozona. First transverse sulcus situated beyond middle of prozona ; interspace between the first and third transverse sulci slightly swollen and with a depression in the middle. Metazona punctate ; shoulders rounded ; posterior margin obtuse-rounded ; median carina distinct. Postero-ventral angle of lateral lobes broadly rounded. Sternum square, punctate anteriorly. Mesosternal and metasternal interspaces twice as wide as long.
Tegmina narrow, their length six times their maximum width ; intercalary vein almost straight, granulate, approaching the media apically ; radius-sector with two branches. Wings elongate, colourless ; venation sparse.
Coloration : Generally pale ochraceous with whitish areas. Antennae with white and brown rings. Tegmina transparent with indistinct brown spots and bands. Inner side of hind femur yellowish ; hind tibia bluish.
Measurements (mm) : Body ♀ 22.0–24.0, ♂ 12.0–15.0 ; tegmina ♀ 17.0–22.0, ♂ 13.0–15.0.
Distribution : North Africa, Israel, Arabia, Mesopotamia.
Israel : Central Negev (17).

Sphingonotus hierichonicus Uvarov, 1923
Fig. 205

Type Locality : 'Palestine'.

Sphingonotus hierichonicus Uvarov B. P., 1923, in : Buxton P. A. & B. P. Uvarov, *Bull. Soc. R. ent. Égypte*, p. 198, figs. 4, 5.
Sphingonotus hierichonicus —. Bodenheimer F. S., 1935, *Arch. Naturgesch.*, 4 (2) : 194.
Sphingonotus hierichonicus —. Mishchenko L. L., 1936, *Eos, Madr.*, 12 (1–2) : 78, 104.

Small, smooth. Head hypognathous, markedly projecting above pronotum. Eyes oval, protruding both laterally and dorsally ; vertical diameter longer than subocular

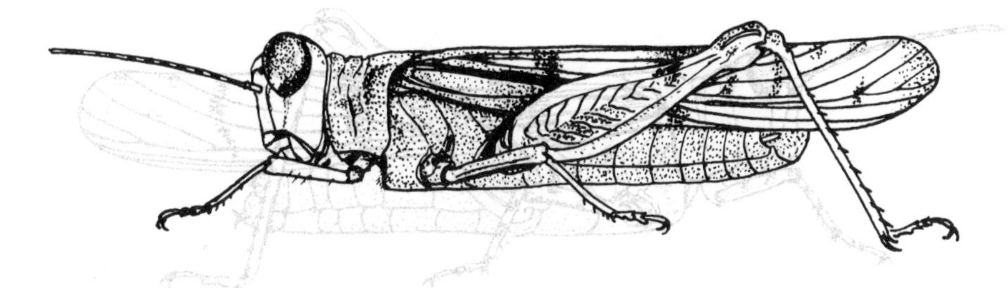

Fig. 205 : *Sphingonotus hierichonicus* Uvarov, 1923, male

groove ; horizontal diameter nearly twice as long as interocular space. Fastigium of vertex almost vertical ; foveolae minute, almost obliterated. Antennae nearly 1.5 times as long as head and pronotum together. Pronotum slightly saddle-shaped, constricted anteriorly ; first transverse sulcus situated in middle of prozona ; a pair of tubercles, connected by a low ridge, between second and third sulci. Metazona obtuse-angled posteriorly. Lateral lobes of prozona with or without processes before the ventro-posterior angle.

Tegmina short, extending to middle of hind tibia, usually transparent ; venation sparse. Intercalary vein straight or slightly sinuate ; radius-sector usually with two branches. Hind tibia broad, distinctly narrowing towards apex.

Coloration : Generally ochraceous-brown, sometimes reddish with white parts. Base of tegmina dark, sometimes with a single oblique band ; apical end transparent with diffuse spots. Wings bluish, sometimes with traces of a brown band. Inner side of hind femur yellowish with a single brown band extending also onto outer dorsal side. Hind tibia greyish ; inner base dark.

Measurements (mm) : Body ♀ 20.0, ♂ 14.0–15.5 ; tegmina ♀ 19.5, ♂ 16.0–17.5.

Distribution : Endemic in Israel; Dead Sea Area (13), 'Arava Valley (14).

Sphingonotus laxus Uvarov, 1952

Fig. 206

Type Locality : 'South Arabia'.

Sphingonotus laxus Uvarov B. P., 1952, *J. Linn. Soc.* (Zool.), 42 (284) : 184.

Small. Head slightly opisthognathous, raised above pronotum. Face punctate ; frontal ridge raised to median ocellus, flattened, with margins diverging ventrally. Vertex pronounced, foveolae obliterated, slightly visible from above ; fastigium concave, rhomboidal, granulate, with raised margins. Eyes oval ; horizontal diameter longer than subocular groove ; minimum interocular distance less than half the horizontal diameter. Pronotum slightly saddle-shaped, especially in female. Metazona 1.2–1.5

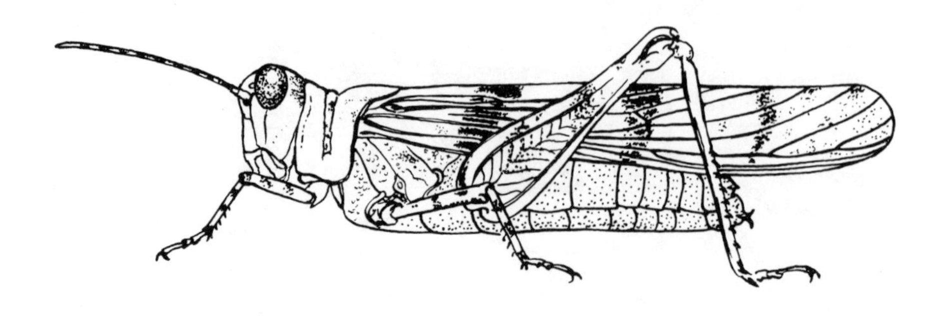

Fig. 206: *Sphingonotus laxus* Uvarov, 1952, female

times longer than prozona, densely granulate and punctate; shoulders slightly raised, obtuse-angled posteriorly. First and second transverse sulci are fused and situated beyond middle of prozona. Postero-ventral angle of lateral lobes rounded. Sternum square or slightly wider than long, punctate anteriorly. Mesosternal interspace slightly swollen, 1.5 times as wide as long; margins diverging. Metasternal interspace nearly quadrangular, 1.2 times as wide as long. Tegmina with straight intercalary vein; radius-sector with one or two branches.
Coloration: Pale brown with two oblique brown bands on tegmina, the anterior one with straight margins, extending to base of hind femur. Wings coloured. Inner surface of hind femur yellowish-grey with dark band in the posterior third, extending also on outer and dorsal surface. Lobes of hind knee dark grey; hind tibia bluish-grey.
Measurements (mm): Body ♀ 18.5, ♂ 12.0; tegmina ♀ 18.0, ♂ 13.0.
Distribution: Israel, Saudi Arabia.
Israel: Southern Negev (16; Wadi Radadi).

Genus MORPHACRIS Walker, 1870

Catalogue of the specimens of Dermaptera Saltatoria in the Collection of the British Museum, London, Part IV, pp. 721, 790

Cosmorhyssa Stål C., 1873, *Recensio Orthopterorum*, Stockholm, 1: 116, 121.

Type Species: *Morphacris adusta* Walker, 1870 (nymph).
Diagnosis: Medium sized. Head slightly opisthognathous or hypognathous. Eyes oval; frons concave. Antennae filiform, as long as head and pronotum together.

Morphacris

Pronotum with one distinct transverse sulcus and numerous longitudinal carinulae on both sides of the median carina (Fig. 208); posterior margin acute.
Distribution : Africa, Madagascar, Arabia, Syria.
One species in our region.

Morphacris fasciata (Thunberg, 1815)
Figs. 207–209 ; Plate V : 1

Type Locality : 'South Russia' (Uppsala Museum).

Gryllus fasciatus Thunberg C. P., 1815, *Mém. Acad. Sci. St.-Pétersb.*, 5 : 230.

Morphacris fasciata ab. **sulcata** (Thunberg, 1815)

Gryllus sulcatus Thunberg C. P., 1815, *Mém. Acad. Sci. St.-Pétersb.*, 5 : 234.
Cosmorhyssa sulcata —. Stål C., 1873, *Recensio Orthopterorum*, Stockholm, 1 : 122.
Morphacris sulcata —. Kirby W. F., 1910, *A synonymic catalogue of the Orthoptera*, Vol. III, Orthoptera Saltatoria, Part II, London, p. 219.
Morphacris fasciata ab. *sulcata* —. Uvarov B. P., 1921, *Ann. Mag. nat. Hist.* (9) 7 : 489.
Morphacris fasciata sulcata —. Buxton P. A. & B. P. Uvarov, 1923, *Bull. Soc. R. ent. Égypte*, p. 192.
Morphacris fasciata sulcata —. Bodenheimer F. S., 1935, *Arch. Naturgesch.*, 4 (2) : 188.

Medium sized, strong. Head usually hypognathous in females and slightly opistognathous in males, Antennae filiform (Fig. 207). Pronotum with raised median carina, intersected by one transverse sulcus situated anterior to the middle and numerous longitudinal carinulae on the whole disc (Fig. 208). Posterior margin of pronotum pointed. Tegmina rounded apically, with dense venation ; intercalary vein straight ; radius-sector with four branches.
Coloration : Generally dark grey or brown. Wings yellow with a dark band in the middle, with an anterior branch towards the base (Fig. 209). Tegmina lighter than body, with brown blotches. Lateral lobes of pronotum feature a white line and a shiny black band extending towards the jaws.
Measurements (mm) : Body ♀ 22.0–26.0, ♂ 14.5–16.0 ; tegmina ♀ 22.0–26.0, ♂ 16.0–17.5.
Distribution ; Africa, Madagascar, Syria, Israel.
Israel : Upper and Lower Galilee (1, 2), Jordan Valley (7), Coastal Plain (4, 8, 9), Judean Hills (11), Northern Negev (15).
One of the most common and euryoecous species in the region ; found in very diverse habitats, in which low Gramineae grow. Very dense populations were sometimes found, reaching 10–18 animals per 1 m^2.
These grasshoppers are very active during the season of reproduction, rising frequently into the air and emitting short chirping sounds.
Hoppers of this species are distinguished from those of other species by the numerous carinulae on the pronotum. They are found throughout the year.

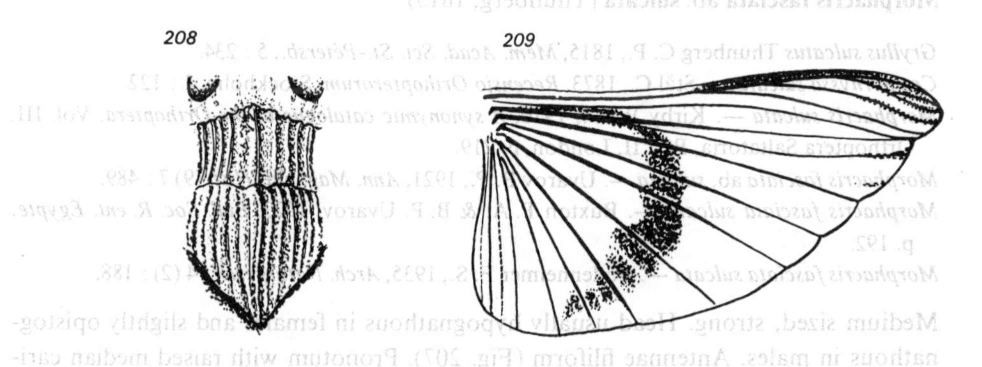

Figs. 207–209 : *Morphacris fasciata* (Thunberg, 1815)
207. female ; 208. pronotum ; 209. wing

Genus OEDIPODA Latreille, 1829

In : Cuvier G., *Règne Animal*, 2nd ed., *Insectes*, 5 : 188

Ctypohippus Fieber F. X., 1852, in : Kelch A., *Grundlage zur Kenntnis der Orthopteren Oberschlesiens, etc.*, Ratibor, Bögner, p. 2.

Oedipoda —. Innes W., 1929, *Mém. Soc. R. ent. Égypte*, 3 (2) : 40, 42.

Type Species : *Gryllus (Locusta) caerulescens* Linnaeus, 1758.

Diagnosis : Medium sized, strong, granulate. Head hypognathous with a slight projection between the oval eyes. Foveolae triangular, small (Fig. 215). Pronotum with distinctly raised longitudinal carina, intersected by posterior transverse sulcus ; metazona longer than prozona, with a sharp posterior angle and short lateral carinae ; ventral margin of lateral lobe with rounded posterior angle.

Tegmina with dense venation, extending beyond hind femur ; radius-sector with two or three branches ; median field with straight intercalary vein. Dorsal carina of hind femur notched behind the middle, forming a prominent ledge (Fig. 212). Arolium short, about one-third the length of the claws.
Coloration : Usually dark brown ; transverse dark bands present on the tegmina (Fig. 211). Wings brightly coloured, with a dark band and a radial branch towards the base (Fig. 211). Inner and ventral surface of femur bluish-black (Fig. 212).
About 15 species distributed around the Mediterranean, east to Kashmir, Siberia and China. Most of them typical geophilous species, inhabitants of open areas with sparse vegetation.
Three species in Israel, differing mostly by the coloration of their wings.

Key to the Species of Oedipoda in Israel

1. Basal part of wings blue or yellow — 2
– Basal part of wings red (Fig. 211). — **O. miniata** (Pallas)
2. Wings yellow ; the dark band with a branch reaching the base of the wing (Fig. 213). — **O. aurea** Uvarov
– Wings blue ; the branch of the dark band shorter, not reaching the wing's base (Fig. 214). — **O. caerulescens** (L.)

Oedipoda miniata (Pallas, 1771)
Figs. 210–212 ; colour plate : 5

Type Locality : 'South Russia'.

Gryllus miniatus Pallas P. S., 1771, *Reisen durch verschiedene Provinzen des russischen Reiches in den Jahren 1768–1774*, 1 : 467.
Gryllus salinus Gmelin J. F., 1790, *Systema Naturae*, revised. 13th ed., Leipzig, 1 (4) : 2083.
Acrydium salinum —. Fischer de Waldheim G., 1820, *Entomographie de la Russie. Orthoptères de la Russie*, p. 39, pl. 1, fig. 3.
Acrydium germanicum Costa O. G., 1836, *Fauna del regno di Napoli. Ortotteri*, Naples, p. 17, pl. 2, fig. 4A, b–d.
Oedipoda gratiosa Serville J. G. A., 1838, *Histoire naturelle des Insectes*, in : Roret, *Collection des Suites à Buffon. Orthoptères*, Paris, p. 727.
Oedipoda salina —. Eversmann E., 1848, *Additamenta quaedam levia ad Fischeri de Waldheim. Orthoptera rossica*, Moscow, p. 9.
Oedipoda fasciata var. *c* Fischer L. H., 1853, *Orthoptera Europaea*, Leipzig, p. 413.
Oedipoda miniata —. Brunner von Wattenwyl C., 1882, *Prodromus der europäischen Orthopteren*, Leipzig, p. 162.
Oedipoda miniata —. Giglio-Tos E., 1893, *Boll. Musei Zool. Anat. comp. R. Univ. Torino*, 8 (164) : 6.
Oedipoda miniata —. Krauss H. A., 1909, in : Kneucker A., *Verh. naturw. Ver. Karlsruhe*, 21 : 57.

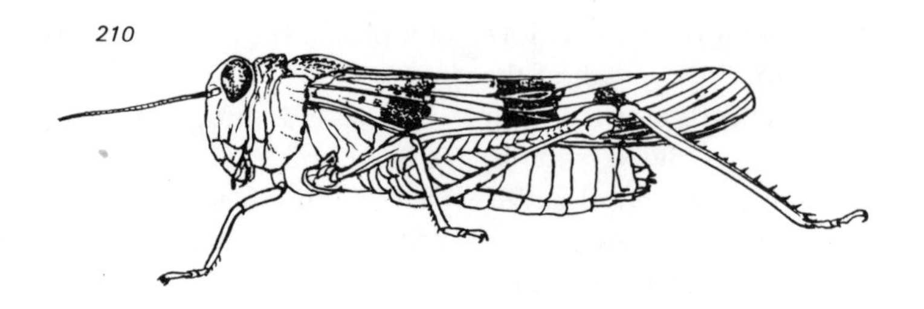

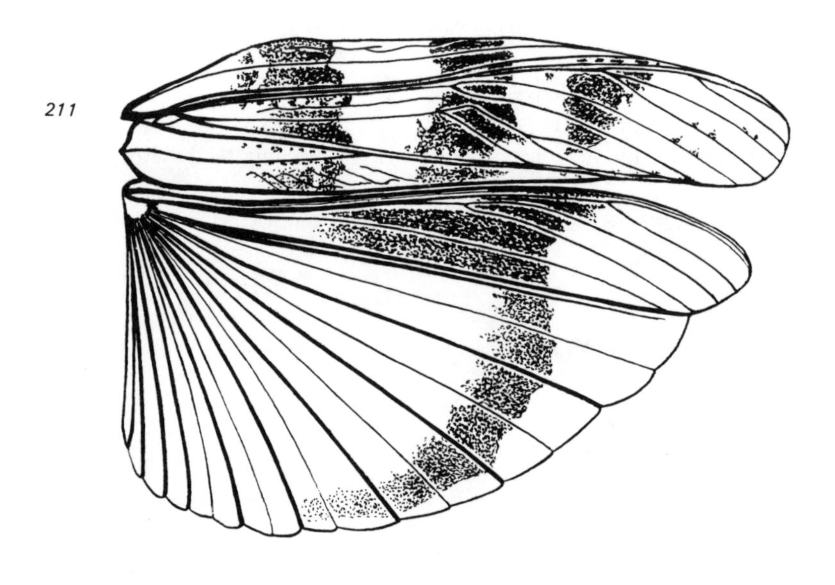

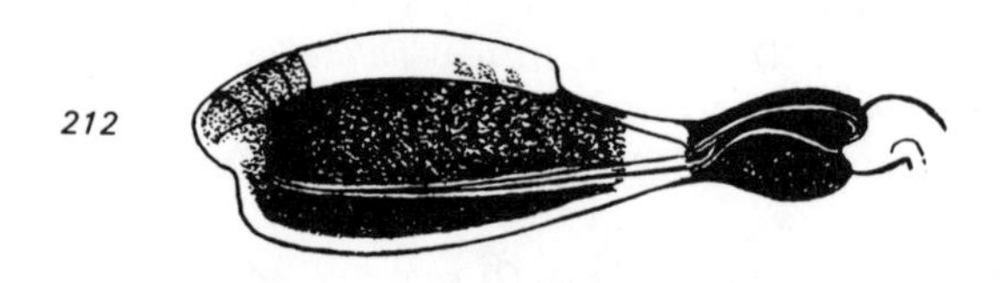

Figs. 210–212 : *Oedipoda miniata* (Pallas, 1771)
210. female ; 211. wings ; 212. hind femur, inner side

Oedipoda miniata —. Navas L., 1911, *Revista Montserratina*, Barcelona, p. 1.
Oedipoda miniata —. Buxton P. A. & B. P. Uvarov, 1923, *Bull. Soc. R. ent. Égypte*, p. 193.
Oedipoda miniata —. Innes W., 1929, *Mém. Soc. R. ent. Égypte*, 3 (2) : 42.
Oedipoda miniata —. Bodenheimer F. S., 1935, *Animal Life in Palestine*, Jerusalem, pp. 28b, 86, 88, 311, 320, 323.
Oedipoda miniata —. Bodenheimer F. S., 1935, *Arch. Naturgesch.*, 4 (2): 190.
Oedipoda miniata —. Bei-Bienko G. Ya. & L. L. Mishchenko, 1951, *Locusts and Grasshoppers of the U. S. S. R. and Adjacent Countries*, Moscow, II : 592 [in Russian].
Oedipoda miniata —. Ramme W., 1951, *Mitt. zool. Mus. Berl.*, 27 : 425.

Dark brown or ochraceous. Pronotum granulate with longitudinal rugulae on the lateral lobes. Tegmina with dark cross-bands, the first between the base of the hind legs, the second in the middle, and an additional small one close to the posterior third (Fig. 211). Inner surface of hind femur dark bluish with a white ring beyond the middle (Fig. 212). Inner surface of hind tibia white-sulphurous ; spines dark.
Measurements (mm) : Body ♀ 20.0–27.0, ♂ 17.0–21.0 ; tegmina ♀ 23.0–30.0, ♂ 17.0–23.0.
Distribution : North Africa, Southern and Central Europe, Israel, Urals, Siberia.
Israel : Lower and Upper Galilee (1, 2), Carmel (3, 5), Coastal Plain (4, 8, 9), Judean Hills (6, 7, 10), Jordan Valley (7), Dead Sea Area (13), Northern Negev (15).
One of the most important species of the summer fauna. A typical geophile, cryptic coloured ; usually found in areas of sparse ephemeral plants, on which it feeds. Hatching occurs from February to July. Females with ripe ovaries observed from March to August ; copulation and oviposition from June to August. Adults were observed to overwinter.

Oedipoda aurea Uvarov, 1923
Fig. 213

Type Locality : 'Palestine'.

Oedipoda miniata var. *flava* Saussure H. de, 1884, *Mém. Soc. Phys. Hist. nat. Genève*, 28 (9) : 149.
Oedipoda aurea Uvarov B. P., 1923, *Entomologist's mon. Mag.*, 3rd ser., 11, pp. 33–35.
Oedipoda aurea —. Bodenheimer F. S., 1935, *Arch. Naturgesch.*, 4 (2) : 190.
Oedipoda aurea —. Bei-Bienko G. Ya. & L. L. Mishchenko, 1951, *Locusts and Grasshoppers of the U. S. S. R. and Adjacent Countries*, Moscow, II : 592 [in Russian].
Oedipoda aurea —. Ramme W., 1951, *Mitt. zool. Mus. Berl.*, 27 : 425.

Body ochraceous or brown ; basal part of tegmina dark, with one prominent square band, sometimes marked by light fields on both sides. Wings golden-yellow in basal part. The base of hind tibia with a whitish ring.
Measurements (mm) : Body ♀ 22.0–26.0, ♂ 18.0–20.0 ; tegmina ♀ 22.0–24.0, ♂ 19.0–20.0.
Distribution : Asia Minor, Syria, Israel.
Israel : Lower and Upper Galilee (1, 2), Carmel (3, 5), Judean Hills (6, 7, 11, 12).

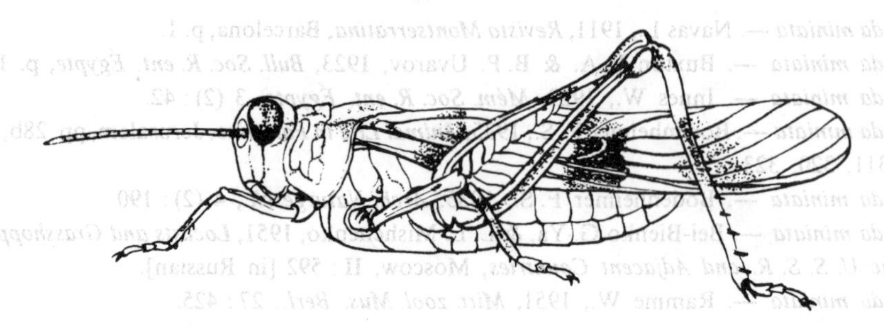

Fig. 213 : *Oedipoda aurea* Uvarov, 1923, male

O. aurea is a typical highland species, found on dry mountain slopes. On Mount Hermon it is found at altitudes up to 1,500 m. Hoppers observed in February and adults from May to December. More hygrophilous than *O. miniata*, this species is usually found in bushy areas.

Oedipoda caerulescens (Linnaeus, 1758)

Figs. 214, 215

Type Locality : 'Habitat in Meridionalibus'.

Gryllus Locusta caerulescens Linnaeus C., 1758, *Systema Naturae*, 10th ed., 1 : 432.

Acrydium coeruleipenne De Geer C., 1773, *Mémoires pour servir à l'histoire des insectes*, Stockholm, 3 : 473.

Ctypohippus coerulescens —. Seoane V. L., 1878, *Stettin ent. Ztg.*, 39 : 371.

Oedipoda coerulescens —. Brunner von Wattenwyl C., 1882, *Prodromus der europäischen Orthopteren*, Leipzig, pp. 159, 164.

Oedipoda coerulescens —. Bodenheimer F. S., 1935, *Arch. Naturgesch.*, 4 (2) : 191.

Oedipoda coerulescens —. Bei-Bienko G. Ya. & L. L. Mishchenko, 1951, *Locusts and Grasshoppers of the U. S. S. R. and Adjacent Countries*, Moscow, II : 592 [in Russian].

Strong, ochraceous, with indistinct cross-bands on tegmina. Wings bluish with a wide black band branching at the wing base. Hind tibia usually with prominent black and white basal part.

Measurements (mm) : Body ♀ 22.0–28.0, ♂ 15.0–21.0 ; tegmina ♀ 22.0–26.0, ♂ 16.0–26.0.

Distribution : North Africa through Asia Minor to Iran, Central and Northern Europe to China.

Israel : Upper and Lower Galilee (1, 2), Carmel (3), Judean Hills (11).

Mostly inhabits high mountain levels and valleys and occurs generally on dark basalts ; in the north of Israel found together with *O. aurea.* Adults observed from March to November.

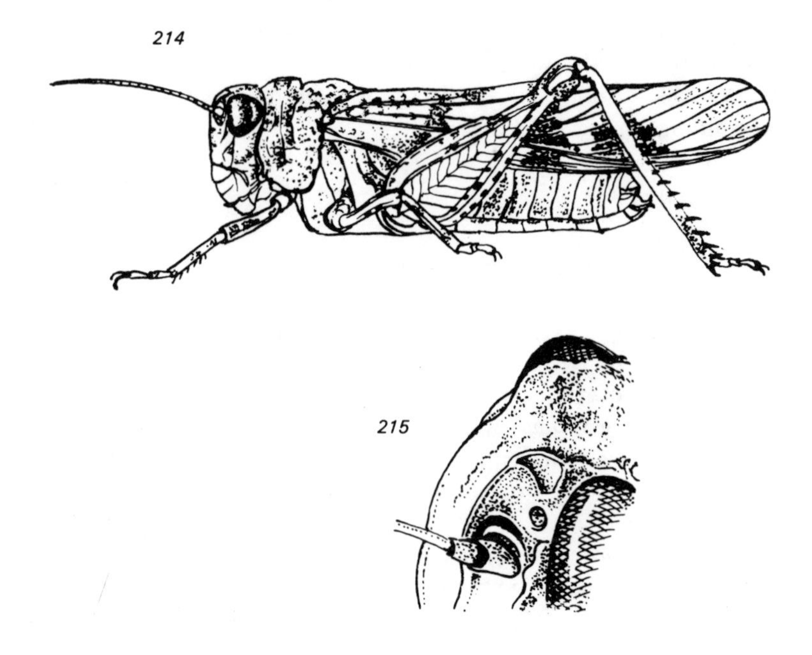

Figs. 214–215 : *Oedipoda caerulescens* (Linnaeus, 1758)
214. female ; 215. head, latero-dorsal

Genus AIOLOPUS Fieber, 1853
Lotos, 3 : 100

Epacromia Fischer L. H., 1853, *Orthoptera Europaea*, Leipzig, pp. 296, 360.
Aiolopus —. Innes W., 1929, *Mém. Soc. R. ent. Égypte*, 3 (2) : 11, 29.
Aiolopus —. Hollis D., 1968, *Bull. Br. Mus. nat. Hist. (Ent.)*, 22 (7) : 314.

Type Species : *Gryllus thalassinus* Fabricius, 1781.
Diagnosis : Medium sized, smooth, compressed laterally. Head slightly opisthognathous ; face smooth ; fastigium of vertex pentagonal ; frons oblique. Fastigial foveolae elongate, narrowing anteriorly, continuous in front (Fig. 218). Pronotum elongate, saddle-shaped, slightly roof-shaped. Prozona narrower than metazona. Median carina indistinct, slightly intersected by the posterior transverse sulcus. Lateral carinae absent or spurious. Posterior margin of metazona obtuse-angled ; margins of lateral lobes rounded ventrally.
Tegmina and wings developed. Intercalary vein in median field straight, serrate.

Wings colourless; outer margin sometimes greyish. Inner side of hind femur light, often reddish with dark brown bands. Posterior part of hind tibia usually reddish. Distributed in the tropical and subtropical regions of the Old World, penetrating into the Indian Ocean areas. Ecologically connected with humid areas of grassy vegetation, feeding and hiding among Gramineae.
Three species in our region.

Key to the Species of Aiolopus in Israel

1. Hind tibia as long as or slightly shorter than hind femur, with 9–12 outer spines and 10–13 inner ones. Pronotum flat dorsally with traces of lateral carinae — 2
– Hind tibia much shorter than hind femur, with nine outer and 10 inner spines. Pronotum usually slightly raised along median carina. — **A. simulatrix simulatrix** (Walker)
2. Anterior margin of tegmina with a distinct white band between two dark ones. Pronotum without distinct constriction in prozona. Length of hind femur 3.0–3.2 times its maximum width (Fig. 216). — **A. strepens** (Latreille)
– Tegmina without white band on anterior margin. Pronotum distinctly narrower in prozona, diverging posteriorly. Length of hind femur 4.0–4.2 times its maximum width (Fig. 217). — **A. thalassinus thalassinus** (Fabricius)

Aiolopus simulatrix simulatrix (Walker, 1870)
Colour plate: 3

Type Locality: South Hindustan (British Museum).

Epacromia simulatrix Walker F., 1870, *Catalogue of the specimens of Dermaptera Saltatoria in the Collection of the British Museum*, London, Part IV, p. 773.
Heteropternis savignyi Krauss H. A., 1890, *Verh. zool.-bot. Ges. Wien*, 40: 262.
Epacromia affinis Bolivar I., 1902, *Annls Soc. ent. Fr.*, 70: 600.
Epacromia strepens —. Krauss H. A., 1909, in: Kneucker A., *Verh. naturw. Ver. Karlsruhe*, 21: 36.
Acrotylus simulatrix —. Kirby W. F., 1910, *A synonymic catalogue of the Orthoptera*, Vol. III, Orthoptera Saltatoria, Part II, London, p. 267.
Aeolopus laticosta Bolivar I., 1912, *Trans. Linn. Soc. Lond.* (Zool.), 15: 270.
Aeolopus affinis —. Kirby W. F., 1914, *The Fauna of British India, including Ceylon and Burma. Orthoptera (Acrididae)*, London, p. 122.
Aeolopus strepens deserticola Uvarov B. P., 1922, *J. Bombay Nat. Hist. Soc.*, 28: 358.
Aiolopus strepens affinis —. Buxton P. A. & B. P. Uvarov, 1923, *Bull. Soc. R. ent. Égypte*, p. 191.
Heteropternis savignyi —. Innes W., 1929, *Mém. Soc. R. ent. Égypte*, 3 (2): 91.
Aiolopus affinis —. Bodenheimer F. S., 1935, *Animal Life in Palestine*, Jerusalem, pp. 86, 88, 311, 320, 323.
Aiolopus affinis —. Bodenheimer F. S., 1935, *Arch. Naturgesch.*, 4 (2): 186.
Aiolopus savignyi —. Uvarov B. P., 1942, *Trans. Am. ent. Soc.*, 67: 338.
Aiolopus savignyi —. Ramme W., 1951, *Mitt. zool. Mus. Berl.*, 27: 425.

Medium sized. Head opisthognathous, slightly swollen, raised above pronotum. Face densely pitted; frontal ridge flat, its margins narrowing below fastigium. Fastigial foveolae rectangular, usually with well pronounced margins. Eyes oval. Antennae as long as or longer than head and pronotum together. Pronotum with raised median carina, especially in prozona, with three transverse sulci, the posterior one anterior to its middle.
Tegmina long, reaching beyond middle of the hind tibia; intercalary vein extending along most of the median field. Length of hind femur 3.0–3.5 times its maximum width. Hind tibia shorter than hind femur.
Coloration: Ochraceous, pale or dark, rarely with green or blackish markings. Tegmina with a dark band marginate by lighter areas; apex mottled, transparent. Wings transparent. Inner side of hind femur light with two dark bands; hind knee dark. Hind tibia black at base; middle area with a broad grey ring; apical third pale reddish.
Measurements (mm): Body ♀ 21.5–31.0, ♂ 17.0–25.5; tegmina ♀ 19.0–27.2, ♂ 16.5–22.0.
Distribution: Egypt, Somalia, Uganda, Israel, Central Asia.
Israel: Upper Galilee (1), Jordan Valley (7), Central and Southern Coastal Plain (8, 9), Judean Hills (11).
Associated with areas of low sparse vegetation.

Aiolopus strepens (Latreille, 1804)
Fig. 216

Neotype Locality: Les Eyzies, Dordogne, France.

Acrydium strepens Latreille P. A., 1804, *Histoire naturelle générale et particulière des Crustacées et des Insectes. Orthoptera, Acrididae*, 3: 154.
Gryllus prasinus Thunberg C. P., 1815, *Mém. Acad. Sci. St.-Pétersb.*, 5: 239.
Oedipoda thalassina (Fabricius). Serville J. G. A., 1838, *Histoire naturelle des Insectes*, in: Roret, *Collection des Suites à Buffon. Orthoptères*, Paris, p. 140 (partim).
Aiolopus strepens —. Fieber F. X., 1853, *Lotos*, 3: 100.
Epacromia strepens —. Bolivar I., 1876, *An. Soc. esp. Hist. Nat.*, 5: 348.
Aiolopus strepens —. Innes W., 1929, *Mém. Soc. R. ent. Égypte*, 3 (2): 30.
Aiolopus strepens —. Bodenheimer F. S., 1935, *Arch. Naturgesch.*, 4 (2): 186.
Aiolopus strepens —. Bei-Bienko G. Ya. & L. L. Mishchenko, 1951, *Locusts and Grasshoppers of the U. S. S. R. and Adjacent Countries*, Moscow, II: 569 [in Russian].

Medium sized, robust. Head opisthognathous, in male raised above pronotum; fastigial foveolae trapezoid. Antennae as long as or shorter than head and pronotum together. Eyes oval, their vertical diameter almost twice as long as the horizontal diameter. Pronotum not constricted anteriorly; metazona more than 1.5 times as long as prozona, posterior margin obtuse-angled.

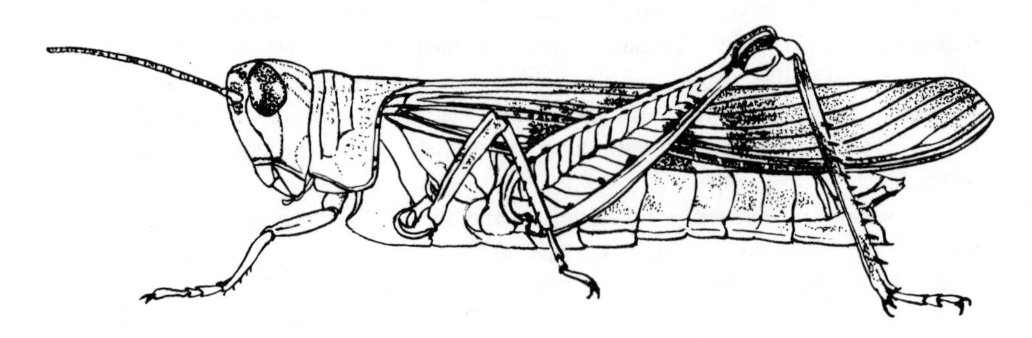

Fig. 216 : *Aiolopus strepens* (Latreille, 1804), female

Tegmina short and broad, slightly extending beyond the knees. Length of hind femur 3.0–3.4 times its width. Hind tibia as long as or shorter than hind femur, with 10 outer and 10–11 inner spines.

Coloration : Brown or ochraceous brown, rarely with greenish parts. Tegmina with a middle transverse dark band marginate by two regular pale transparent areas, the proximal one sometimes obliterated, leaving only a light triangular spot. Inner side of hind femur pale, sometimes reddish ventrally, with dark bands situated dorsally and sometimes also on the outer surface. Hind tibia reddish distally, ochraceous or pinkish at the base, with a black ring in the middle.

Measurements (mm) : Body ♀ 21.7–32.0, ♂ 17.0–32.0 ; tegmina ♀ 20.0–30.5, ♂ 16.5–24.0.

Distribution : Europe, North Africa, Middle East, Central Asia.

Israel : Upper Galilee (1), Jordan Valley (7), Coastal Plain (4, 8, 9), 'Arava Valley (14), Judean Hills (11).

Aiolopus thalassinus thalassinus (Fabricius, 1781)

Figs. 217, 218

Neotype Locality : Locarno, Switzerland.

Gryllus thalassinus Fabricius J. C., 1781, *Species Insectorum*, 1 : 367.

Acrydium thalassinum —. Olivier G. A., 1791, *Encyclopédie Méthodique, Histoire Naturelle*, 6 : 225.

Gryllus prasinus Thunberg C. P., 1815, *Mém. Acad. Sci. St.-Pétersb.*, 5 : 239.

Oedipoda thalassina —. Serville J. G. A., 1831, *Annls Sci. nat. (Zool.)*, 22 : 288.

Acridium grossum Costa O. G., 1836, *Fauna del regno di Napoli. Ortotteri*, Naples, p. 25, pl. 3, figs. 4a–4d.

Gomphocerus thalassinus —. Burmeister H., 1838, *Handbuch der Entomologie*, Berlin, 2 (2) : 647.

Acridium laetum Brullé A., 1840, Orthoptera, in : Webb P. B. & S. Berthelot, *Histoire naturelle des Iles des Canaries*, 2 (2) : 77, pl. 5, figs. 10, 10a.

Epacromia thalassina —. Fischer L. H., 1853, *Orthoptera Europaea*, Leipzig, p. 361, pl. 17, figs. 14, 14a, b.
Aiolopus pulverentulus Fieber F. X., 1854, *Lotos*, 4 : 179.
Epacromia angustifemur Ghiliani V., 1869, *Bull. Soc. ent. ital.*, 1 : 179.
Ochrophlebia savignyi Krauss. Werner F., 1905, *Sber. Akad. Wiss. Wien*, 114 (1) : 423.
Aiolopus thalassinus —. Karny H., 1907, *Sber. Akad. Wiss. Wien*, 116 : 359.
Aelopus thalassinus —. Kirby W. F., 1910, *A synonymic catalogue of the Orthoptera*, Vol. III, Orthoptera Saltatoria, Part II, London, p. 190.
Aiolopus thalassinus —. Buxton P. A. & B. P. Uvarov, 1923, *Bull. Soc. R. ent. Égype*, p. 191.
Aiolopus thalassinus —. Innes W., 1929, *Mém. Soc. ent. Égypte*, 3 (2) : 30, 32.
Aiolopus thalassinus —. Bodenheimer F. S., 1935, *Animal Life in Palestine*, Jerusalem, pp. 320, 323.
Aiolopus thalassinus —. Bodenheimer F. S., 1935, *Arch. Naturgesch.*, 4 (2) : 185.
Aiolopus thalassinus —. Ramme W., 1951, *Mitt. zool. Mus. Berl.*, 27 : 425.
Aiolopus acutus Uvarov B. P., 1953, *Publções cult. Co. Diam. Angola*, no. 21 : 111, figs. 129–131.

Medium sized or small, smooth. Head distinctly opisthognathous. Antennae in male as long as or longer than head and pronotum together , in female shorter. Fastigium pentagonal, marginate ; fastigial foveolae trapezoid, narrowing apically, ventral margin sometimes granulate (Fig. 218). Pronotum slightly saddle-shaped, constricted in prozona, gradually widening in metazona ; posterior margin obtuse-angled.
Tegmina extend beyond hind knees ; radius-sector with two or three branches ; intercalary vein straight. Length of hind femur at least four times its maximum width. Hind tibia with 10 outer and 10–11 inner spines.
Coloration : Ochraceous or brown, with green parts. Head and pronotum with or without a longitudinal median pale stripe and a pale transverse pattern on the dorsum. Tegmina without pale transverse bands or only with signs of them close to anterior margin. Hind femur usually with two dorsal dark spots, its inner part yellowish or reddish with dark cross-bands. Hind tibia reddish or red, pale at base, with a black ring on knee.
Measurements (mm) : Body ♀ 19.5–29.0, ♂ 15.0–22.0 ; tegmina ♀ 18.0–27.2, ♂ 18.0–21.3.
One of the most variable species of *Aiolopus.*
Distribution : South Africa to North Africa, Southern Europe, Mediterranean, India, Southern China.
Israel : Upper and Lower Galilee (1, 2), Jordan Valley (7), Central and Southern Coastal Plain (8, 9), Judean Hills (11), Dead Sea Area (13), Northern and Central Negev (15, 17).
Occurs along water bodies, in grassy areas with green, soft vegetation, especially in the northern regions. Cultivation and irrigation of desert areas enlarge the area of distribution of this species.

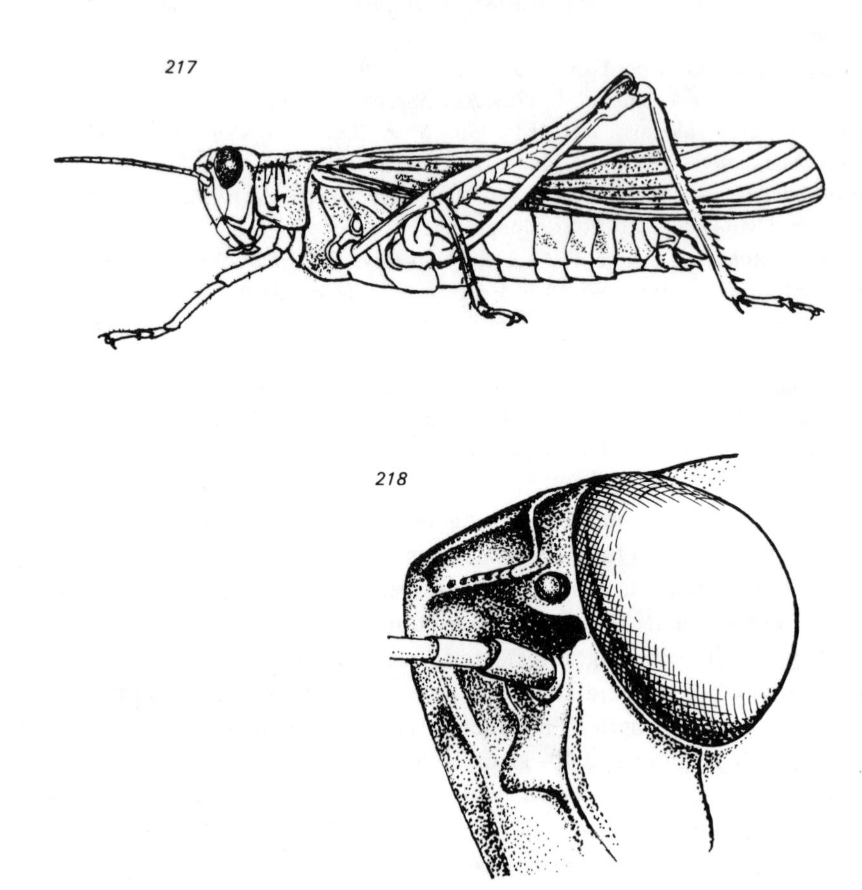

Figs. 217–218 : *Aiolopus thalassinus thalassinus* (Fabricius, 1781)
217. female ; 218. head, lateral view

Genus MIOSCIRTUS Saussure, 1888

Mém. Soc. Phys. Hist. nat. Genève, 30 (1) : 18

Scintharista (Mioscirtus) Finot A., 1895, *Annls Soc. ent. Fr.*, 64 : 439, 448.

Type Species : *Oedipoda wagneri* Eversmann, 1859.
Diagnosis : Medium sized. Head small, hypognathous, not projecting above pronotum. Vertex concave, with sharp margins merging between the eyes. Metazona slightly larger than prozona, with obtuse posterior angle. Median carina distinct along its entire length, slightly raised in prozona. Basal parts of wings yellow or rose, with an

incomplete black band (Fig. 220). Hind femur thin , with two dark spots on its upper outer surface, light spots with two dark bands (one incomplete) on inner surface. A single species distributed from China throughout Southern Russia and Central Asia to Israel ; one subspecies in our region.

Mioscirtus wagneri rogenhoferi Saussure, 1888
Figs. 219, 220

Type Locality : 'South-East Russia'.

Mioscirtus wagneri rogenhoferi Saussure H. de, 1888, *Mém. Soc. Phys. Hist. nat. Genève*, 30 (1) : 1–182.
Mioscirtus wagneri rogenhoferi —. Buxton P. A. & B. P. Uvarov, 1923, *Bull. Soc. R. ent. Égypte*, p. 191.
Mioscirtus wagneri rogenhoferi —. Bodenheimer F. S., 1935, *Arch. Naturgesch.*, 4 (2) : 187.
Mioscirtus wagneri rogenhoferi —. Bei-Bienko G. Ya. & L. L. Mishchenko, 1951, *Locusts and Grasshoppers of the U. S. S. R. and Adjacent Countries*, Moscow, II : 588 [in Russian].

Medium sized. Head small, hypognathous or slightly opisthognathous, not raised above pronotum. Face sparsely punctate ; frontal ridge concave, slightly widening towards clypeus. Fastigium of vertex oblong, concave with sharp margins. Foveolae obliterated or only slightly visible. Antennae filiform, longer than head and pronotum together. Pronotum raised along the median carina, with three transverse sulci, of which the first and third intersect the median carina. Lateral carinae obliterated in prozona, forming shoulders in metazona. Lateral lobes of prozona longer than high. Sternum longer than wide ; mesosternal interspace square in female, widening posteriorly in male.
Base of tegmina densely reticulate ; intercalary vein straight ; radius-sector with two or three branches. Wings with a black band branching towards the base (Fig. 220).
Coloration : Generally pale brown. Two dark areas on margin of tegmina intersected by a trapezoid light band. Inner part of wings yellow or sometimes rose (in females of ab. *varenzovi* Zub.). Inner side of hind femur light with two dark bands that continue on the dorsal surface. Hind tibia with dark ring in the middle.
Measurements (mm) : Body ♀ 23.0–27.0, ♂ 16.0–17.0 ; tegmina ♀ 21.0–28.0, ♂ 13.0–17.0.
Distribution : Central Asia, Israel.
Israel : Jordan Valley (7), Dead Sea Area (13).
Phytophilous, inhabits arid regions, usually in areas with low dry plants.

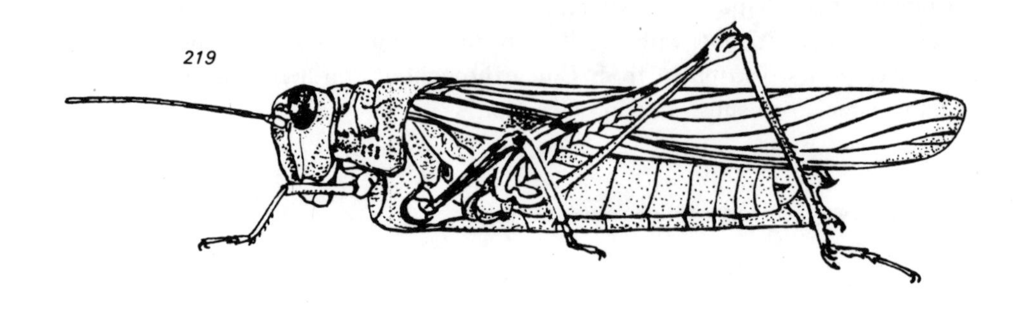

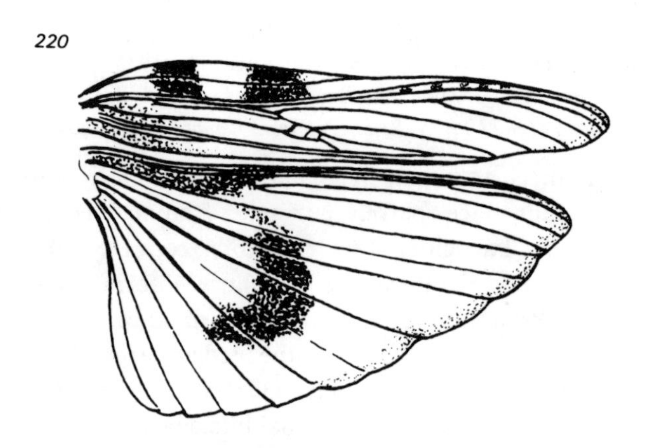

Figs. 219–220 : *Mioscirtus wagneri rogenhoferi* Saussure, 1888
219. female ; 220. wings

Genus HILETHERA Uvarov, 1923

Entomologist's mon. Mag., 59 : 82

Type Species : *Hilethera hierichonica* Uvarov, 1923.

Diagnosis : Small, robust. Foveolae marginate, extending from eyes to apex of fastigium. Antennae short. Median carina of pronotum distinct, intersected by one transverse sulcus. Hind femur very short ; inner side black with one or two light bands (Fig. 223).

Five species known, distributed from China to Africa. One species in Israel.

Hilethera hierichonica Uvarov, 1923

Figs. 221–223 ; Plate IV : 7

Type Locality : Jericho.

Hilethera hierichonica Uvarov B. P., 1923, *Entomologist's mon. Mag.*, 9 : 83.
Hilethera hierichonica —. Buxton P. A. & B. P. Uvarov, 1923, *Bull. Soc. R. ent. Égypte*, p. 191.
Hilethera hierichonica —. Bodenheimer F. S., 1935, *Arch. Naturgesch.*, 4 (2) : 186.

Small, granulate. Head slightly opisthognathous. Frontal ridge flat, concave around the median ocellus, slightly widening towards clypeus. Foveolae triangular, with raised margins, foveolae extending anteriorly towards apex of fastigium. Vertex wide, concave. Antennae shorter than, as long as, or slightly longer than head and pronotum together ; most segments wider than long. Pronotum right-angled posteriorly ; lateral and median carinae distinct along their entire length. Median carina before middle of pronotum intersected by the third sulcus (Fig. 222) ; lateral carinae intersected by all three sulci.

Tegmina wide, their length four times their maximum width. Intercalary vein long and straight, granulate and raised above surface ; radius-sector with two branches. Hind femur thick ; inner side black with a light apical band (Fig. 223). Hind tibia shorter than femur, black with a white ring.

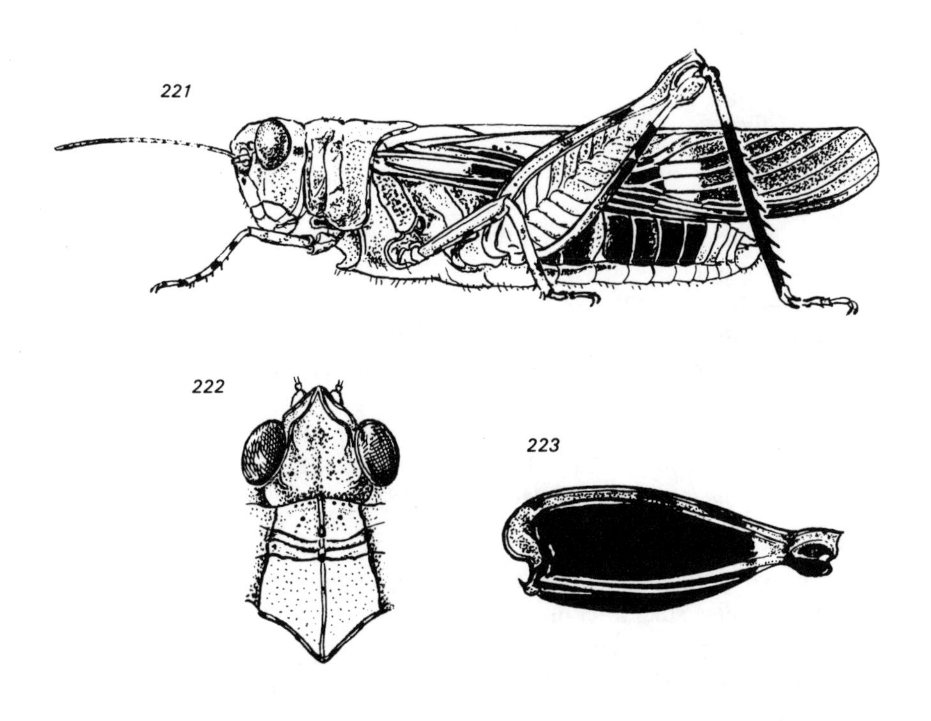

Figs. 221–223 : *Hilethera hierichonica* Uvarov, 1923
221. male ; 222. head and pronotum ; 223. hind femur, inner side

Coloration : Generally dark brown with typical black abdominal tergites (Fig. 221). Tegmina brown with paler transverse fields.
Measurements (mm) : Body ♀ 19.0, ♂ 14.5 ; tegmina ♀ 16.5, ♂ 14.0.
Distribution : Endemic in Israel.
Israel : Southern Jordan Valley (7), Dead Sea Area (13).
Very rare ; inhabits areas with very sparse vegetation.

Genus PYRGODERA Fischer-Waldheim, 1846

Orthoptera Imperii Russici, Moscow, p. 272

Type species : *Pyrgodera armata* Fischer-Waldheim, 1846.
Diagnosis : Large and robust, compressed laterally. Head small, covered above by high, helmet-shaped pronotum (Fig. 224). Wings coloured, with a black band.
Distribution : South-Central Russia, Central Asia, Transcaucasus, Syria, Israel.
Monotypic.

Pyrgodera armata Fischer-Waldheim, 1846

Frontispiece, Fig. 224

Type Locality : Tiflis.

Pyrgodera armata Fischer de Waldheim G., 1846, *Orthoptera Imperii Russici*, Moscow, p. 273, tabl. 21, figs. 1, 2.
Pyrgodera cristata Eversmann. Brunner von Wattenwyl C., 1882, *Prodromus der europäischen Orthopteren*, Leipzig, p. 174.
Pyrgodera cristata —. Navas L., 1911, *Revista Montserratina*, Barcelona, p. 2.
Pyrgoderea armata —. Buxton P. A. & B. P. Uvarov, 1923, *Bull. Soc. R. ent. Égypte*, p. 191.
Pyrgodera armata —. Bodenheimer F. S., 1935, *Animal Life in Palestine*, Jerusalem, pp. 28b, 320.
Pyrgodera armata —. Bodenheimer F. S., 1935, *Arch. Naturgesch.*, 4 (2) : 185.
Pyrgodera armata —. Bei-Bienko G. Ya. & L. L. Mishchenko, 1951, *Locusts and Grasshoppers of the U. S. S. R. and Adjacent Countries*, Moscow, II : 583 [in Russian].

Smooth, laterally compressed. Head hypognathous ; vertex flat ; foveolae triangular. Pronotum very high along median carina, with sharp angles at posterior and anterior margins, with transverse sulci laterally.
Coloration : Generally brown or pale ; hoppers brown or green. Tegmina narrow at base, opaque with two dark bands. Wings red, with a wide black band sending a short branch towards base.
Measurements (mm) : Body ♀ 35.0–40.0, ♂ 20.0–25.0 ; tegmina ♀ 33.0–38.0, ♂ 27.0–32.0.

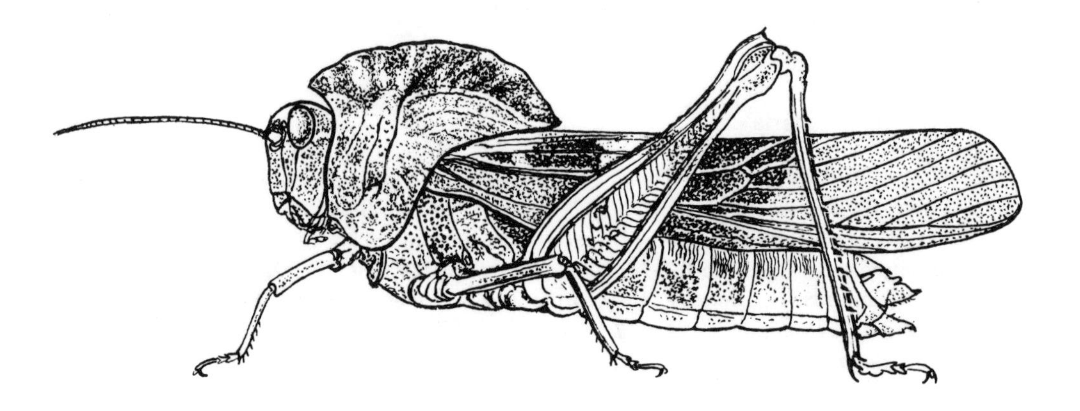

Fig. 224 : *Pyrgodera armata* Fischer-Waldheim, 1846, female

Distribution : Southern U.S.S.R., Lebanon, Israel.
Israel : Jordan Valley (7), Judean Hills (11, 12), Northern Negev (15).
The population in Israel is the southernmost.
Adults and hoppers found in March and April, inhabiting areas with dense vegetation.

Genus LOCUSTA Linnaeus, 1758

Systema Naturae, 10th ed., 1 : 431

Gryllus Locusta Linnaeus C., 1758, *Systema Naturae*, 10th ed., 1 : 431.
Oedipus Berthold A. A., 1827, *Latreille's natürliche Familiën des Thierreichs*, Weimar, p. 411.
Pachytylus Fieber F. X., 1852, in : Kelch A., *Grundlage zur Kenntnis der Orthopteren Oberschlesiens, etc.*, Ratibor, Bögner, p. 5.
Locusta —. Kirby W. F., 1910, *A synonymic catalogue of the Orthoptera*, Vol. III, Orthoptera Saltatoria, Part II, London, p. 228.
Locusta —. Innes W., 1929, *Mém. Soc. R. ent. Égypte*, 3 (2) : 41, 89.

Type Species : *Gryllus (Locusta) migratorius* Linnaeus, 1758.
Diagnosis : Large ; head hypognathous. Frontal ridge flat ; vertex rounded. Fastigium with median carina. Median carina of pronotum raised, straight or slightly curved, intersected in the middle by third transverse sulcus ; posterior and anterior margins acute (Fig. 226). Sternum densely covered by hairs.
Coloration : Brown ; head and pronotum often green. Tegmina with large brown spots, transparent at apex. Wings transparent, colourless ; veins dark.
Distribution : Asia, New Zealand, Australia, Africa.
One species, *Locusta migratoria*, which occurs in two phases : gregarious and solitary. Only the latter occurs in our region.

Locusta migratoria (Linnaeus, 1758) ph. **solitaria**
Figs. 225–227 ; Plate V : 3

Type Locality : 'Tataria'.

Gryllus (Locusta) migratorius Linnaeus C., 1758, *Systema Naturae*, 10th ed., 1 : 432.
Gryllus (Locusta) danicus Linnaeus C., 1767, *Systema Naturae*, 12th ed., p. 702.
Gryllus cinerascens Fabricius J. C., 1781, *Species Insectorum*, 1 : 369.
Oedipoda migratoria —. Serville J. G. A., 1831, *Annls Sci. nat. (Zool.)*, 22 : 288.
Acridium migratorium —. Brullé A., 1840, Orthoptera, in : Webb P. B. & S. Berthelot, *Histoire naturelle des Iles Canaries*, 2 (2) : 77.
Pachytylus cinerascens —. Fieber F. X., 1853, *Lotos*, 3 : 121.
Pachytus migratorius —. Lucas H., 1862, Orthoptères, in : Maillard L., *Notes sur l'île de Réunion*, Annexe I, Paris, p. 23.
Pachytylus migratorius —. Saussure H. de, 1884, *Mém. Soc. Phys. Hist. nat. Genève*, 28 (9) : 119, 120.
Acridium tataricum Hart C. H., 1891, *Fauna and Flora of Sinai, Petra and Wadi Arabah*, London, p. 183.
Pachytylus migratorius —. Giglio-Tos E., 1893, *Boll. Musei Zool. Anat. comp. R. Univ. Torino*, 8 (164) : 6.
Pachytylus danicus —. Bolivar I., 1895, *Annls Soc. entr. Fr.*, 64 : 378.
Locusta danica —. Kirby W. F., 1910, *A synonymic catalogue of the Ortoptera*, Vol. III, Orthoptera Saltatoria, Part II, London, p. 230.
Locusta migratoria ph. *danica* —. Buxton P. A. & B. P. Uvarov, 1923, *Bull. Soc. R. ent. Égypte*, p. 193.
Locusta danica —. Innes W., 1929, *Mém. Soc. R. ent. Égypte*, 3 (2) : 89.
Locusta migratoria —. Bodenheimer F. S., 1930, *Die Schädlingsfauna Palästinas*, Berlin, pp. 71, 105, 325, 344, 350.
Locusta migratoria —. Bodenheimer F. S., 1935, *Animal Life in Palestine*, Jerusalem, pp. 38, 320, 323.
Locusta migratoria —. Bodenheimer F. S., 1935, *Arch. Naturgesch.*, 4 (2) : 189.
Locusta migratoria —. Ramme W., 1951, *Mitt. zool. Mus. Berl.*, 27 : 425.

Large, smooth. Head hypognathous, ventrally rounded. Face smooth ; frontal ridge with low parallel margins, concave around median ocellus. Median carina of vertex in female swollen, in male slightly concave. Interocular distance narrower than vertical diameter of eye. Foveolae absent. Antennae shorter than head and pronotum together. Pronotum roof-shaped, raised along the median carinae, intersected by one transverse sulcus ; posterior margin right-angled ; anterior margin angular (Fig. 226). Sternum hairy ; mesosternal interspace 1.5 times longer than wide.
Tegmina wide, their length five times their maximum width. Radius-sector with three or four branches; intercalary vein straight, in male densely granulate (Fig. 227).
Coloration : Brown with green body parts. Tegmina with dense brown blotches at the base and diffuse blotches apically. Wings shiny greenish or yellowish, sometimes with brown lines on apex.

Measurements (mm) : Body ♀ 45.0–55.0, ♂ 35.0–50.0 ; tegmina ♀ 49.0–61.0, ♂43.5–56.0.

Although the distribution of *L. migratoria* is almost world-wide, the migratory phases are restricted in their distribution. The Caucasus and Central Asia, especially along the Amu-Darya and Syr-Darya rivers, are their largest breeding regions. They also reproduce in China, Malaysia, Madagascar and Central and Western Africa. They usually breed in swampy regions, along rivers and lakes, and with the application of ameliorated techniques and agricultural development, the distribution of migratory phases of *L. migratoria* is becoming more and more restricted. Although only the solitary phase occurs in our region, it may form dense persistent local populations which cause damage to fields.

Usually found on the borders of humid areas rich in green vegetation. During summer it appears in large numbers in fields and irrigated plantations. Hoppers brown or green ; distinguished by the roof-shaped pronotum, intersected by a single sulcus.

Distribution, Israel : Upper and Lower Galilee (1, 2), Hula Valley (18), Jordan Valley (7), Coastal Plain (4, 8), Judean Hills (11), Dead Sea Area (13), Northern and Central Negev (15, 17).

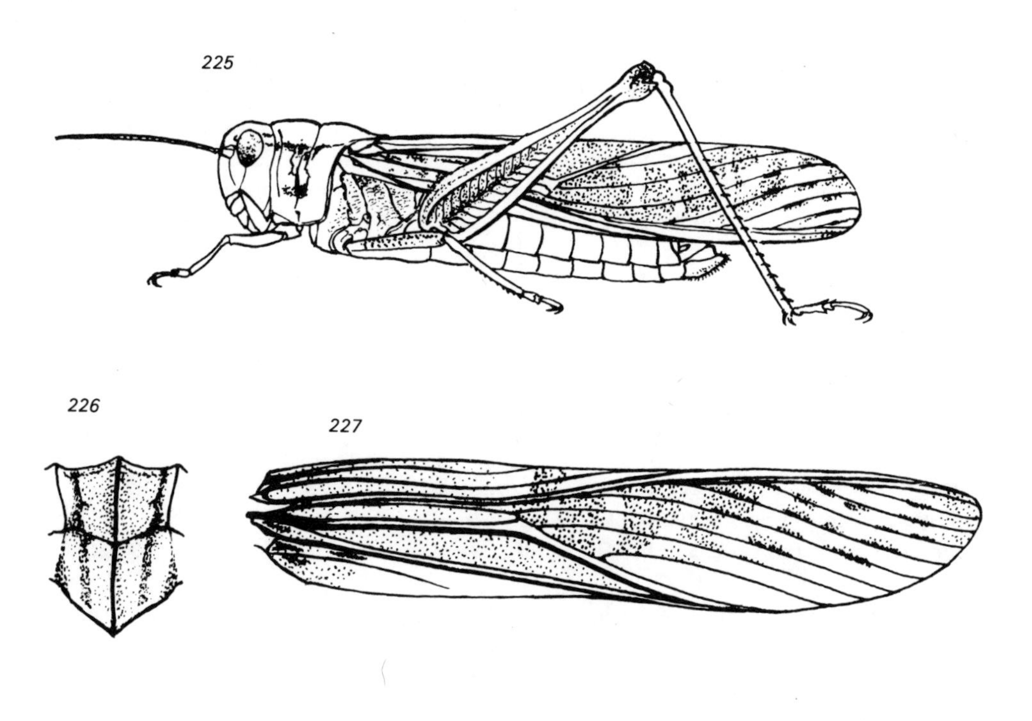

Figs. 225–227 : *Locusta migratoria* (Linnaeus, 1758) ph. *solitaria*
225. male ; 226. pronotum ; 227. tegmen

Genus OEDALEUS Fieber, 1853

Lotos, 3 : 126

Oedipoda (Oedaleus) Fieber F. X., 1853, *Lotos*, 3 : 126.
Pachytylus (Oedaleus) Stål C., 1873, *Recensio Orthopterorum*, Stockholm, 1 : 123.
Oedaleus —. Saussure H. de, 1884, *Mém. Soc. Phys. Hist. nat. Genève*, 28 (9) : 50, 108, 110, 115.

Type Species : *Acrydium nigrofasciatum* De Geer, 1773.
Diagnosis : Medium sized, strong, smooth. Head hypognathous or slightly opithognathous. Vertex oblique ; frontal ridge flat or slightly concave ; foveolae triangular. Pronotum usually with a pale X-shaped marking (Fig. 230) ; median carina slightly raised along its entire length, intersected near the middle by a single transverse sulcus. Shallow constriction present on both sides of prozona. Posterior margin of lateral lobe rounded.
Cubital field of tegmina wider than median one, with square cellules on the apical half. Black band on wing (Fig. 231).
Twenty species, distributed in Africa, Australia and one in Europe. Two species in our region.

Key to the Species of Oedaleus in Israel

1. Inner surface of hind femur whitish, sometimes with dark bands. Hind tibia whitish-grey. Median carina of pronotum straight; posterior margin of pronotum obtuse-angled (Fig. 228). **O. senegalensis** (Krauss)
- Inner surface of hind femur black, with two pale bands. Hind tibia red or reddish-yellow, with a white band near knee. Median carina of pronotum raised (Fig. 230); posterior margin of pronotum right-angled. **O. decorus** (Germar)

Oedaleus senegalensis (Krauss, 1877)

Fig. 228

Type Locality : St. Louis, Senegal (Vienna Museum).

Pachytylus senegalensis Krauss H. A., 1877, *Sber. Akad. Wiss. Wien*, 76 (1) : 56, pl. 1, figs. 9, 9a.
Ctypohippus arenivolans Butler A. G., 1881, *Proc. zool. Soc. Lond.*, 1881 : 85.
Oedaleus senegalensis —. Saussure H. de, 1884, *Mém. Soc. Phys. Hist. nat. Genève*, 28 (9) : 110, 117.
Oedaleus senegalensis —. Bodenheimer F. S., 1935, *Animal Life in Palestine*, Jerusalem, pp. 28b, 40.
Oedaleus senegalensis —. Bodenheimer F. S., 1935, *Arch. Naturgesch.*, 4 (2) : 188.

Medium sized, smooth. Head slightly opisthognathous. Antennae long, in female 1.5 times as long as, in male nearly twice as long as head and pronotum together;

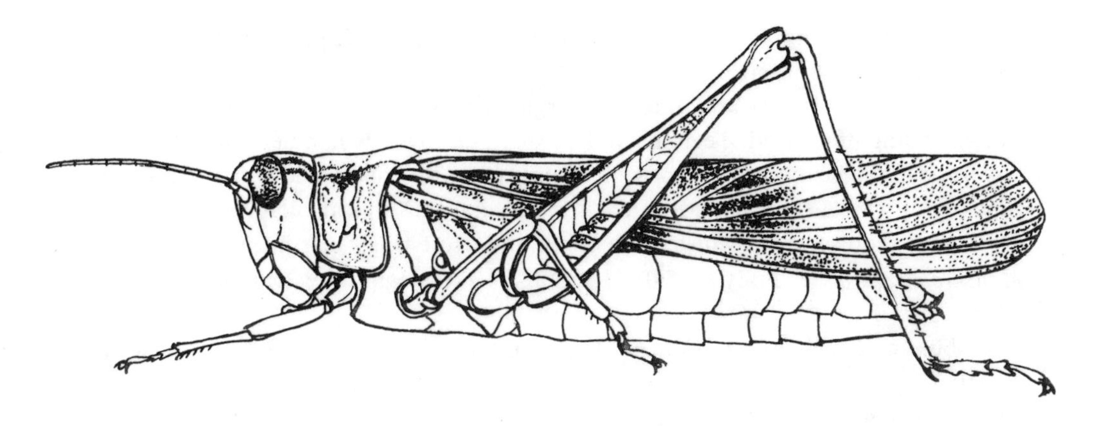

Fig. 228 : *Oedaleus senegalensis* (Krauss, 1877), female

segments 2–3 times longer than wide. Pronotum obtuse-angled posteriorly, compressed in the middle ; single transverse sulcus sometimes intersecting the straight median carina. X-shaped marking on pronotum distinct or obliterated in the middle, extending anteriorly on head, towards the eyes. Tegmina long, extending beyond middle of hind tibia, transparent at apex ; intercalary vein straight ; median field densely veined.
Coloration : Pale or greenish with dark cross-bands on tegmina. Wings with a narrow, sometimes diffuse dark band and yellowish, transparent base. Inner side of hind femur pale yellow ; hind tibia pale or pinkish towards apex.
Measurements (mm) : Body ♀ 22.0–32.0, ♂ 10.0–15.0 ; tegmina ♀ 24.0–34.0, ♂ 18.0–25.0.
Distribution : Senegal, East Africa, through Arabia, Central Asia, Kashmir and Punjab.
Israel : Jordan Valley (7), Northern Negev (15).
Found from April to June in areas with low vegetation.

Oedaleus decorus (Germar, 1826)
Figs. 229–231 ; Plate V : 5

Type Locality : Podolia, Southern U. S. S. R.

Acrydium decorum Germar E. F., 1826, *Fauna Insectorum Europae*, 12 : pl. 17.
Oedaleus nigrofasciatus —. Navas L., 1911, *Revista Monteserratina*, Barcelona, p. 2.
Oedaleus decorus —. Buxton P. A. & B. P. Uvarov, 1923, *Bull. Soc. R. ent. Égypte*, p. 192.
Oedaleus decorus —. Bodenheimer F. S., 1935, *Arch. Naturgesch.*, 4 (2) : 188.
Oedaleus decorus —. Ramme W., 1951, *Mitt. zool. Mus. Berl.*, 27 : 425.

Medium sized, strong. Head hypognathous. Vertex flat; fastigium with median carina (Fig. 230). Pronotum roof-shaped along the median carina, strongly compressed in the middle; posterior and anterior margins angular. X-shaped marking distinct, not extending onto head. Tegmina wide, extending towards middle of hind tibia.
Coloration: Green or pale; bands on tegmina and on outer and inner surface of hind femur. Wings with a wide, dark band and with apical blotches (Fig. 231). Hind tibia reddish or brownish, with black and white rings near knee.
Measurements (mm): Body ♀ 25.0–43.0, ♂ 18.0–31.0; tegmina ♀ 25.0–40.0, ♂ 16.0–33.0.
Distribution: North Africa, Central Asia, European U. S. S. R. and Transcaucasus. Israel: Central Coastal Plain (4, 8), Judean Hills (6, 10, 11).
Inhabits dry grasslands in the northern regions of its distribution. In our region it is hygrophilous, inhabiting wet grasses along streams and lakes.

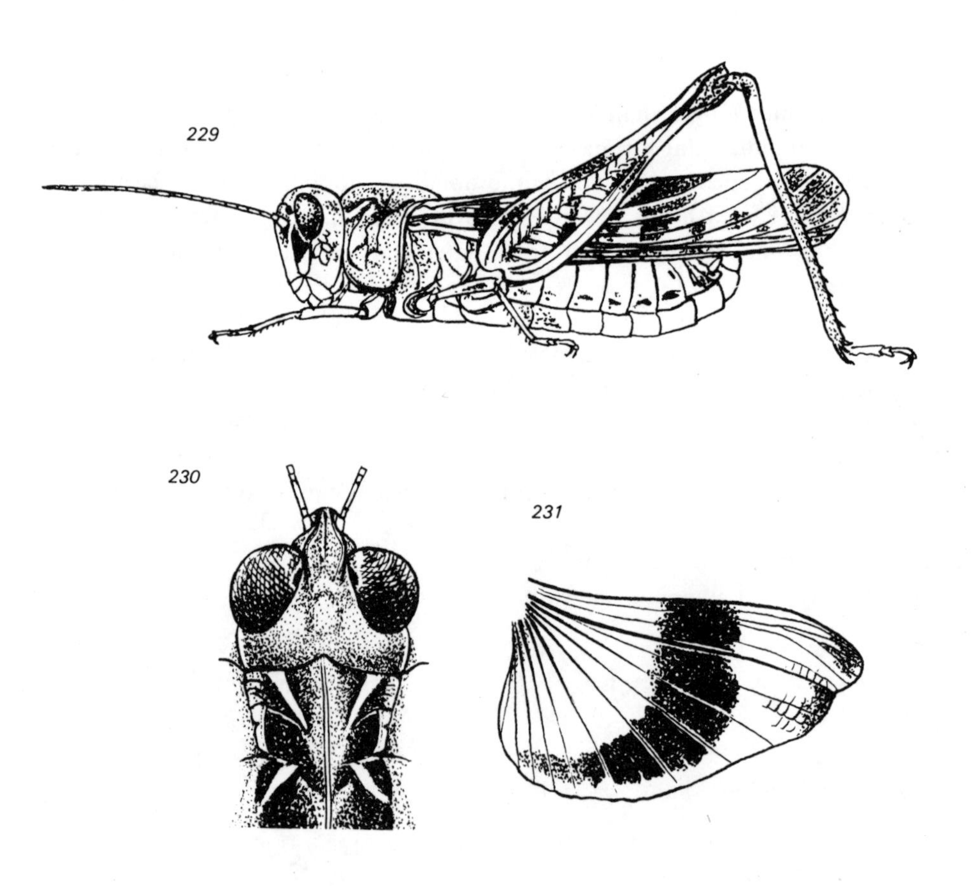

Figs. 229–231: *Oedaleus decorus* (Germar, 1826)
229. male; 230. head and pronotum; 231. wing

Genus SCINTHARISTA Saussure, 1884

Mém. Soc. Phys. Hist. nat. Genève, 28 (9) : 51, 121

Quiroguesia Bolivar I., 1886, *An. Soc. esp. Hist. nat.*, 15 : 515.
Scintharista —. Innes W., 1929, *Mém. Soc. R. ent. Égypte*, 3 (2) : 40, 43.

Type Species : *Scintharista brunneri* Saussure, 1884.
Diagnosis : Strong, smooth. Head large, hypognathous. Vertex slightly depressed ; frontal ridge flat, on median ocellus concave. Eyes oval. Antennae filiform, longer than head and pronotum together. Pronotum angular posteriorly ; median carina raised, straight, intersected by one transverse sulcus (Fig. 232). Tegmina leathery, with dense, irregular venation. Wings with a black band.
Three species, in South Africa, on Canary Islands, Spain, Morocco to India, Taiwan.

Scintharista notabilis (Walker, 1870)

Type Locality : Tenerife, Canary Islands (British Museum).

Oedipoda notabilis Walker F., 1870, *Catalogue of the specimens of Dermaptera Saltatoria in the Collection of the British Museum*, Part IV, p. 745.

Divided into several subspecies, one of which occurs in our region.

Scintharista notabilis blanchardiana (Saussure, 1888)

Fig. 232 ; Plate IV : 8

Type Locality : 'Persia'.

Quiroguesia notabilis blanchardiana Saussure H. de, 1888, *Mem. Soc. Phys. Hist, nat. Genève*, 30 (1) : 35.
Scintharista notabilis brunneri —. Bodenheimer F. S., 1935, *Arch. Naturgesch.*, 4 (2) : 189.
Scintharista notabilis blanchardiana —. Uvarov B. P., 1941, *Proc. R. ent. Soc. Lond.* B. 10 : 95, pl. 1, figs. C, D.

Strong, smooth. Head large, hypognathous or slightly opisthognathous. Eyes oval, shorter than subocular groove. Antennae longer than head and pronotum together. Pronotum raised along the median carina, intersected by the transverse sulcus beyond the middle ; anterior margin slightly projecting over the head ; posterior margin angular. Lateral lobes much higher than wide ; posterior angles rounded. Tegmina wide ; dorsal parts with very dense, irregular venation ; apical half transparent ; radius-sector with five branches (Fig. 232).
Coloration : Ochraceous-brown. Wings with a dark semi-circular band extending along median line from posterior to anterior margin ; apex dark. Inner margin of wings bluish ; the part enclosed by the black band reddish in males, yellow or yel-

lowish-green in females. Inner side of hind femur bluish; hind tibia orange-red.
Measurements (mm): Body ♀ 39.0–42.0, ♂ 26.0–28.0; tegmina ♀ 38.0–41.0, ♂ 25.0–28.0.
Distribution: Arabia, Sinai to Syria.
Israel: Judean Hills (6, 7), Dead Sea Area (13), Southern Negev (16, 17), Sinai (21, 22).
Inhabits bare, rocky mountain slopes, camouflaged by its dark and speckled colour. Adults found from April to July; hoppers observed in March–April.

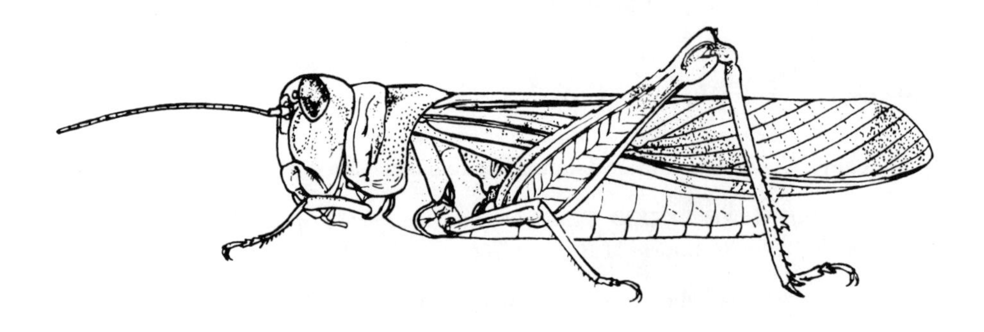

Fig. 232: *Scintharista notabilis blanchardiana* (Saussure, 1888), female

REFERENCES

Ander, K. (1949) 'Die boreoalpinen Orthopteren Europas', *Opusc. ent.*, 14 : 89–104.

Bei-Bienko, G. Ya. (1933) 'Records and description of some Orthoptera from U.S.S.R.', *Bol. Soc. esp. Hist. Nat.*, 33 : 317–341.

Bei-Bienko, G. Ya. & L. L. Mishchenko (1951) *Locusts and Grasshoppers of the U. S. S. R. and Adjacent Countries*, Part I & II, Moscow [in Russian].
Part I (*Opred. Faune SSSR.*, no. 38), pp. 1–378 ; Part II (*ibid.*, no. 40), pp. 379–667. [English translation : Israel Program for Scientific Translations, Jerusalem. IPST Cat. Nos. 834, 835 ; published 1963, 1964].

Beier, M. (1955) Buch 6. Embioidea and Orthopteroidea, in : Weber, H. (ed.), *Dr. H. Bronns Klassen und Ordnungen des Tierreichs*, Band 5 : Arthropoda, Abt. 3 ; Insecta. Leipzig, 118 pp.

Blondheim, S. A. (1978) 'Patterns of reproductive isolation between the sibling grasshopper species *Dociostaurus curvicercus* Uvarov and *D. genei* (Ocskay) (Orthoptera : Acrididae : Gomphocerinae)', Ph.D. Thesis, The Hebrew University of Jerusalem, 113 pp.

Blondheim, S. A. & A. S. Shulov (1972) 'Acoustic communication and differences in the biology of two sibling species of grasshoppers *Acrotylus insubricus* (Scopoli) and *A. patruelis* (Herrich-Schäffer),' *Ann. ent. Soc. Am.*, 65 : 17–24.

Bodenheimer, F. S. (1930) *Die Schädlingsfauna Palästinas*, Berlin, 438 pp.

— (1935a) *Animal Life in Palestine*, Jerusalem, 506 pp.

— (1935b) 'Ökologisch-zoogeographische Untersuchungen über die Orthopterenfauna Palästinas', *Arch. Naturgesch. (N.F.)*, 4 (1) : 88–142; 4 (2) : 145–216.

Broza, M. & M. P. Pener (1969) 'Hormonal control of the reproductive diapause in the grasshopper *Oedipoda miniata*', *Experientia*, 25 : 414–415.

— (1972) 'The effect of corpora allata on mating behaviour and reproductive diapause in adult males of the grasshopper *Oedipoda miniata*', *Acrida*, 1 : 79–96.

Buxton, P. A. & B. P. Uvarov (1923) 'A contribution to our knowledge of Orthoptera in Palestine', *Bull. Soc. R. ent. Égypte*, pp. 167–214.

Dirsh, V. M. (1957) 'The spermatheca as a taxonomic character in Acridoidea (Orthoptera)', *Proc. R. ent. Soc. Lond.*, (A), 32 : 107–114.

— (1958) 'Acridological Notes,' *Tijdschr. Ent.*, 101 : 51–63.

— (1961) 'A preliminary revision of the families and subfamilies of Acridoidea (Orthoptera, Insecta)', *Bull. Br. Mus. nat. Hist. Ent.*, 10 (9) : 349–419.

— (1965) *The African Genera of Acridoidea*, Anti-Locust Research Centre, University Press, Cambridge, 579 pp.

— (1968) 'The post-embryonic ontogeny of Acridomorpha', *Eos, Madr.*, 33 (3–4) : 413–514.

— (1973) 'Genital organs in Acridomorphoidea (Insecta) as a taxonomic character', *A. f. Syst. v. Ecol.*, 11 (2) : 133–154.

— (1974) *Genus Schistocerca (Acridomorpha, Insecta)* (Series Entomologica, E. Schimitschek, ed.), W. Junk, The Hague, 238 pp.

— (1975) *Classification of the Acridomorphoid Insects*, E. W. Classey, Farington, vii + 170 pp.

Euw. J. V., L. Fishelson, J. A. Parsons, T. Reichstein & M. Rothschild (1967) 'Cardenolides (heart poisons) in a grasshopper feeding on milkweeds', *Nature, Lond.*, 214 (5083) : 35–39.

Fishelson, L. (1960) 'The biology and behaviour of *Poekilocerus bufonius* Klug, with special reference to the repellent gland (Orthoptera, Acrididae)', *Eos, Madr.*, 36 : 41–62.

— (1969) 'Two new species of the genus *Chorthippus* (Acridinae) from Israel', *Israel J. Ent.*, 4 : 235–242.

Jago, N. D. (1977) 'Revision of the genus *Ochrilidia* Stål, 1873, with comments on the genera *Sporobolius* Uvarov, 1941, and *Platypternodes* I. Bolivar, 1908, (Orthoptera, Acrididae, Gomphocerinae)', *Acrida*, 6 : 163–217.

Johnston, H. B. (1956) *Annotated Catalogue of African Grasshoppers*, Anti-Locust Research Centre, University Press, Cambridge, xxii + 833 pp.
— (1968) *Annotated Catalogue of African Grasshoppers: Supplement*, Anti-Locust Research Centre, University Press, Cambridge, xiv + 448 pp.
Krauss, H. A. (1909) 'Dermaptera and Orthoptera aus Ägypten, der Halbinsel Sinai, Palästina und Syrien', in: Kneucker, A., 'Zoologische Ergebnisse zweier in den Jahren 1902 und 1904 durch die Sinai-Halbinsel unternommener botanischer Studienreisen', *Verh. naturw. Ver. Karlsruhe*, 21: 35–43, 99–119.
Mishchenko, L. L. (1936) 'Revision of the Palaearctic species of the genus *Sphingonotus* Fieber (Orth. Acrid.)', *Eos, Madr.*, 12 (1–2): 65–192.
— (1952) Insects — Orthoptera IV. No. 2, Locusts and Grasshoppers (Catantopinae), *Fauna SSSR*, no. 54, 610 pp. [in Russian].
[English translation: Israel Program for Scientific Translations, Jerusalem, IPST Cat. no. 883, published 1965].
Orshan, L. & M. P. Pener (1979) 'Termination and reinduction of reproductive diapause by photoperiod and temperature in males of the grasshopper, *Oedipoda miniata*'. *Physiol. Ent.*, 4: 55–61.
Pener, M. P. (1966) 'Note on the Ecology of two Orthoptera of Ethiopian origin new to Israel', *Israel J. Ent.*, 1: 191–192.
— (1968) 'The effect of corpora allata on sexual behaviour and "adult diapause" in males of the Red Locust', *Entomologia exp. appl.*, 11: 94–100.
Pener, M. P. & L. Orshan (1980) 'Reversible reproductive diapause and intermediate states between diapause and full reproductive activity in male *Oedipoda miniata* grasshoppers', *Physiol. Ent.*, 5: 417–426.
Pener, M. P. & A. Shulov (1960) 'The biology of *Calliptamus palaestinensis* Bdhmr. with special reference to the development of its eggs, *Bull. Res. Coun. Israel* (B), 9: 131–156.
Pener, M. P., R. Troetschler, S. Friedman-Cohen, I. Zeldes, S. G. Nasser & G. B. Staal (1981) 'Comparative studies on the effects of precocenes in various grasshopper and locust species', in: *Regulation of Insect Development and Behaviour* (eds. F. Sehnal et al.), Wroclaw Technical University Press, Part I, pp. 357–375.
Rowell, C. H. F. (1961) 'The structure and function of the prothoracic spine of the Desert Locust, *Schistocerca gregaria* Forskål', *J. exp. Biol.*, 38: 457–469.
Shulov, A. (1952) 'Observations on the behaviour and the egg development of *Tmethis pulchripennis asiaticus* Uvarov', *Bull. Res. Coun. Israel* (B), 2 (3): 249–254.
Shulov, A. & M. P. Pener (1961) 'Environmental factors in interruption of development of Acrididae eggs', in: *Cryptobiotic Stages in Biological Systems* (eds. N. Grossowicz et al.), Elsevier, Amsterdam & London, pp. 144–153.
Uvarov, B. P. (1928) *Locusts and Grasshoppers. A Handbook for their Study and Control*, Imperial Bureau of Entomology, London, xii + 352 pp.
— (1948) 'Andalusian Orthoptera described by Rambur', *Eos, Madr.*, 24: 369–390.
— (1966) *Grasshoppers and Locusts. A Handbook of General Acridology*, Anti-Locust Research Centre, University Press, Cambridge, Vol. 1, xi + 481 pp.
— (1977) *Grasshoppers and Locusts, A Handbook of General Acridology*, Centre for Overseas Pest Research, University Press, Cambridge, Vol. 2, ix + 613 pp.
Waloff, N. (1950) 'The egg pods of British short-horned grasshoppers (Acrididae)', *Proc. R. ent. Soc. Lond.*, (A) 25: 115–126.

INDEX

Valid names in roman type. Synonymies in italics. Page numbers in bold type indicate principal reference.

Index

Index

Index

Index

PLATES

PLATE I

1 *Pareuprepocnemis syriaca* Brunner-Wattenwyl, 1893
2 *Heteracris littoralis similis* (Brunner-Wattenwyl, 1861)
3 *Pamphagulus bodenheimeri* Uvarov, 1929
4 *Tenuitarsus angustus* (Blanchard, 1837)
5 *Poekilocerus bufonius* (Klug, 1832)
6 *Chrotogonus homalodemus homalodemus* (Blanchard, 1836)
7 *Pyrgomorpha conica* (Olivier, 1791)
8 *Macrolepta laevigata* (Werner, 1914)

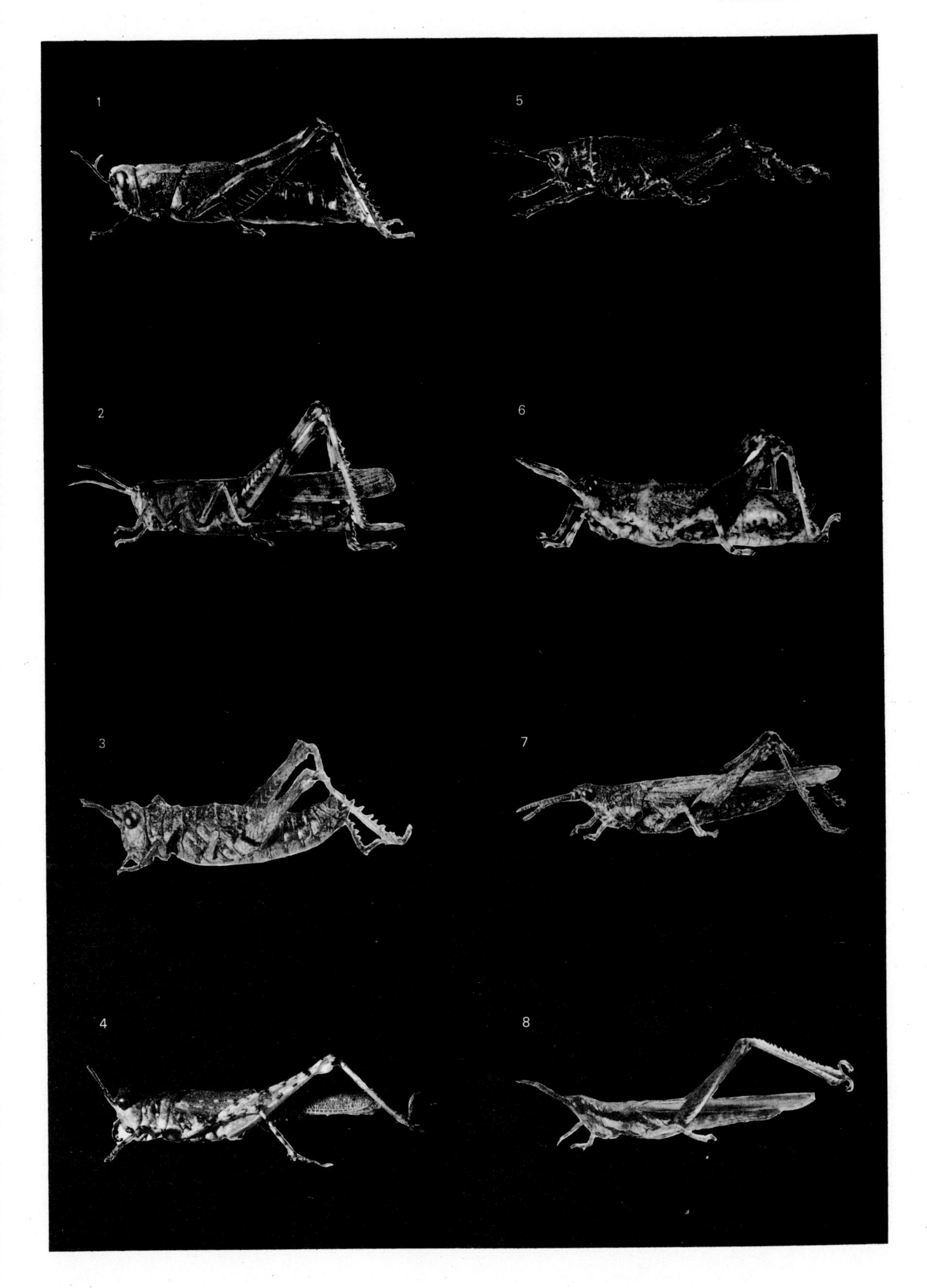
1
2
3
4
5
6
7
8

PLATE II

1 *Schistocerca gregaria* (Forskål, 1775)
2 *Sphodromerus serapis* (Serville, 1838)
3 *Cyclopternacris cincticollis* (Walker, 1870)
4 *Pezotettix curvicerca* Uvarov, 1934
5 *Dericorys albidula* Serville, 1838
6 *Eremotmethis carinatus* (Fabricius, 1775)
7 *Tmethis pulchripennis asiaticus* Uvarov, 1943
8 *Utubius syriacus syriacus* (Bolivar, 1893)

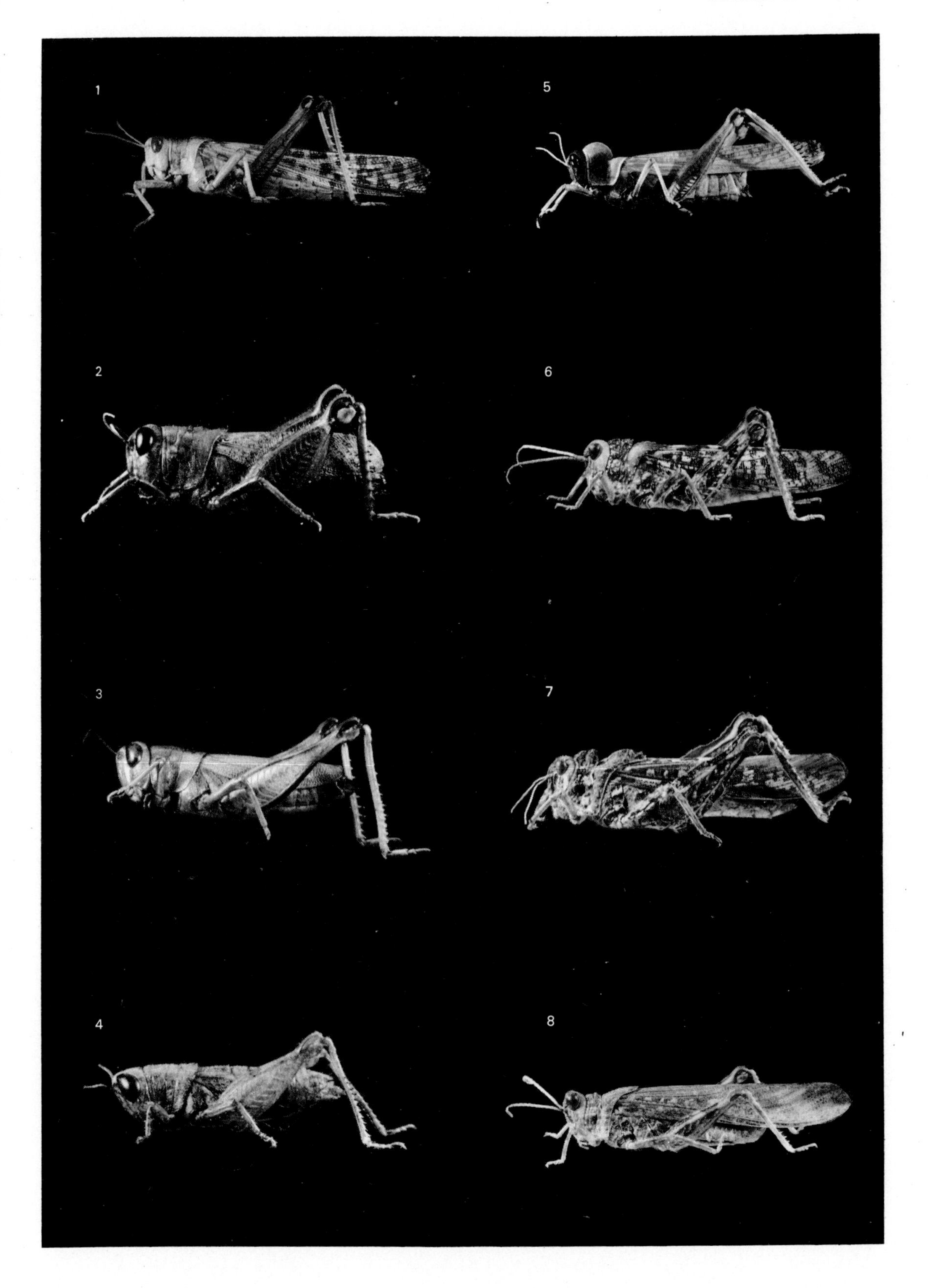
1
2
3
4
5
6
7
8

PLATE III

1 *Acinipe zebratus* (Brunner-Wattenwyl, 1882)
2 *Acinipe hebraeus* (Uvarov, 1942)
3 *Truxalis procera* Klug, 1830
4 *Acrida bicolor* (Thunberg, 1815)
5 *Duroniella laticornis* (Krauss, 1909)
6 *Stenohippus bonneti orientalis* Uvarov, 1933
7 *Notostaurus anatolicus* (Krauss, 1896)
8 *Dociostaurus hauensteini hauensteini* (Bolivar, 1893)

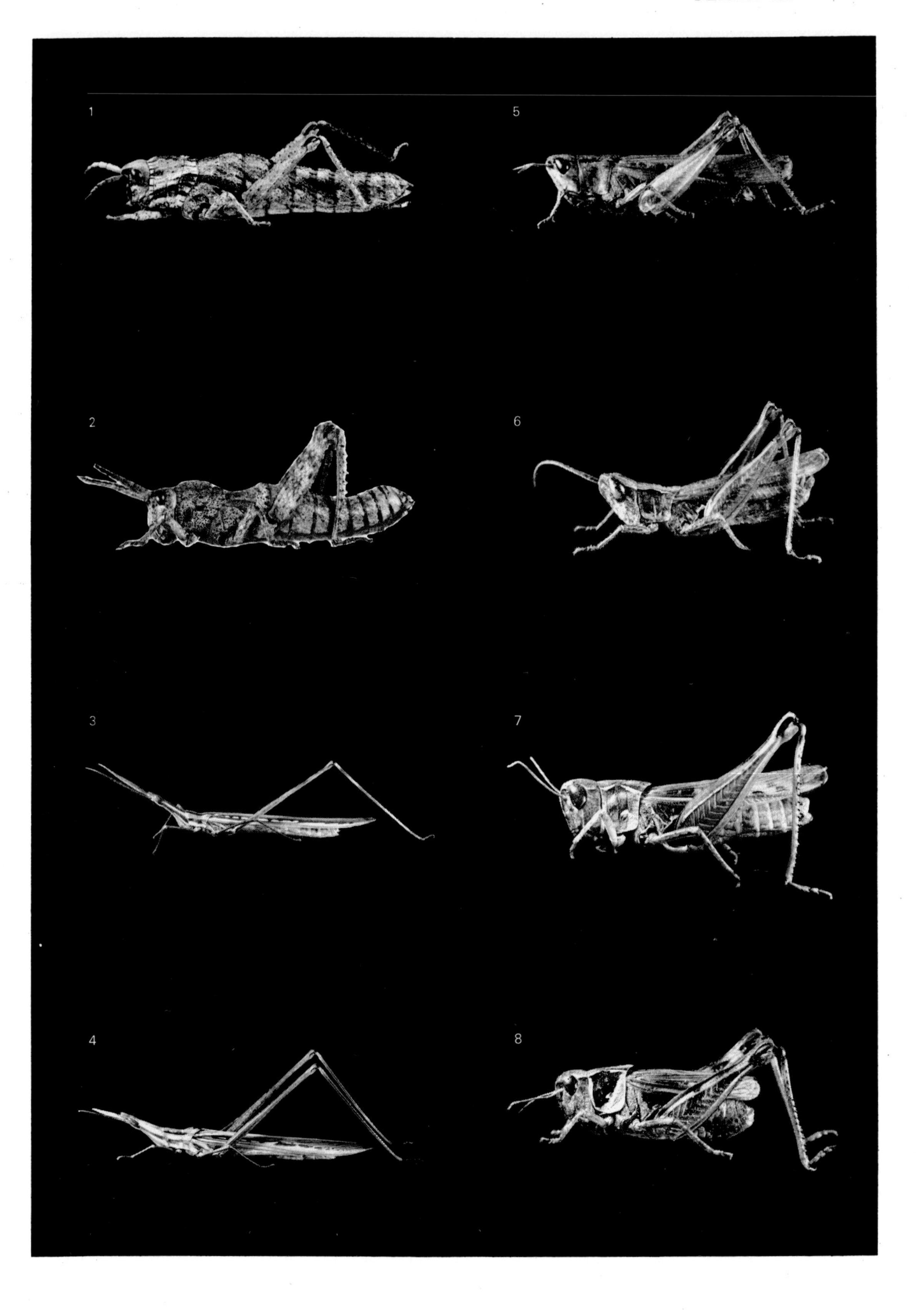
1
2
3
4
5
6
7
8

1 *Anacridium aegyptium aegyptium* (Linnaeus, 1764)

2 *Tropidopola longicornis syriaca* (Walker, 1871)

3 *Aiolopus simulatrix simulatrix* (Walker, 1870)

4 *Pamphagulus bodenheimeri* Uvarov, 1929

5 *Oedipoda miniata* (Pallas, 1771)

6 *Acrotylus patruelis* (Herrich-Schäffer, 1838)

7 *Tmethis pulchripennis asiaticus* Uvarov, 1943 – in copulation

8 *Hyalorrhipis calcarata* (Vosseler, 1902)

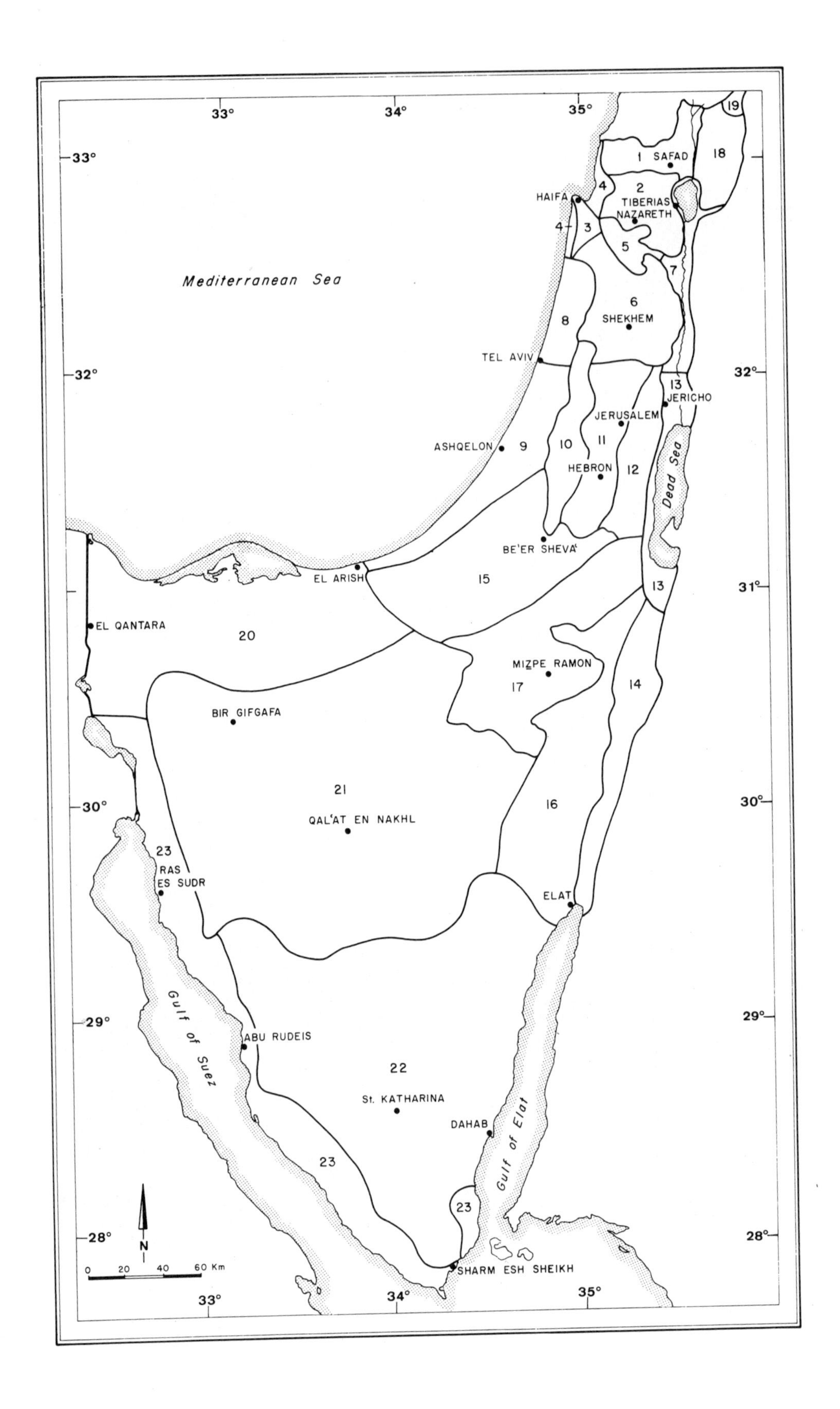

33°
34°
35°
Mediterranean Sea
1 SAFAD
2 TIBERIAS
NAZARETH
HAIFA
3
4
5
6 SHEKHEM
7
8
TEL AVIV
9
10
11
12
13
JERICHO
JERUSALEM
ASHQELON
HEBRON
Dead Sea
BE'ER SHEVA'
EL ARISH
15
14
16
17
MIZPE RAMON
18
19
20
EL QANTARA
BIR GIFGAFA
21
QAL'AT EN NAKHL
22
23
RAS ES SUDR
ELAT
ABU RUDEIS
St. KATHARINA
DAHAB
Gulf of Suez
Gulf of Elat
SHARM ESH SHEIKH
N
0 20 40 60 Km
32°
31°
30°
29°
28°

Geographical Areas in Israel and Sinai

KEY

1. Upper Galilee
2. Lower Galilee
3. Carmel Ridge
4. Northern Coastal Plain
5. Valley of Yizre'el
6. Samaria
7. Jordan Valley and Southern Golan
8. Central Coastal Plain
9. Southern Coastal Plain
10. Foothills of Judea
11. Judean Hills
12. Judean Desert
13. Dead Sea Area
14. 'Arava Valley
15. Northern Negev
16. Southern Negev
17. Central Negev
18. Golan Heights
19. Mount Hermon
20. Northern Sinai
21. Central Sinai Foothills
22. Sinai Mountains
23. Southwestern Sinai

נדפס בישראל במפעלי דפוס כתר, ירושלים
תשמ"ו

כתבי האקדמיה הלאומית הישראלית למדעים

החטיבה למדעי־הטבע

החי של ארץ־ישראל

חרקים 3 — חגבאים: חגבים

(ORTHOPTERA : ACRIDOIDEA)

מאת

לב פישלזון

ירושלים תשמ״ו